Robert J. Howlett (Ed.)

Innovation through Knowledge Transfer 2010

T0189448

Smart Innovation, Systems and Technologies 9

Editors-in-Chief

Prof. Robert J. Howlett
KES International
PO Box 2115
Shoreham-by-sea
BN43 9AF
UK
E-mail: rjhowlett@kesinternational.org

Prof. Lakhmi C. Jain
School of Electrical and Information Engineering
University of South Australia
Adelaide, Mawson Lakes Campus
South Australia SA 5095
Australia
E-mail: Lakhmi.jain@unisa.edu.au

Further volumes of this series can be found on our homepage: springer.com

Vol. 1. Toyoaki Nishida, Lakhmi C. Jain, and Colette Faucher (Eds.)
Modeling Machine Emotions for Realizing Intelligence, 2010
ISBN 978-3-642-12603-1

Vol. 2. George A. Tsihrintzis, Maria Virvou, and Lakhmi C. Jain (Eds.)
Multimedia Services in Intelligent Environments –
Software Development Challenges and Solutions, 2010
ISBN 978-3-642-13354-1

Vol. 3. George A. Tsihrintzis and Lakhmi C. Jain (Eds.)
Multimedia Services in Intelligent Environments –
Integrated Systems, 2010
ISBN 978-3-642-13395-4

Vol. 4. Gloria Phillips-Wren, Lakhmi C. Jain,
Kazumi Nakamatsu, and Robert J. Howlett (Eds.)
Advances in Intelligent Decision Technologies –
Proceedings of the Second KES International
Symposium IDT 2010, 2010
ISBN 978-3-642-14615-2

Vol. 5. Robert J. Howlett (Ed.)
Innovation through Knowledge Transfer, 2010
ISBN 978-3-642-14593-3

Vol. 6. George A. Tsihrintzis, Ernesto Damiani,
Maria Virvou, Robert J. Howlett,
and Lakhmi C. Jain (Eds.)
Intelligent Interactive Multimedia Systems
and Services, 2010
ISBN 978-3-642-14618-3

Vol. 7. Robert J. Howlett, Lakhmi C. Jain, and
Shaun H. Lee (Eds.)
Sustainability in Energy and Buildings, 2010
ISBN 978-3-642-17386-8

Vol. 8. Ioannis Hatzilygeroudis and Jim Prentzas (Eds.)
Combinations of Intelligent Methods and Applications, 2010
ISBN 978-3-642-19617-1

Vol. 9. Robert J. Howlett (Ed.)
Innovation through Knowledge Transfer 2010, 2011
ISBN 978-3-642-20507-1

Robert J. Howlett (Ed.)

Innovation through Knowledge Transfer 2010

 Springer

Prof. Robert J. Howlett
KES International
P.O. Box 2115
Shoreham-by-sea, BN43 9AF
UK
Tel.: +44 2081 330306
Mob.: +44 7905 987544
E-mail: rjhowlett@kesinternational.org
      rjhowlett@smartcentre.org

ISBN 978-3-642-26820-5   ISBN 978-3-642-20508-8 (eBook)

DOI 10.1007/978-3-642-20508-8

Smart Innovation, Systems and Technologies   ISSN 2190-3018

Typesetting: Scientific Publishing Services Pvt. Ltd., Chennai, India.

Printed on acid-free paper

9 8 7 6 5 4 3 2 1

springer.com

InnovationKT Preface

I am pleased to extend a warm welcome to the proceedings of the Second International Conference on Innovation through Knowledge Transfer, Innovation KT'2010, organised jointly by KES International and the Institute of Knowledge Transfer, and sponsored by the University of Wolverhampton.

Featuring world-class invited speakers and contributions from a range of backgrounds and countries, the InnovationKT'2010 Conference provided an excellent opportunity to disseminate, share and discuss the impact of university-business interaction through knowledge transfer in all its forms.

This was the second conference in the InnovationKT series, following on from the inaugural event at Kingston in 2009. There were two main motivations in initiating the Innovation through Knowledge Transfer conference series. The first was to provide a chance for publication on a subject where few opportunities exist already. While there would be advantages to learning of the experiences gained through knowledge transfer projects, the stories to be told often do not fit the profile of papers accepted for conferences and journals, which are focussed more on research. The successes of knowledge transfer therefore often go unreported and this conference provided an opportunity to remedy that deficiency

The second motivation was to foster the development of a community from the diverse range of individuals practicing knowledge transfer. I believe that the delegates of the conference are drawn from an interesting community of practice. Those who are able to offer papers and presentations on the joint and related subjects of innovation and knowledge transfer are not all from an academic background. Certainly academics can provide welcome and insightful contributions, but there is expertise, knowledge, skills, and experience of significant importance, to be drawn from the considerable number of knowledge transfer professionals. These people can relate lessons learned, best practice, what works and what does not, from experience gained through setting up and running real knowledge transfer projects. InnovationKT'2010 has succeeded in bringing together contributions from both the academic and practitioner sections of the knowledge transfer community.

The conference called for both short papers and full papers. Full papers of 10 pages in length, written in a conventional academic style, were presented orally at the conference, and appear in these conference proceedings published by Springer-Verlag as book chapters in the KES Smart Innovation, Systems and Technology series. In addition a summary of each full paper was published in the conference digest. Short papers of one or two pages in length were presented orally at the conference and published in the conference digest, but not in the conference proceedings. The programme contained seven invited keynote talks, 40 oral presentations grouped into eight sessions, and one interactive workshop. The proceedings contain 29 chapters drawn from this material. There were 91 registered delegates drawn from 10 countries of the world, showing that there was truly international participation.

Thanks are due to the many people who worked towards making the conference a success. I would particularly like to thank the Honorary Conference Chair, Professor Ian Oakes from the University of Wolverhampton, for enthusiastically embracing the event and sponsoring it. I would extend my appreciation to the Honorary Conference Series Chairs, Sir Brian Fender of the IKT and Dr Iain Gray of the TSB, for their support. I would also like to thank the invited keynote speakers, the members of the International Programme Committee, and all others who contributed to the organisation of the event.

I hope you find the InnovationKT'2010 proceedings an interesting and useful volume. I hope and intend that future conferences in the InnovationKT series will continue to serve the knowledge transfer community and act as a focus for its development.

Robert J. Howlett
Executive Chair, KES International
InnovationKT'2010 General Chair

Organisation

Honorary Conference Chair

Professor Ian Oakes
Pro-Vice Chancellor Research and Enterprise
University of Wolverhampton, UK

Honorary Conference Series Co-chairs

Sir Brian Fender CMG
Chairman and President of the Institute of Knowledge Transfer

Dr. Iain Gray
Chief Executive, Technology Strategy Board

General Conference Chair

Professor Robert J Howlett
Executive Chair, KES International
Bournemouth University, UK

IKT Liaison Chair

Mr. Russ Hepworth-Sawyer
Institute of Knowledge Transfer

Conference Administration

Peter Cushion, Shaun Lee, Alastair Stewart, Nadia Zernina, Claire Passmore
KES International

Innovation through Knowledge Transfer is organised and managed by **KES International** in partnership with **the Institute of Knowledge Transfer**.

International Programme Committee

Name	Affiliation
Prof. Raffaele de Amicis	Centre for Advanced Computer Graphics Technologies - GRAPHITECH, Italy
Dr. Geoff Archer	Teeside University, UK
Dr. David Brown	Inst. of Industrial Research, Portsmouth University, UK
Prof. Francisco V. Cipolla-Ficarra	ALAIPO and AINCI (Spain / Italy)
John Corlett	Oxford Brookes University, UK
Prof. Sir Brian Fender	Institute of Knowledge Transfer, UK
Dr. Paola Di Maio	University of Strathclyde, Scotland, UK
Dr. Paul Donachy	The Queens University of Belfast, UK
Ms. Kim Dovell	Institute of Knowledge Transfer, UK
Ms. Charlene Edwards	Kingston University, UK
Ms. Carolin A. Fiechter, Dipl.-Kffr.	Universität der Bundeswehr München, Germany
Mr. Marc Fleetham	University of Wolverhampton, UK
Dr. Philip Graham	Executive Director AURIL, UK
Ms. Sue Gunn	City University, UK
Mr. Iain Gray	Technology Strategy Board, UK
Prof. Christos Grecos	University of the West of Scotland, UK
Mr. Mike Hall	Association of Universities in the East of England, UK
Prof. Owen Hanson	University of Middlesex, UK
Prof. Ileana Hamburg	University of Applied Sciences Gelsenkirchen, Germany
Mr. Russ Hepworth-Sawyer	Institute of Knowledge Transfer, UK
Prof. Robert J.Howlett	Bournemouth University / KES International, UK
Prof. Noel Lindsay	University of Adelaide, Australia
Prof. Eva-Maria Kern	Universität der Bundeswehr München, Germany
Dr. Jens Lønholdt	Technical University of Denmark
Prof. Ignac Lovrec	University of Zagreb, Croatia
Ms. Debbie Lock	Kingston University, London, UK
Olivera Marjanovic	University of Sydney, Australia
Dr.-Ing. Maik Maurer	Technische Universität München, Germany
Mr. Martin May	Aston University, UK
Prof. Maurice Mulvenna	TRAIL Living Lab, University of Ulster, UK
Prof. Ian Oakes	Pro-Vice Chancellor Research and Enterprise, University of Wolverhampton, UK
Mr. Hamed Rahimi Nohooji	Iran University of Science and Technology, Tehran, Iran
Dr. Vladimir Stantchev	Berlin Institute of Technology, Germany
Ms. Val Wooff	Durham University, UK
Dr. Cecilia Zanni-Merk	INSA-Strasbourg, France
Shangming Zhou	Swansea University, Wales, UK

Keynote Invited Speakers

Sir Brian Fender CMG MInstKT
Chairman and President of the Institute of Knowledge Transfer

Welcome and Opening Remarks

Professor Ian Oakes
Pro Vice-Chancellor Research and Enterprise
University of Wolverhampton, UK

The Role of University – Business Collaboration in Influencing Regional Innovation

Abstract. The capability to produce and use knowledge through strong systems of innovation is now regarded by many as critical to the success of countries, regions, firms and individuals. In the UK, Higher Education Institutions are widely seen as key contributors to regional economic development and a fundamental part of the knowledge economy.

This presentation will investigate the relationship between knowledge, innovation and competitiveness in a regional context and explore the contributions made by universities in supporting regional innovation systems including an examination of the most common models of university-business partnership in use. It will review the role played by the UK Government in encouraging universities to respond to the needs of business and the wider community through 'third stream' funding programmes and examines the appropriateness of the metrics used to evaluate the effectiveness of this type of activity.

Finally the presentation draws some conclusions on the effectiveness of 'third stream' activities undertaken by UK universities and attempts to demonstrate how research intensive and non-research intensive universities can undertake differential yet complementary roles in supporting regional economic development through 'third stream' activities in the future.

Biography. Professor Oakes is responsible for promoting the University's research agenda and developing the growing knowledge transfer arena at regional, national and international levels.

He was educated at the Universities of Aston and Bath and has held a number of senior management posts in higher education. He has been involved in an extensive programme of technology transfer activities, both national and transnational, operating across a range of sectors and has led the development of a number of initiatives focusing specifically on the transfer of technology from academia to both large and small firms.

He has published widely in the field of innovation and technology transfer in the small firm manufacturing sector.

Dr. Nathalie Gartiser & Dr. Jean Renaud
Institut National des Sciences Appliquees (INSA)
Strasbourg, France

Knowledge Transfer in France – From Academic Research to Companies: Organization and Research Examples

Abstract. The French system of academic research is based on an important transfer system from universities to companies. Based on different organizations and helped by different transfer tools, one important political aim is to develop the fertilization of the industrial world by academic knowledge.

The valorization system is mainly based on two dimensions. The first one is based in universities and academic schools with the aims to help laboratories to identify appropriate knowledge and relevant partners to realize transfers from the academic world to the industrial word. The second dimension is based on public organizations, focused mainly on SMEs. It aims to increase dialogue between partners and to accompany the partners in connecting them, to identify the expertise and to help the partners in the first steps of negotiation and eventually contractualization.

After presenting the general mechanism of knowledge transfer between the academic research and the industry in France, and giving some examples of organizations and tools, we will give some examples of study and research partnerships with the aim to illustrate this way of doing.

Biographies

Nathalie Gartiser is Assistant professor in business sciences at INSA Strasbourg - Graduate School of Science and Technology (France). Dr. Gartiser has been working on organization and industrial innovation management for 10 years. As master degree in innovative design, she has also developed research on problem solving in non technical fields during the last years. Her recent research on this topic has been developed on the Field of Environment and Land Use Planning. Involved in entrepreneurship activities on INSA Enterprises department, she is familiar with valorization activities and knowledge transfer between INSA Strasbourg and industrial partners.

Jean Renaud is a Professor of Innovation and Conception at INSA Strasbourg - Graduate School of Science and Technology (France). He holds a PHD degree in Industrial Engineering. His research focuses on knowledge management and multi-criteria analysis. Dr. Renaud currently serves as an innovation expert in French firms and heads a French national association on project management.

Dr. Iain Gray CEng
Chief Executive
Technology Strategy Board
Swindon, UK

Connect and Catalyse to Stimulate Innovation

Abstract. In the dictionary definition, a catalyst is something that acts as the stimulus in bringing about or hastening a result; it is something which modifies and increases the rate of a reaction.

Since it was created just three years ago, the Technology Strategy Board has established a key position within the UK as a true catalyst for innovation and knowledge exchange; it has demonstrated that funding alone is not sufficient to facilitate true engagement between different communities, whether business, academia or government, to achieve measurable, sustainable outcomes but that, by recognising the barriers to collaboration and devising the appropriate mechanisms for overcoming them, challenges can be met with truly innovative solutions and remarkable results can be achieved.

By drawing upon examples from the Technology Strategy Board's portfolio, Iain Gray will illustrate some of the mechanisms which have been successfully employed to stimulate and enhance collaboration between businesses and academia across the UK, to stimulate and support innovation, bring about strategic commercial developments and to address some of the major societal challenges of our time.

Biography. Iain Gray joined the Technology Strategy Board as Chief Executive in 2007, following its establishment as an executive non-departmental public body.

Prior to joining the Technology Strategy Board, Iain was Managing Director and General Manager of Airbus UK, whose Bristol operation he joined when it was still part of British Aerospace.

Iain Gray completed his early education in Aberdeen, culminating in an Engineering Science honours degree at Aberdeen University. In addition, he gained a Masters of Philosophy at Southampton University in 1989 and has received Honorary Doctorates from Bath, Bristol and Aberdeen Universities in 2005, 2006 and 2007 respectively.

Iain is a Chartered Engineer, a Fellow of the Royal Academy of Engineers, a Fellow of the Royal Aeronautical Society and in 2007 was awarded the Royal Aeronautical Society Gold Medal. He is Chairman of the Business and Industry Panel of The Engineering and Technology Board (ETB), a Governor of the University of the West of England, a Board Member of SEMTA and a Board Member of Energy Technologies Institute.

As Chief Executive of the Technology Strategy Board, Iain is the operational head of the new organisation as it assumes its leading role in driving the UK's technology and innovation strategy.

Iain is married to Rhona and has four children.

Dr. Jarmila Davies CEng
Programmes Development Manager
Department for the Economy and Transport
Welsh Assembly Government, Cardiff, UK

Breaking Barriers and Building Collaborations: Knowledge Transfer Development in Wales

Abstract. Knowledge transfer and innovation is high on the list of priorities for the Welsh Assembly Government (WAG). Creation of a dedicated support for KT dates back to 1997, when following the consultation paper 'An economic strategy for Wales', it become clear that an impartial facility for brokering KT opportunities should be established. The presentation will describe a chronological development of processes that grew from a small group of enthusiastic KT practitioners to a multimillion programme delivering versatile support for knowledge transfer activities in Wales.

Know-How Wales (KHW) launched in 1999 was a free all Wales business support service bringing businesses in Wales closer together with Institutions of Further and Higher education and acted as a gateway to knowledge transfer provision between the two.

A first of the EU funding in 2001 enabled the launch of the Knowledge Exploitation Fund (KEF) that dealt with supporting 3rd mission and capacity building for KT delivery within academia. The KEF funding laid the foundations to a 'KTP Mentoring project for the FE sector' aiming to encourage the spirit of collaboration between HEIs and FEIs.

The second tranche of the EU funding secured in 2007 enabled KT community in Wales to continue and strengthen collaborative activities and embed the spirit of CPD, innovation and enterprise.

Biography. Dr. Jarmila Davies is a Programme Development Manager at the Department for the Economy and Transport of the Welsh Assembly Government.

Having graduated in Civil and Structural Engineering at Prague University Jarmila carried out research for the degrees of MSc at Cardiff University and PhD at the University of Glamorgan. She then pursued a successful career in higher education at the university where she led research programmes of international standing. Being a Chartered Civil and Structural Engineer, she gained considerable experience of collaboration projects working with the construction, manufacturing and engineering industries including a broad range of SMEs in Wales.

Jarmila has played prominent roles in the development of lifelong learning programmes for Welsh engineers and the promotion of the public understanding of science and engineering.

She is committed to establishing new forms of interface between businesses and academia and developing relationship and knowledge management as vital tools in the knowledge transfer process. She is a Fellow of the Institution of Civil Engineers, a Member of the European Federation of Engineering Associations, Honorary Fellow of the Chamber of Czech Engineers and a Member of the Institute of Knowledge Transfer and serves on several Boards concerned with education and promoting the public understanding of science and engineering.

Mr. Michael Smith
Senior Innovation Manager
MidTECH - NHS Innovations West Midlands

The Innovation Management and Knowledge Transfer Process across NHS Trusts

Abstract. Knowledge Transfer across NHS Trusts is slowly gathering momentum. The NHS are increasingly becoming aware of the importance of their IP and their relationships with academic institutions in IP creation.

MidTECH have been working with these Trusts trying to establish a culture where the protection of ideas is a high priority. This has come up against some resistance within the healthcare system but in roads have been made. MidTECH have adopted a system whereby projects are turned over very quickly and a priority is given to "quick-wins". This is showing Trusts that achieving a return from their IP is possible and case studies are feeding more ideas. This rapid turnaround has required an internal change in IP project management. Target-driven, internal competition, bonus schemes and a "hands-off" approach to the technology have all contributed to our model.

This presentation will look at that system and also look generally at how the NHS structure is changing and how that impacts on innovation.

Biography. Mike Smith has worked for various NHS Trusts and Universities in the West Midlands region for over five years, developing and commercialising new ideas and products. Previously, he has worked in the private sector licensing software technologies across the U.S. and Europe. Currently, he is the Senior Innovation Manager at MidTECH - NHS Innovations West Midlands and works directly with NHS staff to assist them in protecting and developing their novel ideas and innovation.

Contents

Session C: Knowledge Transfer Models and Frameworks

Session D: Knowledge Transfer Insights

Session E: Knowledge Transfer Partnership Case Studies

Session F: Innovation and Enterprise

Session G: Knowledge Transfer Case Studies

Session H: Knowledge Transfer with the Third and Public Sectors

Session A
Value Creation through Knowledge Processing – Methodologies, Approaches and Case Studies

Can Knowledge Be Transferred?

Richard Ennals, Peter Totterdill, and Robert Parrington

Kingston Business School, Kingston University, Kingston KT2 7LB, UK

Abstract. The paper argues that conventional models of knowledge transfer are confused and mistaken. Books can be transferred between people. Knowledge is more complex. Knowledge transfer is not a linear process managed by administrators. It is a matter of culture change, with knowledge as integral to the culture.

Knowledge is socially constituted, and not simply held by individuals. Explicit knowledge is only the tip of the iceberg. We need to address implicit knowledge, and most importantly, tacit knowledge. Knowledge is acquired through shared experience, typically by involvement in a particular form of life, with distinctive language games.

On this basis, it is important to create environments in which experience can be shared, and where knowledge can be given practical meaning. In the context of innovation, we can seek to develop innovation systems, contexts in which new ideas can be developed and applied.

In the context of the workplace, we need to facilitate dialogue, and partnership arrangements which engage the local actors, as well as the social partners and external research resources.

The paper considers four new structures for work organisation which enable experience to be shared, ideas applied, and knowledge acquired: Students' Quality Circles, Senior Quality Circles, Forum Theatre, and Network Consultancy. Conclusions are presented from a feasibility study project based at Kingston Business School, and conducted in association with the UK Work Organisation Network.

Keywords: consultancy, dialogue, forum theatre, partnership, Quality Circles, tacit knowledge, work organization.

1 Introduction

The option of simply maintaining the status quo in knowledge transfer is not available. Cuts in UK government spending on universities, and likely impending increases in student tuition fees, are changing power relationships and assumptions.

Academics have been talking of "student engagement", much as employers have been talking of "employee engagement". In both cases, "engagement" constitutes de facto compliance with the wishes of those in authority. Students are now taking greater account of their own personal investment in fees, and expecting

R.J. Howlett (Ed.): Innovation through Knowledge Transfer 2010, SIST 9, pp. 3–11.
springerlink.com   © Springer-Verlag Berlin Heidelberg 2011

service from academics. Students regard themselves as the new masters, as the employers. New models are needed (Nahai et al 2011).

The answer has to be to regard the university as a learning community, with learning as a collaborative activity. In the knowledge society, the university is a knowledge workplace (Gibbons et al 1994; Nowotny et al 2001; Fricke and Totterdill 2004). Old hierarchies are being challenged. Recent administrative super-structures, distant from the learning workplace, have often relied on short-term funding, and may vanish. Non academics have chosen to regard themselves as managers, not required to address or understand knowledge issues, but able to make decisions affecting learning and teaching. This position faces challenge.

We need a new set of practical structures, to empower individuals, broaden participation, and extend dialogue. However, we have entered a new age of austerity. We need to engage in change which uses our own resources, in particular human resources. Learning is not simply to be equated with what takes place in the education system, including universities. Universities themselves need to learn. We need to complement a focus on competition with attention to creating collaborative advantage (Normann and Josendal 2009; Ekman et al 2010; Johnsen and Ennals 2011).

2 Knowledge

It is no longer acceptable to rely on a linear approach to knowledge transfer, top down, whereby teachers, as authority figures, pass on their knowledge. This model does not cover all stages of the process, from research and development, through the ordinary users, including from younger and older generations. Different logics are required at various levels, and, most importantly, we need new buffer zones, including varieties of "Quality Circles" (Hutchins 2008; Chapagain 2006). These act as horizontal filters, between the contrasting discourses on each side, enabling different views and perspectives to be contributed. Dissenting views are not just tolerated, but welcomed as essential seasoning.

Current arrangements for learning and teaching in universities are not sustainable. Mass higher education, with reduced resources made available for teaching, mean that the focus needs to change, as credibility evaporates. Large modular courses are impersonal, with no real opportunity for students and academics to interact. Students may fail their assessments, consider the experience as poor value for money, and leave. Financial and academic judgements are coming into conflict.

The balance needs to change, as between theory and practice. Courses with an orientation towards professional career development should be able to draw on practitioner experience, if they are to be seen as a sound investment of time and money. We should aspire to achieve "skill", and not merely "competence". This means recognising the value of experience, skill and tacit knowledge (Göranzon and Josefson 1988; Göranzon 1995; Göranzon et al 2006). Academic and vocational qualifications alone may not be enough.

The new structures which we have been piloting, and which we introduce in this paper, need not necessarily require the abolition of old institutions. They offer an alternative horizontal mode of development, a new internal skeleton. In the context of universities, ideas and methods can often be best conveyed by students, taking ownership of their own learning, and creating new enterprises as Change Agents. The students are registered for several different modules, and need to be able to make sense of the differences.

The process of knowledge development is organic. It needs to be driven by those who are themselves engaged in the learning process, rather than detached administrators. Universities are not in the business of widget production. Quality is to be defined within the culture, empowering participants. It is not primarily a matter for external measurement.

Since the 1648 Peace of Westphalia we have had stable national borders in Europe, matched by clear boundaries between the academic disciplines, each with its own institutions and traditions (Toulmin 2001). Such silos are becoming harder to defend in an era of globalisation, and in a context where there are cross-disciplinary platforms and social networks. Our students, oriented towards future employment, find it hard to respect such apparent fragmentation.

3 Dialogue

The industrialisation of education has led to an emphasis on outputs from research, at the expense of a concern for the process of research. This approach reached notable heights of absurdity with the UK Research Assessment Excellence, in which research activities were measured in terms of publications in particular journals. Research itself dropped out of consideration.

It had been assumed that adoption of modern scientific approaches would result in "one best way". This assumption appears to have been false, as there are divisions across the disciplines, and little direct communication or mutual understanding between technologists and ordinary citizens, particularly from older generations. There is no general agreement on what constitutes evidence, yet there is glib talk of "evidence based decisions" and "evidence based policy". The truth is that policy determines what is to constitute evidence.

We argue for the importance of dialogue, in education, in the workplace, and in wider society. We can learn from the different views which are expressed. Dialogue need not necessarily result in agreement, but should result in increased understanding. In the European Union, there is a central role for Social Dialogue, engaging the Social Partners (employers' organisations and trade unions). Dialogue has an important role to play. If we all agreed on everything, learning would stop (Ennals and Gustavsen 1999; Gustavsen et al 2007; Nolin 2009; Ekman et al 2010).

4 Feasibility Study

The Feasibility Study at Kingston Business School provides a fixed period in which to observe the emergence of the new structures, and to see the scope for linkage. We are using external funds to conduct a local field experiment, involving each of the four areas listed below. Pilot activities are organised and evaluated.

This was designed to present and exemplify possibilities, so that other actors can become engaged as active partners, and take co-ownership.

UKWON (Fricke and Totterdill 2004; Totterdill et al 2011) has operated in this way since 1998, with a series of externally funded projects enabling new approaches and structures to be prototyped. Development has been in association with partners across Europe, who are part of an ongoing collaborative community. This means that consortia to respond to European calls are always ready and willing.

5 Quality Circles

Quality Circles have a role at transitional points, such as at the beginning and end of working life, where logics and discourses suddenly change. Transitions are not always neat and clean, and individuals follow different paths. It can help to add delaying functions, introducing diverse perspectives and experience, through Circle members.

Ishikawa first introduced Quality Circles in the automobile industry in Japan, with the objective of empowering workers who were suffering adverse effects from Taylorist scientific management (Ishikawa 1990). The idea was that the workers should take co-ownership of the process of continuous improvement, and take pride in their own skill. Quality was thus a bottom-up process.

Experience in UK industry (and indeed in education) has typically been very different. Quality is seen as a top-down matter for managers, meeting externally imposed targets, and with use of check lists rather than the reflections of experienced practitioners. BS 5750, ISO 9000, Investors in People: in each case, achieving certification of compliance requires payment to be made to an external consultant, confirming that paperwork is in order. To return to an agenda of empowerment we have had to take a circuitous route.

6 Students' Quality Circles

Indian visitors to Japan in 1992 were impressed by what they saw of Quality Circles, which they associated with the long record of Japanese industrial success. Apart from developing a Quality Movement in India, they also sought to transfer this powerful approach to the new context of Education. The starting point was City Montessori School and Degree College in Lucknow, which now has over 35,000 students. A movement developed which has engaged students in schools across India, and in 24 other countries, under the auspices of the World Council for Total Quality and Excellence in Education.

Transferring the knowledge of Quality Circles was far from simple. Quality Circles moved from industrial settings, involving experienced adult workers, to an educational setting, involving groups of children as young as 8 years old. In many cases, Students' Quality Circles have been an exercise in English language and public speaking, providing the opportunity of engagement in a practical case study. Educational institutions have continued virtually unchanged, with control very much in the hands of teachers, and a context of scientific management. There has been understandable pride in the achievement of the students, but the status quo has not been disrupted.

Following participation by Kingston University staff and student union officers in Students' Quality Circle events in India, Sri Lanka, Pakistan, Mauritius, and Turkey, it was agreed that Kingston University would host the international convention in 2014. It was time to try to transfer knowledge of Quality Circles to Kingston, through new practical activities.

The first Students' Quality Circle at Kingston University, KCircle, came from an undergraduate module in International Human Resource Management (Nahai et al 2011). Students were in part motivated by the opportunity to present at an international conference in India. They learned from the experience, on their return presented to their classmates at Kingston, and then at a Faculty Learning and Teaching Event. As strong final year business students, the KCircle leaders have established their own consultancy company, Change Agents, to operate after graduation, as they start their own working lives.

The students instinctively followed a path consistent with that of UKWON, whose focus has been on workplace innovation. KCircle identified a market for facilitators of change in Higher Education, and recognised that skills can develop based on experience.

During the Feasibility Study project the KCircle / Change Agents presented to full time MBA students, engaging them in the change process. The MBA, around the world, is a relic of an Anglo-American model of business which is now broken. The financial market system collapsed. The case for developing new generations of general managers, as if nothing had happened, may be flawed. The Kingston MBA requires five years of relevant management experience before the course, and reflection on that experience is a key resource. However, many of the general management textbooks are now obsolete. New approaches are needed. Our students will ultimately gain competitive advantage, through experience of creating collaborative advantage. This requires engagement in practice.

7 Senior Quality Circles

The Senior Quality Circle in the Department of Informatics and Operations Management brings together academics from different discipline backgrounds, with varied professional experience, and assorted elderly relatives. It is a repository of wisdom and tacit knowledge, and the core of a daily lunch club at the Kingston Hill campus, which is usefully situated some miles from alternative catering facilities.

A large proportion of the academic staff of the department are now aged over 50: they would be classified as "Seniors" in Norway, where the Centre for Senior Policy has been addressing practical issues of demographic change, and making special provision for the workplace needs of older workers (Ennals and Hilsen 2010). Those who are aged over 55 could be eligible for Voluntary Early Retirement. However, taking such retirement can mean making a complete break with the workplace. Vital human resources are likely to be lost, individually and collectively. This is an international problem (Hilsen and Ennals 2009; Augustinaitis et al 2009; Ennals and Salomon 2011).

A Senior Quality Circle can reflect on and value the experience, skills and tacit knowledge of the members, and provide a supportive transitional environment which can enable smooth transitions at the latter end of working life. Under European Discrimination Directives, mere chronological age is not a reason to be removed from the workplace. There may be continuing contributions, whether part time, full time, or in the form of consultancy. There can also be support for contributions to life outside work, both before and after retirement.

Demographic profiles of academic workforces suggest that a high proportion of academics are now close to retirement. Younger academics may be more likely to have PhDs, but less likely to have professional experience of working life. In a Business School, this has serious implications, for the learning and teaching culture.

8 Forum Theatre

Forum Theatre brings drama into the workplace, exploring relationships in light of external parallels (Fricke and Totterdill 2004). In employment relations, we often talk of the "workplace actors". In Forum Theatre the actors are also researchers, who investigate a case study situation, and develop a piece of drama. This is presented in the workplace, in such a way that workers and managers can respond, relate to the stories and relationships which are being presented, and eventually intervene in the drama, directing proceedings from the audience. Such interventions can lead to ongoing change processes, jointly owned by audience members.

As part of a project "Dramatic Innovation", Kingston Business School will host a production at the Rose Theatre, for a business audience. An earlier presentation will offer the opportunity for MBA and other students to engage. Kingston University are major sponsors of the Rose.

9 Network Consultancy

Network consultancy enables constructive collaboration across institutional and departmental boundaries. It enables individuals to link up to meet needs of third parties, in a context of trust and partnership. This is particularly important in a business environment when things have fallen apart. Discretionary budgets have been reduced. Needs continue. Gaps increase.

UKWON is developing innovative new approaches, building on unique tripartite engagement with trade unions, employers' organisations and government, as well as universities and research organisations. UKWON has recognised that many older workers re-label themselves as consultants on retirement (whether voluntary or otherwise), partly to retain their own self image. The transition from employee to consultant is not always easy. Collaboration may be unfamiliar.

10 Building on What Is Feasible

Following the feasibility study project, next steps will be driven by practical human need, rather than rhetoric.

One current area for potential development is "Assisted Living", where government is concerned to increase the market for technology vendors, with a view to reducing care costs for the increasing elderly population. It is not enough to push a technocentric view. In order to find human centred solutions for elderly users, intermediate structures are needed, as outlined in this paper. The Feasibility Study project could lead to submission of a major funding bid.

There is also a case for testing the feasibility of syntheses of the structures outlined above. We are advocating a bottom-up approach to change, and are thus not obliged to present a single top down structure to be "rolled out".

A physical example can help to illustrate what is possible. The Matara Centre in the Cotswolds can host organisational dialogue processes, as well as weddings and funerals. Decorated in North American Indian style, the "Council Room of the Elders" provides a suitable and evocative environment for dialogue by a particular Senior Quality Circle, for which on-site accommodation is available. The wider theme of East / West fusion inspires creative flair. The Hilarium Room can host Forum Theatre. Supporting networks from academia, workplace innovation and consultancy can add value to and underpin network consultancy. Ongoing mentoring is available following events.

11 Conclusions

Transferring knowledge is more complex than many people have imagined. It is not like "passing the parcel", with a zero sum game. Tacit knowledge is important, but resists easy transfer.

The status quo in education, work and knowledge, is not a sustainable option. Transition points at the start and end of working life have key roles; new structures can be deployed.

Brief histories of the example structures highlighted the complexities involved in moves between countries, sectors and generations. It is not just a matter of "rolling out" change.

Exploration has begun into how some of the particular challenges of demographic change can be addressed. Instead of regarding age as a form of medical problem, it can be seen as providing invaluable resources of experience, skill and tacit knowledge. Having recognised that potential in older people, the benefits of retaining access to such assets become evident.

The Senior Quality Circle has the potential to benefit its members, the organisation in which the members are currently employed, and wider society, for which they can act as a powerful filter for projects concerning intergenerational relations. There will need to be arenas in which such work can be taken forward. The Matara Centre is one potential venue. Poltimore House, near Exeter, is another. There could be a nationwide network.

Showing the feasibility of one or more components does not in itself guarantee the sustainability of a system constructed from such components. Human beings, and the organisations in which they work, have a remarkable capacity to foul things up, with or without the use of computers (Ennals 1995).

Social science researchers have interpreted the world of which they are part: the problem is to change it. It is not sufficient to criticise conventional accounts of knowledge transfer. This paper has introduced key components for a feasible set of alternatives. There is work to be done.

References

Augustinaitis, A., Ennals, R., Malinauskiene, E., Petrauskas, R.: E-Redesigning of Society: towards experiential connectivity of generations in Lithuania. AI & Society 23.1, 41–50 (2009)

Chapagain, D.: Guide to Students Quality Circles. NQPCN, Kathmandu (2006)

Ekman, M., Gustavsen, B., Pålshaugen, O., Asheim, B. (eds.): The Scandinavian Model of Innovation. Palgrave, Basingstoke (2010) (in press)

Ennals, R.: Preventing IT Disasters. Springer, London (1995)

Ennals, R., Gustavsen, B.: Work Organisation and Europe as a Development Coalition. Benjamins, Amsterdam (1999)

Ennals, R., Hilsen, A.-I.: Older Workers: The Jam in the Sandwich. Presented at Older Workers in a Sustainable Society, Oslo (June 2010)

Ennals, R., Salomon, R. (eds.): Older Workers in a Sustainable Society. Peter Lang, Brussels (in preparation, 2011)

Fricke, W., Totterdill, P. (eds.): Action Research in Workplace Innovation and Regional Development. Benjamins, Amsterdam (2004)

Gibbons, M., Limoges, C., Nowotny, H., Schwartzman, S., Scott, P., Trow, M.: The New Production of Knowledge: The Dynamics of Science and Research in Contemporary Societies. Sage, London (1994)

Gustavsen, B., Nyhan, B., Ennals, R. (eds.): Learning together for local innovation: promoting learning regions. Cedefop, Luxembourg (2007)

Göranzon, B. (ed.): Skill, Technology and Enlightenment: On Practical Knowledge. Springer, London (1995)

Göranzon, B., Josefson, I. (eds.): Knowledge, Skill and Artificial Intelligence. Springer, London (1988)

Göranzon, B., Hammarén, M., Ennals, R. (eds.): Dialogue, Skill and Tacit Knowledge. Wiley, Chichester (2006)

Hilsen, A.-I., Ennals, R.: Virtual Links: intergenerational learning and experience sharing across age divides and distances. AI & Society 23.1 , 33–40 (2009)

Hutchins, D.: Hoshin Kanri: the strategic approach to continuous improvement. Gower, Farnham (2008)

Ishikawa, K.: Introduction to Quality Control. Chapman and Hall, London (1990)

Johnsen, H.C.G., Ennals, R. (eds.): Creating Collaborative Advantage. Gower, Farnham (in preparation, 2011)

Nahai, R., Osterberg, S., Ennals, R.: A Perspective from a Students' Quality Circle. In: Columbus, F. (ed.) Higher Education in a State of Crisis, Nova Science, New York (in preparation, 2011)

Nolin, T. (ed.): Handbook of Regional Economics. Nova Science, New York (2009)

Normann, R., Josendal, K. (eds.): National Pilot in Regional Development. Kingston Business School Working Paper (2009)

Nowotny, H., Scott, P., Gibbons, M.: Re-Thinking Science: Knowledge and the Public in an Age of Uncertainty. Polity, Cambridge (2001)

Totterdill, P., Exton, R., Ennals, R.: Workplace Innovation in Europe. Gower, Farnham (in preparation, 2011)

Toulmin, S.: Return to Reason. Harvard University Press, Cambridge (2001)

The Authors

Richard Ennals is Professor of Corporate Responsibility and Working Life at Kingston Business School, Kingston University, and Visiting Professor at Agder University (Norway), Linnaeus University (Sweden), and Mykolas Romeris University (Lithuania)..

Peter Totterdill is Visiting Professor at Kingston Business School, and Joint Chief Executive of the UK Work Organisation Network.

Robert Parrington is Research Associate at Kingston Business School.

Structured Knowledge Transfer for the Implementation of a New Engineering Service Centre in India

Results from a Captive Offshoring Project in the Automotive Supplier Industry

Franz Lehner and Christian Warth

University of Passau, Innstraße 43, 94032 Passau, Germany
franz.lehner@uni-passau.de, christian.warth@gmx.de

Abstract. Organizations are continuously confronted with stress of competition. The search for lower operational costs is no longer limited to the manufacturing and information technology field and has been extended to engineering services as well. For comprehensible reasons more and more tasks in the engineering service sector are shifted towards India. Along with this, international companies plan at least partly to transfer firm-specific knowledge towards India so that knowledge management has become a key success factor for the performance of plants or subsidiaries in India. This contribution focuses on a research project dealing with the knowledge transfer processes of a global automotive tier 1 supplier to its joint venture in Pune, India. Knowledge transfer processes as part of a holistic knowledge management approach were essential for the success of these off-shoring activities. The major goal of this contribution is to show how this offshoring project was carried out from a knowledge management point of view. This provides deeper insights into the course of action related to knowledge transfer processes between the two locations in the US and India. An internally developed knowledge transfer model leveraged a combination of experienced resources from the joint venture, with task based training and documentation of knowledge and practical cross cultural orientation and assimilation of teams to quickly initiate the new operation. Finally the paper will demonstrate how an above average steady state level can be reached by progress tracking and feedback mechanisms. Furthermore the paper will provide a brief overview of the existing theoretical dominant factors of successful knowledge transfer which were distilled out of empirical studies and prior research in this field.

R.J. Howlett (Ed.): Innovation through Knowledge Transfer 2010, SIST 9, pp. 13–22.
springerlink.com   © Springer-Verlag Berlin Heidelberg 2011

1 Motivation, Background and Research Method

More and more tasks are being transferred by manufacturing and service industries to countries like India, China, Malaysia, etc. In the last years India has become the hub of service industries worldwide due to a growing number of highly
educated, young and English speaking people, stable economic conditions and low
labour costs as compared to other industrialized nations. Companies who would
like to benefit from these conditions must transfer at least parts of their activities
to India.

The findings presented in this paper result from a joint project between the
University of Passau and a global automotive tier one supplier. The partner company is a worldwide technological leader in this branch. Confronted with a downward spiral of business prospects the company had to realign its engineering
services by relocating about 30% of its R&D activities to its joint venture in Pune,
India, in order to adjust the cost structures and to react to changing market requirements. An immediate consequence is that firm-specific knowledge has to be
transferred from the US headquarters to India. Therefore knowledge management
(KM) has become a key success factor not only for the overall firm performance
but also for performance of the service unit. The major goal of this contribution is
to show how this offshoring project was carried out from a KM of view. This will
provide deeper insights into the course of action related to knowledge transfer
(KT) processes between the two locations in the US and India. In the first part it
will be demonstrated how the prearrangements for the KT were developed. This
includes the need and the development process of a shared vision as well as knowledge transfer objectives and a strategy. Subsequently the KT process itself will be
illustrated in detail. Finally the paper will demonstrate how an above average
steady state level can be reached by progress tracking and feedback mechanisms.
The findings of this case study are based upon action research methodology and
can be, at least partly, adapted to similar situations.

According to Avison et al. action research combines theory and practice (and
therefore researchers and practitioners) through change and reflection in an immediate problematic situation within a mutually acceptable ethical framework and
can be described as an iterative process involving researchers and practitioners
acting together on a particular cycle of activities, including problem diagnosis, action intervention, and reflecting learning (Avison et al. 1999). McKay & Marshall
(2001) propose that action researchers should consider two parallel and interacting
cycles: the research cycle (which is focused on the scientific goals) and the problem-solving cycle (focused on the problematic situation). Accordingly a pooled
cycle of academic researchers as well as practitioners of the cooperating company
was permanently implemented for this research project (see Warth 2009).

2 Related Work and Main Influence Factors for KT

After analyzing the last fifteen years of research in the area of KT processes 21
quantitative studies were found which scrutinize the key factors influencing KT
within multinational companies or within alliances (see Lehner/Warth 2010).

At this point it is important to note that none of these publications considered here studied the same or a comparable situation so that a deficit in research and a lack of common understanding has to be stated. Because of their empirical approach it was decided to rely on quantitative models as they allow replication to some extend and at least check if the influence of a certain factor is significant. These factors form the basis of improvements in this project which aims to demonstrate that KT processes can be managed successfully by obeying theoretical insights. Table 1 summarize those influence factors and describes them briefly.

Table 1 List of main factors influencing KT processes

Factor	Description	Reference
Sender	Also disseminative capacity; ability and motivation of an employee to share knowledge	Minbaeva (2004)
Tacitness	Implicit and non-codifiability accumulation of skills	Zander/Kogut (1995)
Complexity	Number of critical and interacting elements embraced by an entity or activity	Hayes/Wheelwright (1984)
Specifity	Transaction cost's asset specifity	Reed/DeFillippi (1990)
Teachability	Extent by which know-how can be taught to new workers	Hayes/Wheelwright (1984)
Reciprocity	Sum of a partner's account of the resources committed by itself and its perception on the extent of resources committed by the other party	among others: Williamson (1991)
Codifiability	Extent to which the knowledge has been articulated in documents	Kogut/Zander (1992)
Ambiguity	Extent with which the knowledge can be transported, interpreted and absorbed	among others: Kogut/Zander (1992)
Recipient	Employees' job related abilities and overall competencies, job related motivation, involvement, job satisfaction, absorptive capacity (overall ability and willingness to absorb new knowledge)	Minbaeva et al. (2003)
Learning intent	Degree of desire for internalizing a partner's skills and competencies	among others: Hamel (1991)
Cultural distance	People from members of our corporate global network including our parent tend: 1) to think like us and 2) to behave like us	among others: Lin/Germain (1998)
Relationship	Degree of involvement in MNCs network	./.
Ability-based trust	The focal party's perception of the partner's capabilities, knowledge and skills related to alliance	Mayer/Davis (1999)
Benevolence-based trust	Extent to which the focal party perceived the partner would not intentionally harm its interests	Mayer/Davis (1999)
Integrity-based trust	The focal party's perception regarding partner's fairness, sense of justice, consistency and values	Mayer/Davis (1999)

It has to be added that in this specific project it was not aimed to evaluate or improve KT models but instead to use their practical implications to support management in a specific case of KT. Hence only those factors were used for which a common acceptance can be assumed. The factors listed in table 1 refer to an ideal KT process consisting of four components. These components are: sender, recipient, the knowledge to be transferred, and finally the environment in which the KT process is embedded.

3 Prearrangements for KT

To prepare the organization for the new offshoring model, managers from the two locations met for an initial due diligence activity to understand the work (tasks) done in the different departments, functional roles executing the task at the sending location. The complexity level of these tasks was determined by analysis for intensity of collaboration and domain knowledge. This resulted in a finite set of tasks that were deemed offshore-able.

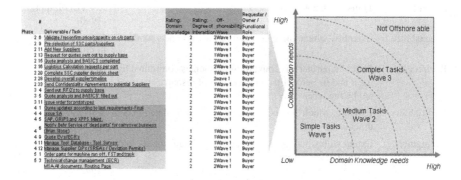

Fig. 1 Due Diligence – Break down of tasks and clustering by waves

Based on this process, the wave 1 functions were determined as design, simulation, quality, process planning across all product lines, wave 2 was determined to be product engineering, program management and cost estimation, wave 3 established scale to the operations. This then was followed by defining job descriptions and resource assignments to the different functional roles. Minimum entry criteria for resources were established in terms of qualifications and foundation knowledge needed before the resources arrived in the US. Trainers from the parent in Stuttgart traveled to train the resources on standard engineering tools. Dedicated training plans were created for each of the individuals by the sending organization based on the minimum entry criteria and focusing dependent on their pre-qualifications.

4 KT Process

Figure 2 shows the overall timeline and transition process for this project.

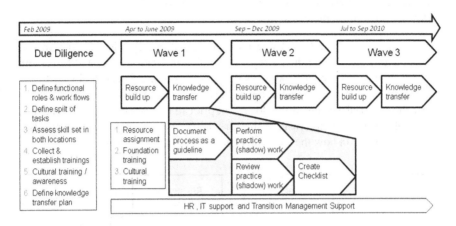

Fig. 2 Overall transition process

A team of transition managers at both locations prepared the sending and receiving organization with an overall plan for the KT, explaining how the process works and expectations from the resources on both sides. Additionally the teams were exposed to cultural orientation trainings. By execution of those one day trainings it was possible to improve the collaboration between the two parties. An Indian employee who works permanently for the sending organization could be enabled to present measures how to improve the virtual teamwork. In doing so the US participants (mostly designers or team leaders) learnt multilayer aspects in team working with their Indians counterparts (e.g. intercultural facets, technical aspects, organizational issues). Those trainings were planned to carry out also at the receiving side from a US expat with similar contents adjusted for the Indian members. Finally a week before the Indian team arrived to the US, physical space for working, IT set up like phones, computers etc was organized. HR teams were prepped to conduct a training program on the lines of the "new hire orientation". Managers from the US were prepped to keep the first day of reporting open to communicate expectations, offer openness to meet the new teams and support to resolve any issues they face during the process.

Most of the technical knowledge was documented and available on-line in an internally development system known as BDS which was globally accessible for all employees. The BDS housed all relevant standards, guidelines, manufacturing requirement and technical specification for all the product lines. However it did not capture nuances that people followed in day to day work or requirement that were unique to meeting the North American market customer needs or manufacturing plants. Examples of such kind of information would include

- Preparing CAD deliverables per OEM standards
- Documenting DFMEA, PFMEA per internal standards or migrating an existing one to an OEM specified format
- Translating DVP&R into test orders in the test request system
- Developing control plans and like wise

It was essential that the knowledge transfer team captures this information and documents this. To enable this, a separate section in BDS was created to capture this kind of information. A process as described in figure 3 was deployed to enable robustness of the documentation. A key point was incorporation of a loop to review the editorial aspects of the document prior to reviews by the US Managers. The collaborative element ensured that the management teams in both organizations were aligned in how the work was currently being performed.

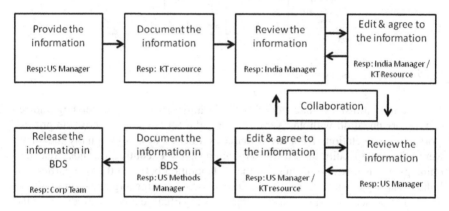

Fig. 3 Documentation of the transferred knowledge

The teams then performed work per the new documented procedure. This is shadow work which was then reviewed by their US counterpart. The review process was captured by means of a checklist. Several iterations of similar tasks were performed to ensure the robustness of the process and checking documentation. This also served to build confidence in the KT resource as well.

Towards the end of each wave each team created a simulated offshore environment, by deliberately moving the KT resources to another building away from the US teams for two weeks. Tasks were provided and additional information passed on using electronic forms of communication such as phones, emails, chat or desktop sharing. The results were reviewed and documentation further strengthened based on the observed failure modes. Another advantage of the "little India environment" was that it gave a firsthand impression to the US resources as to how the business model would impact their day to day work.

5   Steady State of KT

On return to India, the KT team had to create one final deliverable which was called as the "Procedure Manual". The procedure manual was the document that links the process as described in the US with the way it will be actually performed in India. Table 2 shows the key contents of the procedure manual.

Table 2 Content of procedure manuals

Chapter	Key question	Contents
1	How are the tasks requested?	Required inputs, input review, time/cost estimation
2	How are the tasks executed?	Working process, issue and status reporting
3	How are the tasks reviewed?	Checklists, error reporting, fixing errors
4	How are the tasks delivered?	Assumptions, issues, acceptance notes
5	How are the tasks accepted?	Closure, rework, feedback, lessons learned, billing
6	What reference information exists?	Guidelines, standards, expert list, methods, forms

The information in the KT phase at the US covers one aspect of chapter 2 and 3, however recognizing the fact that an entire organization does not turn up for KT, only a small representative team is sent, it is essential that a holistic approach to ensure quality of service is addressed. This deliverable is due back to the sending location within three months of completion of KT, return of the team to India and start of the engineering service.

6   Governance – Ensuring an Effective KT

One of the challenges while executing a knowledge transfer program is to ensure that all tracks (functions) which are off-shored to the engineering service are moving at a steady pace. Any deviation is quickly identified and fixed at the very earliest. The framework is as shown in figure 4.

The monitoring framework focused on the end to end process was established. Successful completion of the pre-arrangement phase was a necessary criterion to migrate to task training and documentation. Likewise successful evaluation of the shadow work with a feedback rating of 3.25 on a scale of 5 was necessary to authorize "Go Live" for that function. This milestone indicated completion of KT and start of payable work from India. Slow movers were identified and Go Live dates adjusted as needed.

Figure 5 shows finally how the KT was tracked from a progress side by linking the monitoring framework to the nominated resources and function areas.

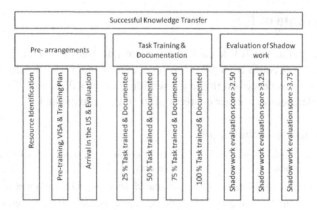

Fig. 4 KT monitoring framework

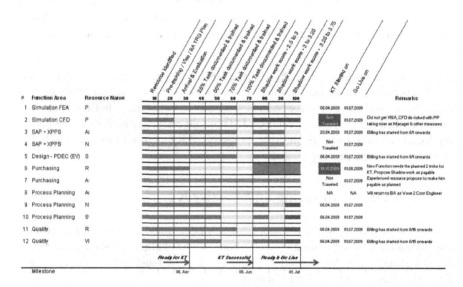

Fig. 5 Sample of KT dashboard.

After completing wave 1 an IT tool was implemented to ensure that there was a standard operation procedure for off-shored tasks to the engineering service. With this tool it was possible to submit a work request, to review and provide feedback on the delivery made by the engineering service, to monitor the overall status of work requests and to log defects on the helpdesk. Furthermore this tool distilled monthly key performance indicators which served as base for measures to improve the capabilities of the service unit. The transition management office utilized this tool to estimate the off-shored hours as well as the billable hours to manage the engineering service also from a financial standpoint.

7 Conclusion and Outlook

The findings of this case study show that theoretical insights from research on KT is a promising way to address the selected problem area and to improve the KT process. The factors identified in the literature review formed the basis of the taken KT measures. By doing so it could be demonstrate that KT processes can be managed successfully by obeying theoretical insights. Due to guiding this research with the methodology action research it is possible to get by now first practical insights into the "mechanics" of KT but further research will be needed to intensify those insights. Following the action research approach the strength of the study can be seen in the possibility to design an integrated KT model which describes the "mechanics" of a KT in the global automotive supplier industry from Germany towards India. Those insights are up to now absent but relevant as this industry is rising. By proceeding so it is moreover possible to feedback into the research cycle how the academic knowledge was applied into practice.

Acknowledgments. The authors would like to thank the anonymous reviewers for helpful comments on the paper.

References

Avison, D., Lau, F., Myers, M., Nielsen, P.: Action Research. Communications of the ACM 42(1), 94–97 (1999)

Hamel, G.: Competition for competence and inter-partner learning with international strategic alliances. Strategic Management Journal 12 (special issue), 83–103 (1991)

Hayes, R., Wheelwright, S.: Restoring our competitive edge – competing through manufacturing. John Wiley and Sons, New York (1984)

Kogut, B., Zander, U.: Knowledge of the firm, combinative capabilities, and the replication of technology. Organization Science 3(2), 383–397 (1992)

Lehner, F., Warth, C.: Knowledge Transfer Processes in the Automotive Supplier Industry – Designing an Integrated Knowledge Transfer Model. In: Moreira, R., Silva, R. (Hrsg.) Proceedings of the 11th European Conference on Knowledge Management, Academic Conferences, Universidade Lusíada de Vila Nova de Famalicão, Famalicão, Portugal, September 2-3, pp. 591–601 (2010)

Lin, X., Germain, R.: Sustaining Satisfactory Joint Venture Relationships: The role of conflict resolution Strategy. Journal of International Business Studies 34(3), 179–196 (1998)

Mayer, R., Davis, J.: The effect of the performance appraisal system on trust for management: a field of quasi-experiment. Journal of Applied Psychology 84(1), 123–136 (1999)

McKay, J., Marshall, P.: The dual imperatives of action research. Information Technology & People 14, 46–59 (2001)

Minbaeva, D., Michailova, S.: KT and Expatriation in Multinational Corporations. Employee Relations 26(6), 663–679 (2004)

Minbaeva, D., Pedersen, T., Björkman, I., Fey, C.F., Park, H.J.: MNC KT, Subsidiary Absorptive Capacity, and HRM. Journal of International Business Studies 34(6), 586–599 (2003)

Reed, D., De Fillippi, R.: Causal Ambiguity, Barriers to Imitation, and Sustainable Competitive Advantage. Academy of Management Review 15(1), 88–102 (1990)

Warth, C.: Aktionsforschung. In: Lehner, F. (Hrsg.) Forschungsstrategien im Wissensmanagement. Passauer Diskussionspapiere, Schriftenreihe Wirtschaftsinformatik, Diskussionsbeitrag W-31-09, pp. 29–36 (2009)

Williamson, O.: Comparative economic organization: the analysis of discrete structural alternatives. Administrative Science Quarterly 36, 269–296 (1991)

Zander, U., Kogut, B.: Knowledge and the Speed of the Transfer and Imitation of Organizational Capabilities: An Empirical Test. Organization Science 6(1), 76–92 (1995)

Supporting Cross-Border Knowledge Transfer through Virtual Teams, communities and ICT Tools

Ileana Hamburg

Institut Arbeit und Technik, FH Gelsenkirchen
hamburg@iat.eu

Abstract. Many multinational organisations support work collaborative practices like virtual functional or project teams within cross-border business. Cross-border knowledge transfer within virtual teams or communities may face an extra challenge of cross-cultural hurdles. In this paper, after a short presentation of virtual teams and communities and the problem of cross-border transfer in this context, some methods and tools for achieving intercultural competence and tools supporting knowledge transfer as well as activities of an on going European innovation transfer project about Lifelong learning in SMEs are given.

1 Introduction

Nowadays also due to the economic situation and globalisation, many multinational organisations which have subsidiaries with staff working in different locations support work collaborative practices like virtual functional or project teams. Cross-border business exploded over the past 20 years and intercultural problems rise in this context.

The topic of knowledge sharing and transfer was researched in this context by some authors [15], [16] but most in conventional face-to-face collaboration forms. The knowledge transfer is considered as an aspect of knowledge management (KM); it is very complex depending on actors, tools and tasks [1]; much knowledge is tacit or hard to articulate [17]. But the cross-border knowledge transfer within virtual teams may face an extra challenge of cross-cultural hurdles. Less support and a not productive and systematic dealing with cultural differences transform this difference into an obstacle instead of a source of synergy and a stimulus for knowledge transfer and mutual learning.

The problem of diversity, which is an important aspect in cross-border business, is not really understood by many company managers and the advantages

R.J. Howlett (Ed.): Innovation through Knowledge Transfer 2010, SIST 9, pp. 23–29.
springerlink.com   © Springer-Verlag Berlin Heidelberg 2011

are not used for individuals and organisations. Common goals in the transfer of cross-border knowledge are not defined, success criteria are not understood, and the achieving of intercultural competence by using intercultural learning is not supported.

Particularly in connection with virtual teams (VT) and other cooperation like virtual communities of practice (VCoPs), which could be powerful environments for knowledge transfer, these aspects have to be researched. Small and medium sized companies (SMEs), which have existence problems, need support in this direction. Collaborative portals for virtual teams and communities and staff Web logs are tools which support cross-border transfer and intercultural competence and can be a big help for SMEs to remain internationally competitive.

In this paper, after a short presentation of virtual teams and communities and the problem of cross-border transfer in this context, some methods and tools for achieving intercultural competence and tools supporting knowledge transfer are given.

2 Virtual Teams and Communities – Knowledge Transfer

One of the examples of approaches for collaborative work in companies refers to virtual teams that are composed of members who could reside in different time zones or countries [11]. Particularly, being supported by the development of new Internet and Web-based technologies, the work and communication could be done at anytime, anywhere, in real or virtual spaces. The team members can see the results of their work, evaluate them and their motivation might increase.

Another form of cooperation particularly used in big multinational companies like Shell and Hewlet Packard are communities of practice (CoPs) [2], [10], [20]. They are networks of individuals who share a domain of interest and knowledge about which they communicate (online in the case of virtual ones – VCoPs [12], [5]).

There are some differences between VCoPs and functional or project virtual teams. A project team has specific objectives, with members working towards formal milestones and deadlines and it is often dissolved once its mission is accomplished. VCoPs membership changes, objectives and needs too; VCoPs can exist as along their members (who are volunteers), are interested to contribute and to gain knowledge and resources.

In a CoP, members share knowledge, which is boundless; they learn how to converse theory into practice. CoPs help participants to bridge the gap between tacit knowledge (How) and explicit knowledge (That) [7]. ICT based procedures have the potential for the combination of synchronous and asynchronous communication and access.

Knowledge to be transferred is embedded in people (Human), processes, relationships (Social), environments (see Figure 1).

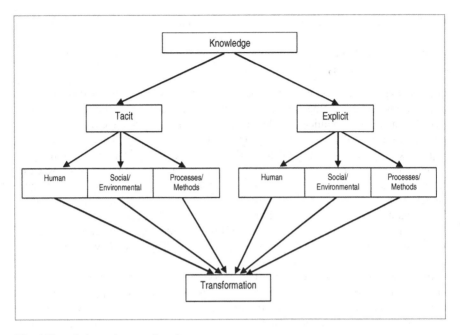

Fig. 1 Knowledge to be transferred

In virtual teams and CoPs knowledge is created when people participate in solving a common problem and exchange the needed knowledge for the problem.

Sharing knowledge makes more sense in the context of a CoP because its members have common interests in learning and exchanging experience in their specific area of activity and this favours reciprocal trust.

We have presented some benefits derived from transferring knowledge teams and in CoPs by a sense of shared interests and an extending/deepening knowledge, which derived from on-going interaction. But cross-border transfer is very complex and there are many factors to be considered and there are barriers which hinder the achieving of such benefits particularly in cross-border business.

The literature generally refers to impediments to coordinate and collaborate within such environments due to divergent nationally based culture attributes, language barriers and new technologies [4]. Culture and cultural differences can have a strong impact, particularly in the case of tacit knowledge such as leadership skills or management know-how. This type of knowledge is very valuable, complex, culturally determined and not easily to be codified.

Gupta and Govindarajan suggest that knowledge transfer within an organisation or more organisations is a function of five forces [8]. We used this model and adapt this in the case of the teams/communities taking into consideration the following factors:

- Value/reputation of the source of knowledge (person, process, etc.)
- Motivational disposition of knowledge sources: politics, rivalry and other barriers may reduce the desire of the source to share its knowledge with others

- The use of clear and rich communication channels, open and frequent communication among members of teams/communities
- Motivational disposition of member to whom the knowledge is directed
- Absorption capacity of these members.

Transferring relevant knowledge costs time and resources; the willingness of teams/communities members to do this could be affected by their national culture.

Referring the communication, the most basic barrier is language. In our projects we tried to use English as basic language for the communities, but due to the low English knowledge of some countries this caused communication problems particularly in the virtual teams/communities (see below). So we decided to increase the face to face sessions to have a clear communication in our VCoP. Another aspect refers to communication channels and methods which are different in different countries (formal, organized, informal, spontaneous, and unplanned).

One important activity for the successful management of knowledge transfer is to define common goals for this process in advance. The goals have to be identified and agreed by all members.

In connection with the use of technology, one barrier for the virtual teams/communities refers to selectivity in the choice of ICT to support them. Virtual teams and VCoPs need to use Internet standard technologies [9]. Our experience and results of other projects show that members have often difficulties with the ICT access and ICT skills referring for example to the use of on-line forums and eLearning. The best software to use is the one the team/community is most familiar with and is most prepared to use.

Other aspects are trust and the depth of relationships. Face-to-face interaction and socialization processes consolidate the relations between members and group membership. Trust is important for knowledge sharing and development in a virtual team or VCoP and this develops primarily through face-to-face interactions.

Another aspect is that because virtual infrastructures can be set up across cultures via the Web, cultural and language differences can change interactions and hinder the flow of CoP activities. The use of technology to bridge geographical gaps can lead to a misinterpretation of messages; cues and feedback are often missing. Crossing virtual boundaries between institutions can involve legal issues like data protection, intellectual property.

3 Intercultural Competence

There are no universal solutions or specific rules for responding to cross-border knowledge transfer in virtual teams/communities.

This is a complicated and sensitive matter. When it is brought up it has the potential to create uneasiness, resentments and arguments amongst participants. The goal of such process is to use the intercultural difference at including knowledge of all teams/community members, to be able to create a culture of open sharing/transferring of knowledge both tacit or explicit and create new one. In our projects before starting the cross-border VCoP we initiated an intensive training week for a knowledge transfer moderator because intercultural differences call for intercultural competences of CoP moderators. The role of the moderator in a CoP with many cultures is a crucial one; she/he should observe conversations, give advice and try to be a mediator between cultures. However the challenge of different cultures also affects the role of participants. They also need a good deal of sensitivity and awareness of participants prejudices (and also their own), to be prepared for collaborating with people from a different background. A week of intercultural learning and dialog for the VCop members was also organized.

While communicative competence is characterized by the negotiation of intended meanings in authentic contexts of language use, intercultural competence has to do with far less negotiable discourse worlds, the "circulation of values and identities across cultures, the inversions, even inventions of meaning, often hidden behind a common illusion of effective communication" [13].

Intercultural competence does not ask to behave like someone else or imitate another culture, but to learn actively about the people you are cooperating/working/learning with for effective collaboration and communication (www.uq.edu.au). One important step in the knowledge process transfer is to develop openness to differences for example understanding that culture is not static, that cultural context is changing. The ability to operate across many types of boundaries, real and virtual ones, is helpful. A key for developing intercultural competences is to respect and understand diversity and grounds of discrimination.

In connection with the used language to tackle interculturality in CoPs, it is important that language that stereotypes or shows bias against groups of people should be avoided. The use of inappropriate language has the potential to damage the credibility of the moderator/trainer and alienate the learners.

Web 2.0 [18] helps students to work in an autonomous way, to work collaboratively, to find, to publish and to share data, information and resources easily. Collaboration is one of the most important factors in knowledge transfer. It also provides on-line spaces to publish and classify contents in different formats. Therefore, it is a way to improve on the different competences that our language curriculum mentions: intercultural competence, communicative competence and audiovisual competence.

Within our Leonardo project, simultaneously with the process of building the VCoP a concept for a portal has been developed to serve as a tool for cross-border knowledge sharing/transfer and intercultural learning within the VcoP.

4 Tools for Knowledge Transfer

Within our project team it was decided to develop a portal with information about the project available for all interested people and having some restricted areas for the VCoP member use only. Figure 2 shows the functionality of the restricted area.

Within the Interface Community the members of the VCoP can communicate directly by using different tools, search information and collaborate. Moderated online intercultural learning sessions are planned. There is a moderator in the VCoP with intercultural competence but some project members would like to achieve also such competence.

The project team encourages the VCoP members to use particularly the Web logs supported by the portal which can contribute to transfer also tacit knowledge.

Web blogs are used till now within big multinational companies but the the experience cannot be applied within SMEs without adaptations.

It is planned within the project to do short studies about the impact the Web blogs created within the project on the process of cross-border knowledge transfer particularly in the SMEs which are members of the VCoP.

The Community Cross border Knowledge Transfer/memory component contain the technologies for the support of knowledge management and storing.

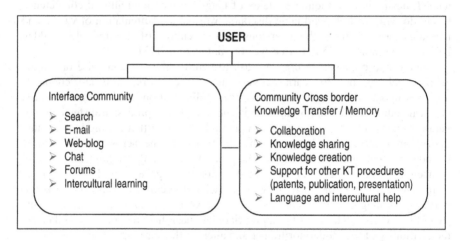

Fig. 2 Restricted area

5 Conclusions

This paper tries to show that virtual teams and virtual communities of practice can have an important role within cross-border knowledge transfer processes. Their popularity is increasing but in the context of SMEs a lot of work should be done to encourage them to join such co-operations. CoPs have the advantage that they are not too closely intertwined with established patterns of companies, particularly when innovative product/processes have to be found.

One important issue in cross-border knowledge transfer is the achieving of intercultural competence: The lack of it can produce frictions in cross-border business that importance arises.

Intercultural portals and Web logs promise qualitative advantages within the knowledge transfer process.

Projects have to be developed and work has to be done to convince SME managers about the advantages of these aspects and environments.

References

[1] Argote, L., Ingram, P.: Knowledge transfer: A basis for competitive advantage in firms. Organizational Behavior and Human Decision Processes 82(1), 150–169 (2000)

[2] Beer, D., Busse, T., Hamburg, I., Mill, U., Paul, H. (eds.): eLearning in European SMEs: observations, analyses & forecasting. Münster, Waxmann (2006)

[3] Brandas, C.: Intercultural Knowledge management Support Systems. In: InterKnow Euroworkshop II, Regensburg, Germany (2003)

[4] David, G., Chand, D., Newell, S., Resende-Santos, J.: Integrated collaboration across distributed sites: the perils of process and the promise of practice. Journal of Information Technology 23, 44–54 (2008)

[5] Diemers, D.: Virtual Knowledge Communities. Erfolgreicher Umgang mit Wissen im digitalen Zeitalter. Dissertation der Universität St. Gallen (2001)

[6] Dougiamas, M.: Moodle: Using Learning Communities to Create an Open Source Course Management System, Perth, Australia (2004)

[7] Duguid, P.: The Art of Knowing: Social and Tacit Dimensions of Knowledge and the Limits of the Community of Practice, pp. 109–118. The Information Society (Taylor & Francis Inc.) (2005)

[8] Gupta, A.K., Govindarajan, V.: Knowledge flows within multinational corporations. Strategic Management Journal 21, 473–496 (2000)

[9] Hall, B.: Learning goes online: how companies can use networks to turn change into a competitive advantage (Cisco Systems: Packet Magazine) (2000)

[10] Hamburg, I., Engert, S., Petschenka, A., Marin, M.: Improving eLearning 2.0-based training strategies on SMEs through communities of practice. In: The International Association of Science and Technology for Development: The Seventh IASTED International Conference on Web-Based Education, Innsbruck, Austria, pp. 200–205 (March 17-19, 2008)

[11] Horwitz, F.M., Bravington, D., Silvis, U.: The promise of virtual teams: identifying key factors in effectiveness and failure. Journal of European Industrial Training 30(6), 472–494 (2006)

[12] Johnson, C.: A survey of current research on online communities of practice. Internet and Higher Education 4, 45–60 (2001)

[13] Kramsch, C.: Context and Culture in Language Teaching. Oxford University Press, Oxford (1993)

[14] Krogh, G., Ichijo, K., Nonaka, I.: Enabling Knowledge Creation. In: How to Unlock the Mystery of Tacit Knowledge and Release the Power of Innovation. Oxford University Press, New York (2000)

[15] Lukas, L.: The role of culture on knowledge transfer: the case of the multinational corporation. The Learning Organization 13(3), 257–275 (2006)

[16] Mäkelä, K.: Essays on international level knowledge sharing within the multinational corporation. Helsinki School of Economics A-227 (2006)

[17] Nonaka, I., Konno, N.: The concept of 'ba': building a foundation for knowledge creation. California Management 40(3), 40–54 (1998)

[18] O'Reilly, T.: What is Web 2.0. Design patterns and Business models for the next generation of Software (2005), http://www.oreillynet.com/lp/a/6228

[19] Stocker, A., Tochtermann, K.: Investigating Weblogs in Small and Medium Enterprises: An exploratory Case Study. In: Proceedings of 11th International Conference on Business Information Systems – BIS 2008 (2nd Workshop on Social Aspects of the Web), Innsbruck (2008)

[20] Wenger, E., McDermott, R., Sydner, W.: Cultivating communities of practice: a guide to managing knowledge. Harvard Business School Press, Boston (2002)

Knowledge Management Can Be Lean: Improving Knowledge Intensive Business Processes

Carolin A. Fiechter[1], Olivera Marjanovic[2], Julia F. Boppert[3], and Eva-Maria Kern[4]

[1] Universität der Bundeswehr München
    carolin.fiechter@unibw.de
[2] University of Sydney, Australia
    olivera.marjanovic@sydney.edu.au
[3] trilogIQa
    julia.boppert@trilogiqa.de
[4] Universität der Bundeswehr München
    eva-maria.kern@unibw.de

Abstract. The main objective of this paper is to analyse the knowledge dimension of a repetitive, but highly complex business process (BP) in a case organisation – a large logistics service provider. More precisely, the paper illustrates an application of a combined Business Process Management (BPM) and Knowledge Management (KM) framework to one of its core BPs and demonstrates a possible approach to analysing "knowledge-intensiveness" of the chosen process. The paper illustrates that in this particular example of a BP, a sustainable source of competitive advantage does not come from process automation, but is related to the experiential knowledge of decision makers, and complexity of their decisions. Also, in order to improve this type of process, our research shows that it is necessary to consider human-centred process knowledge rather than simply focus on process structure, as it was typically done in the past, in the case of highly structured BPs.

Keywords: Knowledge intensive business processes, knowledge intensity, business process analysis, business process improvement, business process management, integrated framework.

R.J. Howlett (Ed.): Innovation through Knowledge Transfer 2010, SIST 9, pp. 31–40.

1 Introduction

The field of Business Process Management (BPM) continues to be considered as top business priority, across industry sectors (Gartner, 2010). In the past the main focus was on process efficiency, typically achieved through workflow automation. However, these days, BPM systems have been widely used and are no longer considered as competitive differentiators. "Emergent work practices are becoming common rather than prescribed projects. Most of the simple tasks have been automated or soon will be" (Davenport and Prusak, 1998).

In their quest for a more sustainable source of competitive advantage, more mature BPM organizations are now turning their attention away from process automation and towards process-related knowledge. In fact, knowledge is now considered an integral part of the BPs and not something to be managed separately. Furthermore, knowledge is deeply embedded in all types of business processes, even in those that up now have been considered highly repetitive, structured and somewhat "mechanistic'.

Indeed, both BPM and Knowledge Management (KM) scientific communities agree that process-related knowledge is relevant for all types of BPs, regardless of their structure and complexity. (cf. (Amelingmeyer, 2004, p.15); (Davenport, 2005, p.6); (Remus, 2002, p.73f.); (Stewart, 1998, p.50)). However, the nature of knowledge and knowledge management differs significantly for different types of processes. Even more, a very comprehensive review of BPM and KM literature confirms that this integration problem is very challenging. "It is still not clear how to integrate knowledge management more thoroughly into business process management... connecting knowledge activities to the core business processes is the second and more effective stage of knowledge management in an organization" (Smith and McKeen, 2004).

The relevant KM literature, offers two views of process-related knowledge, resulting in two very approaches how this knowledge might be managed (Engelhardt, 04). On the one hand, process-related knowledge could be captured in its explicit form and embedded within the process itself. This is typically achieved through rule-based components, used to support highly structured decision making that could be expressed through a set of rules (i.e. explicit knowledge). The second approach recognises the complexity of decision making that in many instances requires experiential knowledge, impossible to reduce to a set of rules. As the most important and the most complex component of process-related knowledge is held by humans, its management need to be human-centred. In other words, it needs to include better support for process participants and facilitate sharing of experiential knowledge among participants.

While the first solution seems to be an adequate way for simple processes, the second solution is suitable for the more complex processes. However, this is very hard to determine a priory, and through an application of traditional BP analysis frameworks that do not consider individual decisions and their knowledge intensity, as discussed in this paper.

The main objective of this paper is to analyse the knowledge dimension of a repetitive, but highly complex business process in a case organisation. The paper illustrates that in this particular example of BP, a sustainable source of competitive advantage does not come from process automation, but is related to the experiential knowledge of decision makers. Also, in order to improve this type of process, our research shows that it is necessary to consider human-centred process knowledge rather than process structure, as it was typically done in the past, in the case of highly structured BPs.

2 Theoretical Background: The Integrated Framework for BPM and KM

This section introduces a theoretical framework for process-related knowledge, previously introduced by (Marjanovic and Freeze, 2011). This integrated framework is designed to combine three influential models from the KM and BPM disciplines, namely, a holistic model of BPM (Harmon, 2007), a model of process/knowledge continuum (Crandall, Klein, and Hoffman 2006) and the reversed knowledge pyramid (Jennex, 2008). The integrated framework is depicted by Table 1. The first column describes the key elements coming from different theoretical frameworks, used to analyse three different types of BPs captured by columns 2, 3 and 4.

For example, the element called "BP Type" comes from the process-knowledge continuum. The same process-knowledge continuum was used to describe the main characteristics of all three types of processes, provide examples and classify the types of workers required (captured by column 1, rows 2, 3, 4). Business examples used include the retail sales BP representing a simple procedural process, the equipment repair BP that is a more complex process, performed by knowledge workers and the new product development as an example of a very complex BP performed by experts.

Data sources, information type, types of process knowledge and knowledge intensity (rows 5, 6, 7, and 8) originally come from the reverse pyramid framework. The same key element "information type" is also considered by the Holistic BPM model, within its BPM systems component. At the same time, the people component of the same framework is predominantly described by the combination of several elements, including "worker types", "types of process knowledge", "knowledge intensity", defining what kind of knowledge is required to perform these processes and to what extent it could be captured in its explicit form. BP modeling and BP improvement methodologies originally come from the process component of the holistic model, while the last two elements come from its strategy component.

Table 1 The framework for BPM and KM integration.

BP Complexity	Simple procedural processes	More complex processes	Very complex processes
Main Characteristics	Step-by-step sequence; Few rules or decision points; Well defined subject matter	Branching sequence; Many rules or decision points; Less defined subject matter	Sequence defined by process; heuristic and guesses; evolving subject matter
Examples	Mfg line; retail sales; book keeping	Equip. Repair; Field sales; Process Analysis	New Product development; S/W system Design; Consulting
Worker Types	Ordinary workers	Knowledge workers	Experts
Data sources	Deterministic	User-selected	Require human-expertise
Information type	Predefined; highly structured; coming from BPM, ERP or Workflow systems	Structured and unstructured; Generally similar system sources	Structured/unstructured; Source cannot be predicted in advance;
Types of process-related knowledge	Predominantly explicit in the form of process models	Explicit – process models, business rules; Experiential – Exceptions, process-related insights	Predominantly experiential: lessons learned; new practices; tips and hints
Knowledge intensity	Knowledge is resident in the process model. Data is captured largely by mechanical sensors	Knowledge-intensive processes that require human expertise for completions. Mix of human and mechanical data collection	New combinations of data and information occur frequently through human interpretation
BP Modeling	Quite detailed	Only High Level	Not advisable
BP improvement methodologies	Traditional ◄——————►	Knowledge-based ◄——————►	Discovery ——————►
BP Automation	Automated with little human interaction. ——————————————►	Human interaction required at key points.	Not possible
Process-related competitive advantage	Process efficiency; standardization to minimize variations	Process effectiveness; knowledge processes designed to leverage human knowledge	Expert's knowledge; competitive advantage not achieved through processes but is linked to expert work outcomes
BP performance monitoring	Measures related to process efficiency and control: cost/ time/ output/ throughput ◄——————►	◄——————►	Measures related to process effectiveness expressed in terms of goals and learning ——————►

This theoretical model is used to guide and inform our exploratory research of the process dimension of a complex BP, found in a case organisation, as described in the next section.

3 The Case Study: Business Processes at a Logistics Service Provider

3.1 Case Study Setting and Data Collection

As already stated, this exploratory study was conducted in a complex case organisation, engaged in ongoing provision of complex logistic services. Our research focused on a set of business processes, ranging from very simple to knowledge-intensive, with different types of knowledge (explicit as well as experiential). This paper focuses on an interesting example of a BP called "reverse logistics", especially on the process step "receipt of goods" This particular process could be classified as highly structured, in terms of its control flows and pre-defined structure. Therefore would be considered as a prime candidate for a traditional BP improvement methodology. However, as this paper illustrates, an in-depth analysis of its process-related knowledge offers new insights and opportunities for a very different approach its improvement.

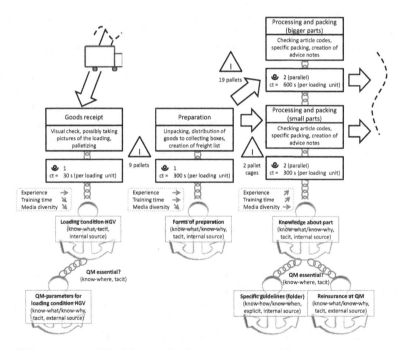

Fig. 1 Value-stream model of the examined process.

In the process called "reverse logistics" the logistics service provider takes delivery of diverse goods (goods receipt) from different European locations, inspects and returns them to his customer's warehouse. Therefore, the parts are unpacked, checked basically as well as particularly based on specific guidelines and finally are either sorted out or repacked with new packaging materials (special processes preparation, processing, packing). In the figure above the first three process steps are presented as a value chain, with the process relevant knowledge (detailed by the dimensions knowledge form, type and source) depicted as an anchor.

In the step "goods receipt" the employee records the arrival of the lorry, carries out a visual examination of the loading condition and – if there is any evidence the loading condition might have led to damages to the goods – takes pictures of the lorry loading. In some cases he also needs to palletise the incoming goods for further processing. If the employee has the feeling (tacit knowledge) that s/he is not able to assess the acceptability of the loading condition, s/he has to come for an expert of the quality management (QM, external source) to decide in his/her place.

In the second step called "preparation", the same employee unpacks the goods, allocates them to different collecting boxes and creates the freight list. To do so, s/he has to decide which part to assign to which form of preparation in the next step based on his experience (tacit knowledge).

In the third step, the "processing and packing", the process flow is divided into two parallel processes, one for small and one for bigger parts. Despite different cycle times (ct), the activities within both process alternatives are similar, starting with checking the article codes, followed by packing the goods based on specific guidelines and ending with the creation of advice notes. Based on his/her accumulated experience, the employee knows these guidelines for the most popular goods (tacit knowledge), for all others s/he has to look them up in a provided folder (explicit knowledge, internal source) or to reinsure his decision at the QM (tacit knowledge, external source).

In order to describe and assess the process, we collected data using different sources (numbers / frequency in brackets): Structured interviews with process owner (2), general observations (3), detailed observation and questioning (2) and semi-structured interviews with operation personnel (5).

3.2 Presentation and Analysis of Data

In order to classify the chosen process and analyse its knowledge dimension, we applied the integrated KM/BPM framework, taking into account different criteria, as described below. A special emphasis was placed on the "knowledge intensiveness", as the key characteristic of process-related knowledge in a BP. More precisely, in order to express a level of knowledge intensity of a given BP, we applied a methodology developed by (Kern and Boppert 2010). This particular methodology includes data collection guided by twelve different items (see table 3 below) founded in the previous research by (Gronau, 03, p.315), (Davenport, 05, p.10) and expert interviews. During data collection phase, experts were asked to rate each item on a 5-step Likert scale, with 1 being the lowest possible and 5 being highest possible value. According to the arrays corresponding to the averages of these twelve values, the processes were classified as less knowledge intensive (values from 1 to 2.3), moderately knowledge intensive (from 2.4 to 3.6) or strongly knowledge intensive (3.7 and higher).

Table 2 Knowledge intensiveness in the single process steps from experts view and assignment to the three rating arrays.

Criteria for knowledge intensiveness	Value for "goods receipt"	Value for "preparation"	Value for "packing"	Value for "processing"
Variability and exceptions	2	3	4	3
Diversity and incertitude of input and output	1	2	3	5
Variety of sources and media	1	1	3	3
Variance and dynamic development	1	2	3	4
Many participants with different expertise	3	1	2	4
High degree of innovation	1	1	3	3
Disposable scope for decision-making and insertion of creativity	2	2	3	4
Complexity of work	1	2	3	4
Degree of singularity (vs. degree of recurrence	1	2	3	2
Business criticality	2	2	4	5
Required experience	1	2	3	4
Training time	1	2	3	4
Average	1.7	2.2	3.7	4.5
Rating	Less knowledge intensive	Less knowledge intensive	Strongly knowledge intensive	Strongly knowledge intensive

This evaluation and classification already reveals one critical aspect of determining the knowledge intensity (and probably also other criteria for process classification): A process consists of different steps that may differ in their characteristics and thus in their requirements in terms of process design and support. By aggregating the single process steps to a single value for the process "reverse logistics", we would calculate a value of 3.0 and hence classify the process as "moderately knowledge intensive" – a category none of the process steps falls in.

It becomes clear than further analysis is required to determine where this "knowledge intensity" does comes from and, most importantly in which form. In order to better understand the nature of the process-related knowledge, we then proceeded to analyse the chosen process, focusing on its knowledge-intensive task "Goods receipt" and using the previously described integrated BPM/KM framework to better understand the nature of knowledge involved. The outcomes are depicted by the following Table 3.

Table 3 Application of the integrated framework on the selected process

Criterion	Cat. 1	Cat. 2	Cat. 3	Description
Main Characteristics	X			Step-by-step sequence, few decision points, well defined subject matter
Worker Types	X			Mostly blue-collar workers
Data sources	X			Deterministic (perception of the truck, rules & manuals, personal experience, known experts)
Information Type		X		Structured and unstructured (rules and experience)
Types of process-related data		X		Predominantly business rules, also experiential lessons
Knowledge Intensity	X			See table 3; the process step requires some human expertise
BP Modelling	X			Detailed modelling is possible; a value stream model is existing
BP Improvement Methodologies				(The intention of the case was to improve the process in terms of knowledge management)
BP Automation		X		Process is not automated; human interaction is required at key points (decision making)
Process-related competitive advantage	X	X		Process efficiency (process time) as well as process effectiveness (minimal or no wrong decisions)
BP performance monitoring	X			Process efficiency control: Time; process effectiveness: number of complaints / disruptions in later processes

4 Application of the Research Results in the Case Organisation

Table 4 shows that the process step predominantly contains the elements of a simple procedural process as well as some of a more complex process: Human interaction and decision are necessary, but the process step itself is not really complex (knowledge-intensive), in terms of its experiential knowledge required. The integrated model suggests using traditional methods for process improvement for some its aspects, rather than the entire process.

This finding was used in the case organisation to enhance its lean management approach to BP improvement and discover further opportunities for process improvement through more appropriate management of some aspects of its process knowledge. More precisely, given the organisation's strategic orientation on, and adoption of the principles of lean management, we analysed the chosen process for waste, as per lean management (traditional approach). Then using the insights related to its process-related knowledge, the process itself was made more "intelligent" through better management of its explicit knowledge component. In practical terms, this means the optimisation by standardisation of the process steps and codification and visualisation of its explicit knowledge, relevant for each step. This also included elimination of unnecessary external information, not required for decision making.

This also meant that redundant knowledge anchors (as seen in the step "goods receipt" in figure 1) should be eliminated. We did so by simply developing and installing explicit pictured decision guidance for the acceptance of trucks, thus avoiding the need for going back on external knowledge sources. By doing so, we could reduce cycle times and process costs.

5 Discussion

Compared to a "simple" lean approach to BP improvement, our approach offers added value achieved through a systematic analysis and integration of its process-related knowledge. Our research also offers new insights on the role of knowledge, perception and decision making on process reliability and efficacy. Thus, it facilitates Business Process Management by providing a framework that supports designing reliable and at the same time streamlined processes.

From the research perspective, the most important finding is related to the nature of BP improvement methodology suitable for this type of processes. While in the past, highly structured processes were typically improved through methodologies that focused on possible automation of control flows, this case illustrates a need for a dual-type BP improvement method, not currently considered by the BPM community. While lean management focused on elimination of waste, a complementary aspect of BP improvement focused on process-related knowledge, as the key source of competitive advantage. More precisely, the case showed a clear link between improved decisions and improved processes, prompting BPM researchers to focus on decision and improved decision-making support as an important aspect of BP improvement. We argue that this could be achieved with or even without technology, through improved knowledge sharing among decision makers. However, further research is required to discover and confirm the most appropriate approaches and strategies for better integration of human-centred KM into more intensive BPs.

6 Conclusion and Outlook

The integrated framework offers a comprehensive approach for analysing the knowledge component of business processes as shown above, without stipulating or prescribing a particular course of action. Thus, it leaves room for improving the processes according to the organisation's strategy, in our case lean production combined with improved decision making. With the methodology for determining the knowledge intensiveness of a product developed by Kern and Boppert (2010), this research offered a possible approach to operationalizing one of the criteria of the integrated model. Further research is needed in order to operationalize the other criteria and confirm their applicability in practice.

Acknowledgements. We would like to thank the case organisation for its support in conducting the case study as well as the anonymous reviewers for helpful comments on the paper.

References

Amelingmeyer, J.: Wissensmanagement – Analyse und Gestaltung der Wissensbasis von Unternehmen. Deutscher Universitäts-Verlag, Munchen (2004)

Crandall, B., Klein, G., Hoffman, R.R.: Working Minds: A Practitioner's Guide to Cognitive Task Analysis. Bradford Books, MIT Press (2006)

Davenport, T.H.: Thinking for a Living – How to Get Better Performance and Results from Knowledge Workers. Harvard Business School Press, Boston (2005)

Davenport, T.H., Prusak, L.: Working Knowledge: How Organisations Manage What They Know. Harvard Business School Press, Boston (1998)

Engelhardt, C., Hall, K., Ortner, J.: Prozesswissen als Erfolgsfaktor. Effiziente Kombination von Prozessmanagement und Wissensmanagement. Deutscher Universitäts-Verlag, Munchen (2004)

Garner: Gartner EXP Worldwide Survey, Gartner Press Release (January 19 , 2010), http://www.gartner.com/it/page.jsp?id=1283413

Gronau, N., Palmer, U., Schulte, K., Winkler, T.: Modellierung von wissens-intensiven Geschäftsprozessen mit der Beschreibungssprache K-Modeler. Professionelles Wissensmanagement – Erfahrungen und Visionen, Nummer 06 (2003)

Harmon, P.: Business Process Change, 2nd edn. Morgan Kaufmann, San Francisco (2007)

Hexelschneider, A., Schlager, R.: Wissensmanagement in der Krise – Krise im Wissensmanagement? Ein Whitepaper WissensmanagementKriseWhitepaper090609 (2009), http://www.scribd.com/doc/16316124/

Jennex, M.E.: Re-Visiting the Knowledge Pyramid. In: Proceedings of the 42nd Hawaii International Conference on System Sciences HICSS-42, Hawaii, USA (2009)

Kern, E.-M., Boppert, J.F.: Value Creation by Knowledge Management – A Case Study at a Logistics Service Provider. In: Proceedings of I-KNOW 2010, Graz, Austria, pp. 218–230 (2010)

Marjanovic, O., Freeze, R.: To Appear in the Proc. of the 44th International Conference on Systems Sciences HICSS-44, Hawaii, IEEE Computer Society, Los Alamitos (2011)

Remus, U.: Prozessorientiertes Wissensmanagement - Konzepte und Modellie¬rung, Dissertation, Universität Regensburg (2002)

Smith, H., McKeen, J.D.: Developments in Practice XII: Knowledge-Enabling Business Processes. Communications of AIS 13, 25–38 (2004)

Stewart, T.A.: Der vierte Produktionsfaktor: Wachstum und Wettbewerbs¬vorteile durch Wissensmanagement, Carl Hanser (1998)

Session B
Strategic and Organisational
Approaches to Knowledge Transfer

A Conceptual Approach towards Understanding Issues in the Third Stream: Conceptions of Valid Knowledge and Transfer in UK Policy

Nicolette Michels

Oxford Brookes University, Business School, Wheatley, Oxford, OX331HX,
United Kingdom
nmichels@brookes.ac.uk

Abstract. The value of academic knowledge and knowledge transfer (KT) as part of the third stream activity within HE (Higher Education) has for some years now been regarded as important for global competitiveness and consequently a key feature of UK HE policy-making. However there remain some issues for achieving a fully-fledged third stream. Few meaningful conclusions exist regarding the issues of mismatch between policy trajectory and achievement. More in-depth understanding of how and if third stream policy is meaningful in the interpretation by various stakeholders, is argued as important for understanding policy implementation issues. This conceptual paper seeks to establish a more meaningful approach to investigating the KT policy domain. Questioning the coherency and clarity of UK policy discourse, the paper asks: how is valid knowledge and knowledge transfer conceptualised? A model for analysis and investigation of such issues is developed. Drawn from conceptions in the academic and wider literature a 'Four Metaphor Framework' categorises valid knowledge and the transfer process as: 'Transfer', 'Exchange', 'Partnership'; 'Beyond a Capitalist Transaction'. The usefulness of the framework is assessed through its application to the discourse of key documents from UK policy. The mixed metaphors revealed in policy discourse are potentially significant in the light of the gap between government aspirations and achievement. For those concerned with the issues of effective design and implementation of KT policy, this paper provides an analytical model for subsequent empirical studies.

1 Introduction

This paper will be of consequence for anyone interested in the nature of knowledge, of academic knowledge and of so-called 'valid knowledge' in the new economy. The study is of particular significance for those who are concerned with the nature and issues surrounding implementation of the Higher Education (HE) third stream agenda. Policy and its associated body of research should be

R.J. Howlett (Ed.): Innovation through Knowledge Transfer 2010, SIST 9, pp. 43–61.
springerlink.com   © Springer-Verlag Berlin Heidelberg 2011

fit-for-purpose. In this respect, this conceptual paper starts by identifying some issues in the third stream policy arena. The UK third stream objective is to 'increase the impact of the HE knowledge base to enhance economic development and the strength and vitality of society' and 'to secure long-term and adequate support for third stream activities as a significant HE function; to integrate third stream activities into every HEI in a sustainable way' (HEFCE, 2008, pg 27). This citation from the Higher Education Funding Council (HEFCE)'s updated 2006-2011 Strategic Plan, illustrates that UK policy is based on the established notion of the value of HE knowledge and its transfer. It also illustrates a policy aim of universal engagement by all Higher Education Institutions (HEIs) in the third stream knowledge transfer (KT) agenda. This policy's trajectory of universal engagement by HEIs in third stream activity is contrasted with evidence of the inconsistent involvement by all stakeholders (across different disciplines, departments, HEIs, businesses, regions and policy instruments). With the third stream considered important for innovation and for economic and social growth, the urgency of the issue increases given the current economic context and impending cuts facing the sector in the Comprehensive Spending Review of October 2010. Further, this paper critiques the range and depth of insight provided by the existing body of research into the issues of KT policy implementation. The need for more meaningful and conceptually coherent investigation into the range of policy instruments, activities, stakeholders and different perspectives is identified.

In light of the context outlined above, this paper focuses on language and narrative. It casts doubt on the clarity of the knowledge transfer discursive domain. Accepting that 'there is a relationship between the type of knowledge and its transfer' (Ozga and Jones, 2006, pg 7), this paper builds on the assumption that in policy there should be a conceptual as well as practical link between what is considered valid knowledge and the associated knowledge transfer process that it aims to facilitate. Further, taking an interpretavist perspective, this paper builds on the understanding that what counts as valuable knowledge is subject to different discursive domains This paper thus foregrounds the potential significance for achievement of policy trajectory and for research into the issues, of the clarity in policy discourse of the so-called shared values that UK policy claims to embody. The link between discourse and engagement is currently not clear but the paper asserts that better identification and understanding of KT's discursive domain(s) could be important for those involved in designing, implementing and participating in the KT agenda. This paper asks the questions: how does UK government policy conceptualise valid knowledge and the knowledge transfer process? Is it coherent? How can we achieve more clarity and consistency in understanding the KT domain?

This paper seeks to develop a conceptual framework to achieve more meaningful insight into the discursive domain(s) of knowledge transfer. A framework for analysis is thus developed embodying four different metaphors of knowledge transfer. The proposed Four Metaphor Framework is shown to draw on the evolving understanding(s) of the past few decades regarding the nature of knowledge, of academic knowledge, and of valid knowledge and knowledge creation in the new economy. Consequently it situates itself and takes in the early disciplinary-based conceptions of academic knowledge (e.g. Biglan, 1973a; 1973b; Becher,

1989) and the later evolutions into multiple, inter and trans-disciplinary knowledge conceptions (e.g. Gibbons et al, 1994; Rhoades, 2007; Godemann, 2008). It also takes in the conceptions contained in explorations and critiques in the wider literature of HE and business with respect to the sometimes divergent views regarding the nature of valid knowledge, knowledge creation and the university-industry knowledge exchange relationship (e.g. Delanty, 2001; Williams, 2007; Etzkowitz, 2008).

As a basis for assessment of this model, the Four Metaphor Framework is applied in this paper to an analysis of KT discourse. Key policy documents issued by the relevant departments and agencies associated with UK university KT and the third stream are the subject of this analysis. The nature and coherency of the conceptions of valid knowledge and knowledge transfer in UK KT policy are critiqued. Further, the implications of this for policy effectiveness, and the possible links between discourse and the issue of engagement are illustrated. The paper concludes with consideration of the usefulness of the framework for supporting the identified need for further empirical investigation of different perspectives and dimensions in the KT policy agenda whether of region, HE institution, business stakeholders, individual, or policy instrument.

2   Issues in the Third Stream

2.1   Valid Knowledge and the Trajectory of Universality of the Third Stream

The central role of universities in knowledge creation and hence in policy discourse is deemed to date back to the agreement of the European Councils of Lisbon and Barcelona in 2000 and 2002 respectively (Mayo, 2009). The notion of university knowledge as a driver of innovation and business production (Nonaka et al, 1995, cited Geuna and Muscio, 2009), and the focus by most national governments on the pivotal role universities can perform for enhancing their country's competitiveness in the 'global knowledge economy', means a well-established marker has been put on the contemporary value of university knowledge. Acknowledging Foucault, Peters suggests that it is now 'impossible to pursue the question of knowledge separately from the question of capital' (Peters, 2001, pg 17). From this has emerged the 'Knowledge Transfer' agenda (or 'third stream') as the 'third mission' (Smith and Taylor, 2009, pg 13) and 'second revolution' for academia (Etzkowitz, 2008, pg 30) alongside and in addition to the 'first mission' of teaching and the 'second mission' and 'first revolution' of research (Geuna and Muscio, 2009, pg 94). The UK government's adoption of this narrative is evident in the series of White Papers, policy statements and initiatives that have emerged since the start of the new millennium. Indeed, the narrative in the 2003 White paper on the future of Higher Education (BIS, 2003) centres on the 'critical role' HE will play to support change in the knowledge economy. Five years later, the positioning of HE's 'leading role' in the 'national innovation ecosystem' remains evident (HEFCE, 2008, pg 28).

Dedication by UK HE policy-makers to the third stream agenda extends to a stated aim of achieving greater economic and social impact of the HE knowledge base through more involvement in third stream by all stakeholders (HEFCE, 2008). A policy trajectory of universal engagement is evident from HEFCE's strategic plan. A contribution from all HEIs is expected. The long-term aim is 'to secure long-term and adequate support for third stream activities as a significant HE function: to integrate third stream activities into every HEI in a sustainable way' (HEFCE, 2008, pg 27). Policy acknowledges differences between types of HEI, types of knowledge and types of sectors, but nevertheless universality of engagement with policy objectives is assumed possible, allowing for 'institutional autonomy within a framework of shared values and goals' (BIS, 2009, pg 1).

2.2 Understanding Inconsistency of Engagement in the Third Stream

Against a policy expectation of greater engagement in knowledge transfer it has been noted that in general, universal engagement in KT is however not materialising (Urwin, 2003; Pilbeam, 2006). 'Supply' and 'demand' issues in the third stream agenda are acknowledged, and deciding its principle responsibility lies supply side (with academia), HEFCE considers the lack of engagement of HEIs and academics to be a key risk to the achievement of a fully-fledged third stream (HEFCE, 2008, pg 27). KT is considered by some to provide financial support as much for HEIs as for industry. In light of the contemporary economic context and drastic changes to HE funding with the October 2010 Comprehensive Spending Review, these facts are of particular concern.

Those of us involved in almost any aspect of the Knowledge Transfer agenda, will know from personal experience that KT is an exciting but thus far not well-established nor easy agenda to be involved in. Whether we are policy agencies working regionally, managers at HEI level trying to facilitate implementation of the agenda, academics engaged in the activity itself of transferring knowledge, or industry-based partners, all can testify to the challenges of KT. The author's own experience in this area has played a not insignificant factor in the pursuit of the issues in this paper. Over ten years personal involvement in various roles and projects under the KT banner has included: consultancy projects; contract research; student in-company projects; careers fairs; industry guest speakers; Knowledge Transfer Partnerships (KTPs); consultancy; coaching and mentoring services; diverse projects funded through the Higher Education Innovation Fund (HEIF). One may note the range of activities which KT can encompass. Further however, the volume and diversity of KT activities taken on by one individual is indicative also of the relatively limited engagement of other academic colleagues who could have but for whatever reason did not take on some of those projects (and who arguably in some cases might have been better suited in terms of relevant academic expertise). Semi-formal involvement at faculty level within one HEI as KT champion to try and engage more colleagues in KT has provided first-hand evidence of lack of numbers of academics engaging as a key feature of the current KT agenda at institutional as much as national level. This involvement has given insight into the activities and attitudes of different institutions, different academic faculty, departments, and disciplines, as well as different policy-agents and businesses. This personal insight confirms that engagement in KT is not non-existent but certainly not universal either.

Investigation of heterogeneous engagement with third stream activities across and within regions, individual institutions and disciplines has been the subject of some investigations. Conclusions confirm inconsistent levels of participation with tentative suggestions for a closer look at factors associated with individual, departmental, disciplinary, institutional and regional differences (Urwin, 2003; Ozga and Jones, 2006; Pilbeam, 2006; Geuna and Muscio, 2009). Analysis of a review of the body of research into the third stream reveals however a tendency of this research to focus on the academic domain (Agrawal, 2001). Further, even with this HEI-focused research, the diverse nature and focus of the studies which take different definitions of the term third stream, makes meaningful comparison and insight difficult.

2.3   The Policy Value Framework and Coherency of Discourse?

If third stream policy is intended to provide a framework for innovation (DIUS, 2008b) as well as of 'shared values and goals' (BIS, 2009, pg 1), then the difficulty one has in actually defining 'the third stream' is significant and arguably precedes the subsequent question about the clarity of definition of this policy's underpinning framework of values. The third mission has emerged as a loosely defined term, a fact which the broader KT policy discursive domain appears to have no trouble with. Various terms are used interchangeably and include 'third stream', 'third mission', 'knowledge transfer', 'knowledge exchange', 'partnership', 'enhanced contribution' to name a few. These appear to encompass a broad but nowhere concretely defined set of activities. The author's own engagement in the diverse set of activities listed earlier, and the difficulty in choosing which term to use throughout this paper, is illustrative of the situation. The possibility of misinterpretation by the reader of the author's meaning when referring to KT and third stream is also illustrative of the issues explored here.

The UK government appears to regard all types of university knowledge as valuable capital as evidenced by the contribution it states it envisages from all HEIs to KT, both research-intensive and non-research-intensive (BIS 2003), and from all academic disciplines ('the full range of HE subjects': HEFCE, 2008, pg 31). However, by apparently encompassing and celebrating heterogeneity, there is also a risk of ambiguity and lack of coherence which Ozga and Jones (2006), citing Allan Luke (2003) suggest is problematic.

Fowler and Lee, draw on Foucault to express their concern with the different interpretations they note within policy of what can be counted or delimited as 'within the true' (Foucault, 1985, pg 7 cited Fowler and Lee, 2007, pg 184). As is explored in more detail later in this paper, their interpretation that some policy discourse and certain policy instruments do not seem to cater for certain types of valid knowledge domains does not sit well with the so-called universality of KT policy. But Fowler and Lee's observations also highlight that what counts as valuable knowledge ('within the true') is subject to different discursive domains. Multiple discursive domains poses potential issues for policy implementation. In the KT domain we have already understood there to be diverse stakeholders and the potential importance of different types and levels of implementation. One may contemplate the link between coherency in policy discourse to issues of convergence, divergence and local inflection of KT policy (Ozga and Jones, 2006).

The observations of Fowler and Lee foreground for this paper the significance in policy-making of clarity, coherency and relevance of what is considered valuable. It goes without saying that accuracy, consistency and alignment of conceptions of types of knowledge and policy measure is important. The coherence and relevance of the value framework underpinning policy are significant if as Ozga and Jones point out, we have come to understand that there is a 'relationship between the kind of knowledge and its transfer' (Ozga and Jones 2006, pg 7). Thus, the significance of different discursive domains and of local inflection and implementation of policy, highlights the potential problem of policy which may not be fit-for-purpose or perceived as such. The Technology Strategy Board (TSB)'s stated view is that innovation is a choice (TSB, 2008). With a significant number of academics apparently choosing to continue to prioritise teaching and research, the concept of a shared third mission, of a shared framework of values and goals, and ultimately the relevance of the notion of a second revolution are brought in to question (Peters, 2001; Geuna and Muscio, 2009). Understanding the different conceptions of knowledge and hence also of knowledge transfer is an important consideration for policy-makers trying to engage and enhance different stakeholders' efforts in the transfer of their knowledge. Conceptualisations of knowledge and of knowledge transfer in the discursive domains of UK knowledge transfer policy are worth further investigation. We must start with gaining greater clarity in understanding the conceptions underpinning policy discourse itself. Specifically: How is knowledge and valid knowledge conceptualised in UK policy discourse? How is the knowledge transfer process conceptualised? What is the overall narrative? Is it coherent? Is it contradictory? Further, can we establish a conceptual framework with which to achieve clarity and consistency of analysis and understanding?

3   Analysing Conceptions of Knowledge and Knowledge Transfer

3.1   Evolving Conceptions of 'Valid Knowledge': Four Metaphors

From the evolving debate over the last few decades regarding the nature of knowledge, of academic knowledge and of valid knowledge in the new economy, two not mutually exclusive strands of conceptualisation have emerged which permeate the KT policy discursive domain regarding 'types of knowledge' and ultimately of 'valid knowledge'. Firstly as triggered by the likes of Biglan (1973a, 1973b) and Becher (1989), is that academic knowledge and its associated impact on attitudes and behaviours may be perceived as distinct and definable along (academic) disciplinary structures. Secondly, explored initially by Gibbons et al (1994) and later by the likes of Godemann (2008) and others, is that 'valid' knowledge is variously regarded as situated and produced: either within the academic domain (largely disciplinary-based); or outside the academic domain ('practical'); or/and as produced through various combinations of knowledge of different types and from different sources (inter-disciplinary, multi-disciplinary; trans-disciplinary). It is recognised that the focus on these explorations of academic knowledge does not pursue the post-modern perspective which would argue for much greater consideration than is possible here of social practice and power dimensions (Trowler, 1998).

Drawing on this evolving body of debate and research, an analytical framework can be established. This framework for analysis of the KT policy discourse is structured around categories of valid knowledge and its transfer which draws inspiration from Godemann (2008), but it is here represented under four metaphors:

- Knowledge Transfer as 'Transfer' of knowledge from expert 'knowledge base' to consumer (largely disciplinary-based)
- Knowledge Transfer as 'Exchange' of knowledge between different parties but ultimately each knowledge domain remaining intact/unintegrated (inter-disciplinary)
- Knowledge Transfer as 'Partnership' or 'Co-creation' recognising equality of validity of different knowledge sources and resulting in 'new knowledge' (multi or trans-disciplinary)
- Knowledge Transfer as 'Beyond a Capitalist Transaction', for the 'Greater Social Good', with wider social resonance

The coherency of third stream policy discourse and the potential implications for KT policy effectiveness may be considered by applying this Four Metaphor Framework to UK policy discourse. The usefulness of the framework itself for investigation of the KT policy domain is thereby also able to be assessed. Thus the proposed Four Metaphor Framework is applied below to analysis of the discourse of key texts emanating from key UK policy agents since the 2002-3 Lisbon and Barcelona Councils referred to earlier: Department for Business, Innovation and Skills (BIS); the Department for Innovation, Universities and Skills (DIUS); the Higher Education Funding Council (HEFCE); the Technology Strategy Board (TSB). 'Policy text' in this paper is analysed but it is based on a non dichotomous approach to policy 'as text' and 'as discourse' (Ozga, 2000). Key documents issued since the White Paper 'The Future of Higher Education' (BIS 2003) are reviewed from the following policy agencies:

4   UK Policy and the Four Metaphors of Knowledge Transfer

4.1   The Value of Academic Knowledge

The influence of a Biglan-inspired, disciplinary-based viewpoint on the conceptions of knowledge, and specifically what is counted as valuable knowledge in the
Foucault sense is evident in KT policy and in stakeholder interpretation and response to this policy. Disciplinary conceptualisation of knowledge is of course
integrated into and underpins much of the way academia conceptualises and structures itself and the suggested influence of disciplinary-based epistemology on attitudes and behaviours (Biglan, 1973b; Becher, 1989; Neumann et al, 2002), may of
course equally extend to associated value judgements about knowledge. Biglan-
type categorisations appear to underpin policy discourse which attempts to deal
with HE knowledge and HEI heterogeneity. Hence we see in policy text (e.g. The
Lambert Report: HMSO, 2003) and from commentators on policy (e.g. Van
Vught, 2009): definition of HE knowledge transfer as 'hard' and 'soft', where
hard KT involves IP, technology transfer and spinouts and soft KT is about networks, sponsored students, contract research, collaborative research and consultancy (BIS, 2003). Significantly Lambert's discourse appears to attach equal value
to both hard and soft types of KT. In terms of 'what' is being transferred or exchanged, the department of Business Innovation and Skills (BIS) appears to differentiate between 'pure' and 'applied' and attaches equal value to 'cutting-edge,
internationally competitive research' on the one hand and 'technologies, knowledge and skills development' on the other (BIS, 2003). Trying to accommodate
and even integrate diversity through acknowledging that different types of knowledge of equal value exist in different types of HEI is reflected in the distinction
between research-intensive and non-research intensive HEIs (BIS, 2003). This
perhaps underpins the later discourse that values and expects each HEI to have its
own distinct third stream mission (BIS, 2008).

But coherency in the implied equality of value attributed in policy discourse to
all types of knowledge (BIS 2003; HEFCE, 2008); and the problems of heterogeneity and range of interpretations about what is valuable and transferable academic knowledge in what Urwin (2003) calls a 'wide-ranging moniker', appears
for some to be easily undermined. A disciplinary-based viewpoint appears to lie
behind many critics of UK KT policy who express concern at bias in policy fa-
vouring one conception of valid academic knowledge over another. In particular
the divide between science and the humanities and between pure and applied disciplines appears inflamed by the KT agenda and by KT policy ambiguity. Fowler
and Lee's (2007) criticism of KT policy's 'intellectual inadequacies', argued in
the context of the pure/applied continuum and evident also in the discussion of the
likes of Ozga and Jones (2006) and Smith and Taylor (2009) in the context of the
soft/hard dimension, largely consider policy discourse to favour the codified,
paradigmatic knowledge of the sciences and pure subjects over the non paradigmatic knowledge of the humanities and arts. Even Gibbons et al (1994), who question the singular value of peer-reviewed, discipline-based, Mode 1 knowledge, are
like other commentators on the role of universities in a new knowledge economy

(e.g. Etzkowitz, 2008), complicit in seeming to perpetuate a science-orientated focus. This science bias stems from their heavy, albeit acknowledged reliance on a sciences context arising within the tech transfer era. Finding little link between mode 2 knowledge and 'the values and practices' of the humanities (Gibbons et al, 1994, pg 90), and Biglan's original presentation of the social sciences as still striving to find themselves a paradigm, does little to help an implicit value-laden narrative of paradigmatic-based knowledge as the only kid on the block.

A policy bias towards certain types of knowledge (here certain disciplines) creates problems. If such a bias exists there is a risk of inappropriate policy structures and measures for recognising and facilitating the transfer of knowledge which is deemed by the disadvantaged knowledge providers as (equally) valuable. Indeed, Urwin's (2003) suggestion that non-research intensive HEIs cannot engage in meaningful knowledge transfer highlights discord with the 'universality' trajectory of policy discourse and signals potential problems of universal engagement either perceived or otherwise at HEI as well as at discipline level. Likewise Furlong and Oancea's (2006) attempt to find meaningful measures of quality for applied as opposed to only pure research, and Hammersley's subsequent critique (2008), reveals discomfort with a system which struggles to convince all stakeholders of its unbiased application to all disciplinary knowledge bases. Policy-makers both by funding the Furlong and Oancea project and in the stated policy aim of wanting to do more with the creative, media and culture industries (HEFCE, 2008), appear to be recognising the problems of unequal efficacy of current policy. Admission that what is currently only able to be defined as 'hidden innovation' (DIUS, 2008b) means that more needs to be understood about 'the meaning of knowledge exchange in these newer areas' (BIS, 2008, pg 31).

Biglan's discovery of the apparent higher levels in applied subjects of 'social' and external 'connectedness' and their greater propensity for engagement in 'service' activities, and the relative ease of 'collaboration' of the hard disciplines, appear to show encouraging similarity of some disciplinary preferences (possibly applied and hard) to the nature of KT activities. A disciplinary-based perspective may suggest that some knowledge types are more suited to KT than others, and appears to emerge in the work of Etzkowitz and Webster. Pilbeam's discipline-based mapping of third stream seeks to test this further (Etzkowitz and Webster, 1998, cited Pilbeam, 2006), but ultimately does not appear to bear out any correlation: a disciplinary link has not proven conclusive.

4.2 The Value of Academic Knowledge and the Transfer Metaphor

The mere difficulty of defining disciplinary categories and boundaries in an academic domain of ever more fragmented and numerous sub-disciplines, may mean disciplinary boundaries are dissolving (Gibbons et al, 1994), but still leads some to attempt to fall back on broader sweeps of categorisation which refer loosely to the 'two great academic cultures' of natural sciences and humanities (Godemann, 2008, pg 627). This attempt to simplify the issue still reflects a tendency for a non homogeneous view of knowledge 'as portrayed by the knower' (Becher, 1989). Different discursive domains which lead to apparently 'competing knowledges' (Fowler and Lee, 2007, pg 182), have created and perpetuated a divisive, oppositional discourse. Thus we have become familiar with not just the division of knowledge as hard/soft and pure/applied, but also as formal/informal (Fowler and Lee, 2007), explicit/tacit (Polanyi, 1958 cited Williams 2007), and the differing associated behaviours of these opposites, presumably equally relevant for KT as for teaching and research as: didactic/interactive (Neumann et al, 2002) and positivist/ phenomenologically-inflected (Fowler and Lee, 2007). Clearly for some academic stakeholders, including and especially perhaps in response to third stream policy discourse, there is an urge to define and assert the merit of what does not seem to be incorporated in another party's (i.e. policy) conception of valid knowledge.

All of the debates charted above trying to find ways to recognise the value of different types of academic knowledge, nevertheless serve to reinforce the conception of the over-riding 'value' widely attributed to university academic knowledge which underpins the KT policy agenda. Policy articulates that 'in a knowledge economy, universities are the most important mechanism we have for generating and preserving, disseminating and transforming knowledge into wider social and economic benefits' (BIS, 2009; pg 2). This narrative, which conceives universities as central in knowledge and knowledge production, is critiqued by some scholars for what they consider to be an inappropriate discourse in KT policy of a conception of 'expert towards consumer' (Fowler and Lee (2007). The term 'knowledge transfer' itself is critiqued for its linearity (Gibbons et al, 1994, pg 9). Talk in policy of demand and supply and the path from laboratory to marketplace (DIUS, 2008b) reinforces this conceptual linearity in policy. The humanities and applied subjects in particular detect in policy a narrative with 'uni-directional movement' which belies a 'positivist stance' and hence what they regard as 'legitimisation of formal, evidence-based scientific knowledge' (Fowler and Lee, 2007, pg 182). Calls for recognition in policy of the value of the social and interactive dimensions of their subject areas' contributions might suggest reasons for lack of engagement due to policy mismatch. The discomfort of the humanities and applied subject domains with linear, paradigmatic conceptualisation of knowledge and transfer in policy discourse mirrors or perhaps signals the changing understanding about the nature of valid knowledge itself.

4.3   The Value of Inter-disciplinary Knowledge

In the post-modern era, the relevance and over-simplicity of Biglan's disciplinary categorisation of academic knowledge and what Trowler categorises as the associated epistemological essentialist approaches have been acknowledged for some time (Becher, 1989; Trowler, 1998). The fragmentation and proliferation of multiple sub-disciplines as referred to previously, are indicative of tension about relevance of conceptualisations of certain knowledge bases and discursive domains for differing contexts. In the social-constructivist era, we would expect and indeed now see the impact on policy conceptions of valid knowledge of the post-modernist attitude to identity which Delanty charts as moving 'in favour of multiplicity and heterogeneity' (Delanty, 2001, pg 4). Becher of course acknowledged at the time the changing nature of knowledge and the limitations of categorisation, and not surprisingly therefore criticisms of policy as out-of-touch which stem from a discipline-based, bi-polar and categorisation-based perspective are arguably themselves likewise dated. In actual fact policy critics drawing on examples which appear disciplinary-based (e.g. arts and humanities and applied healthcare), are arguing for the recognition of the existence and value of knowledge which is not currently conceived perhaps as academic knowledge at all, embodying a conceptualisation of knowledge which is more 'socially-distributed' (Gibbons et al, 1994, Pg 17). Here they illustrate if not entirely divorced from association with a disciplinary and oppositional stance, a discursive domain which recognises and values knowledge beyond traditional boundaries. Importantly, although Gibbons et al struggled to find links between the values and practices of the humanities and their mode 2 knowledge, their conceptions of valid knowledge and hence knowledge transfer envisaged an activity crossing university boundaries which found resonance with humanities scholars.

Interestingly however, in calls for representations in policy of the value of knowledge transgressing (but in some cases also mirroring those of) the traditional academic categorisations, an oppositional, 'bi-polar' (Williams, 2007) conceptualisation of knowledge in discourse appears to continue. So we see knowledge categorised as: propositional/procedural (Schon 1963, cited Williams, 2007); just in case/just in time (Moe et al Lau, 1999, cited Williams 2007); tacit/explicit; contemplative/performative (Barnett 2000); scientific knowledge/natural knowledge; lay/professional (Delanty, 2001); academic/practical (Godemann, 2008). But for our argument, the question about post-modern conceptualisations of knowledge is whether the multiple knowledge sources are of equal value. The so-called broken unity of knowledge, the post-modern heterogeneity of knowledge (Delanty, 2001) and the value of 'heterogeneous teams' (Weinert, 1998, cited Godemann, 2008, pg 635) which are all part of the new knowledge economy value system, appears to envisage and recognise 'new sites of knowledge production' (Delanty, 2001, pg 103) with boundaries between university and the outside world becoming what Rip calls 'porous' (Rip 2002a, cited Lazzaretti and Tavoletti, 2005, pg 491).

4.4   The Value of Inter-disciplinary Knowledge and the Exchange Metaphor

The move away from a conceptualisation of valid knowledge as 'univalent' towards 'polyvalent' (Etzkowitz, 2008, pg 32), and Gibbons al's Mode 2 or transdisciplinary conception of this new knowledge, results in a critique of the 'transfer' metaphor which is deemed indicative of a conceptual separation of production and solution. This it is argued is an out-moded view of valid knowledge based on an out-moded view of innovation as linear (Gibbons et al, 1994, pg 9). With 'the locus of added-value in innovation having shifted' (Gibbons et al, 1994. pg 46), mode 2 knowledge is deemed to feature heterogeneity and diversity in the group,

social accountability to multiple stakeholders, reflexivity and quality control beyond the academic peer group, knowledge produced in practice and in context not
from an already existing structure, and disseminated through practice. In this case
the 'Exchange' rather than 'Transfer' metaphor would seem more appropriate for
policy trajectory.

Policy discourse appears to acknowledge the innovation process as being more
'open', of the research base sitting apparently 'along side other sources of knowledge like large companies, SMEs and users' (BIS, 2009, pg 6), and its policy role
as bringing everyone together (DIUS, 2008b). Noting that 'not all knowledge can
be codified and innovators are helped by interaction' (DIUS, 2008b, pg 8), the
notion of blurred boundaries seems to exist. But the ambiguity and mixed messages of policy discourse remain evident in the varied use (as referred to earlier) of
'transfer' and 'exchange'. 'Interdisciplinary knowledge' as defined by Godemann
distinguishes this conceptualisation of combined knowledge from multiple sources
as ultimately reinforcing or building upon the existing knowledge domains: 'an
attempt to counteract specialisation in the academic system, whilst nevertheless
remaining loyal to the disciplinary structures' (Godemann, 2008. pg 628). Knowledge 'Exchange' in this case remains a useful metaphor. BIS's assertion that there
is 'no question of compromising pure research' (BIS 2009, pg 10) suggests even
with knowledge exchange and lowered boundaries, in policy conceptions the academic knowledge base remains largely untouched by exchange with other knowledge sources.

4.5 The Value of Trans-disciplinary Knowledge

The democratisation of knowledge (Delanty, 2001) where 'post-disciplinarity
takes over', (Turner, 1999, cited Delanty, 2001, pg 3), is particularly relevant perhaps for the concept of HE KT. Changes in conceptions of what is valid now arguably lead to an undermining of the notion of what Williams calls 'the university
as the sole authority in creation, validation and dissemination of knowledge' (Williams, 2007, pg 514). The argued increasing dominance of the market in the HE-
industry-market power nexus, the decline of the donnish dominion and value of
subject-based knowledge, it is argued lead to an end of the singular value of peer
review with a move to validity as utility (Williams, 2007).

Whether the inclusion in policy discourse of acknowledgement of different
types of knowledge really encompasses and values equally practitioner knowledge
for example is not clear and of course this because of the question it raises in this
case about the contribution and value of HE. The stated aim to establish multi-
disciplinary research centres (BIS, 2009) may be the nearest we get to policy apparently according equality of value of different knowledge types but ultimately
appears to continue to retain an academic-orientated discourse. Equality of value
of all types of knowledge (university and non university) is understandably perhaps not readily evident in the discursive domain of HE policy because of the implications of what Williams calls 'the end of the monopoly of universities on valid
knowledge' (Williams, 2007, pg 514). However, Williams' interpretation and focus on the power of the market arguably misses a key point about developments in

new knowledge conceptions. Trans-disciplinary knowledge is conceived to be about more than the existence of multiple sources of equally valid knowledge. Godemann's (2008) distinction between inter-disciplinary and trans-disciplinary knowledge, conceives trans-disciplinary knowledge as more than that which Gibbons et al conceived of as trans-disciplinary (theirs being multi-disciplinary knowledge through application and solution to 'external' problems). Godemann conceives trans-disciplinary knowledge to be that which has value because it has *integrated* knowledge from the diverse sources, 'freed' from the interests and methodologies of the discipline, and resulting in the creation of new knowledge beyond the contributing knowledge base(s). Godemann's association of trans-disciplinary knowledge with a totally new, transcendental 'vantage point' (pg 628) beyond academic methods and structures, seems to expand beyond the fragmented nature of context-specific trans-disciplinary activities (Delanty, 2001) and the transience noted by Gibbons et al in such time-limited problem-solving activities.

4.6   The Value of Trans-disciplinary Knowledge and the Partnership Metaphor

Godemann's assertion of the importance in trans-disciplinary knowledge of arriving at a 'meta knowledge of common ground' suggests the group that is engaging in knowledge sharing, exchanging and creating, decides what counts as 'within the true'. Godemann's emphasis on assimilation by all stakeholders of knowledge from each other and 'common group ground' (pg 632) echoes the 'intertwined' boundaries and link to 'meta innovation' of Etzkowitz (2008, pg 145). No single, agreed definition of mode 3 knowledge has thus far emerged. Whilst the educational development context of Rhoades and Slaughter's mode 3 knowledge (Rhoades, 2007) did recognise changing modes of production and a combination (matrix) of different knowledge contributors including non academics, their conceptualisation appears more multi-disciplinary then trans-disciplinary in orientation with its focus on a matrix of separate expert sources internal to the HEI (albeit including non academic sources and in response to external factors). One suggestion from Ray and Little (2001) in the management literature for mode 3 to be defined by 'group tacit knowledge' arguably has more resonance with the notions of common ground, integration and new vantage point. In light of this, policy discourse which declares a stated aim of consensus between 3 parties (BIS, 2009) appears appropriate, and 'Partnership' or perhaps 'Co-creation' better metaphors and an advancement on 'Exchange' for trans-disciplinary knowledge.

Godemann's assertion of the importance in trans-disciplinary knowledge of each contributor knowing and defining what it is contributing in order to be able to arrive at the 'meta knowledge of common ground' may be significant. In this context what is the university's 'capital'? What is its valuable role? Whilst Delanty (2001) defines the role and valid contribution of the university in this new era as the dispenser of credentials, and arbiter of cultural capital and ultimately acting as a site where knowledge and culture interconnect which celebrates dissensus rather than consensus, the lack of clarity in policy discourse as to the nature of the contribution of university knowledge in the KT process is significant. Godemann would presumably argue that all parties, including the academic stakeholders need to understand their role and contribution and hence if not articulated already, to go and find out in order be able to play a meaningful part in establishing the common ground. The lack of clear and definitive statement about this in policy discourse is of course noteworthy.

Looking to practice, Sylva et al offer an example of collaborative research which challenges the idea of knowledge as owned by any one party (as exemplified for example by patents in the 'old way' of university KT), and posits what they call the 'new way' of 'partnership working' which they suggest see an 'unusual equality' of practice versus research (Sylva et al, 2007, pg 159). Appearing to echo Godemann, assimilation of each partner's viewpoint and joint ownership are noted by that research team, but also the role of the university-partner to be one of providing academic 'integrity'. In contrast to Williams (2007), this appears to attribute to the university a specific, defined, valid contribution to the knowledge-creating equation. As has been noted, Godemann's new mutually arrived-at 'vantage point' has some similarities with Gibbons et al's association of trans-disciplinary knowledge in that it conceptualises no requirement to return to the discipline for validation and the significance for this type of knowledge production of a socially-distributed network. Although we cannot not pursue this in this paper, acknowledging the inherent implications of the significance of the power dimension in achieving such a trans-disciplinary group situation would resonate with the social-constructivist and post-epistemological perspectives of the postmodern era (Trowler, 1998).

4.7   Value beyond an Economic, Capitalist Transaction: The Social Good Metaphor

Significantly perhaps, Godemann's trans-disciplinary knowledge goes beyond the discussion of Gibbons et al on 'instrumentalisation', and also envisages a broader relevance than Peter Williams' conceptualisation of validation of knowledge by the 'the user' which seemed to reduce the value and validity of knowledge to a single (industry consumer) discursive domain. Further, Godemann's conceptualisation echoes but extends the call from the humanities scholars for policy to recognise the wider societal dimension of HE knowledge transfer (Ozga and Jones, 2006), linked back perhaps to Stehr's reference to the same (Stehr 94 cited Peters, 2001, pg 5). Developments in understanding knowledge which leads Ozga, and Jones, (2006) to focus on lack of recognition in policy of 'sticky knowledge', supports and dovetails with another suggestion for mode 3 put forward by Huff and Huff (2001, pg 49) as concerned with 'the human agenda' (i.e beyond economic/business). Interestingly, Rhoades' (2005) call for a more inclusive, democratic academic republic in HE governance, which envisages external 'partnerships' including with non profit-making bodies for 'community development' and 'the public good' starts to resonate with these other conceptualisations of the HE-industry nexus as something more than a capitalist transaction.

4.8 Mixed Metaphors

The presentation of the dual ideologies of economic and social benefit as 'unproblematic co-habitees' in official discourse has been noted (Fanghanel, 2007). In policy discourse things are not clear-cut in spite of this stated dual purpose of social and economic contribution. For example, 'third stream' may denote 'income stream', and consequently income generation for the HEI. Certainly this economic dimension seems for some to be interpreted as a defining and possibly definitive feature of KT (as for Pilbeam, 2006 and Etzkowitz, 2008). But this commercial dimension is not overtly articulated in HEFCE's 2006-2011 KT strategy discourse (HEFCE 2008). Indeed a stated focus on 'social impact' dominates in that policy text. But by contrast, a later KT-focused section of policy text from BIS which refers to the context of new funding challenges for HE does adopt some elements of the commercial narrative, albeit indirectly rather than overtly' (The Future of HE in the Knowledge Economy', BIS, 2009). The ambiguity and inconsistency of the policy narrative across two different documents, issued by two different policy agencies may point to the significance of the different meso levels of interpretation and implementation. However, either way, overall, the metaphor of 'the greater good' beyond the commercial transaction seems difficult to define and locate in KT policy discourse.

Heterogeneity leading to mixed metaphors and mixed messages continue to confound universality and coherency of UK KT policy but importantly also the ability to conduct meaningful analysis. For example, Pilbeam's (2006) interesting attempt, referred to earlier, to analyse KT across and within institutions and academic disciplines takes an economic interpretation, with a particular definition of third stream based on income (research income from government, charity and business), whilst Francis-Smythe (2008) looking at barriers to engagement in KT, by including Knowledge Transfer Partnerships, includes some HEI-income generation dimension but does not clarify what else a 'responsibility for KT' of the respondents actually involves (which might for example focus more on relationship-building activities than income generation). Similarly, in their work looking at KT across HEIs in one region (Scotland) through a comparison of web-based information, Ozga and Jones (2006) acknowledge the limitations of such research based on a 'currently visible institutional view of the KT agenda'. But even so, the relevance of their research difficult to place given the fact that it probably incorporates many different non comparable types of KT activity. The co-habitation in KT policy discourse and in the research and critique of this policy agenda of the social and economic ideologies is far from unproblematic.

5 Conclusion

UK knowledge transfer policy appears to aim for universal engagement in the third stream. Inconsistent engagement within and across different disciplines, departments, HEIs, regions and policy instruments is confounding the second revolution. If government plays a 'critical role' for providing a 'framework' for innovation but at the same time expects and values the heterogeneity of

HEI-specific missions, then policy which supports 'institutional autonomy' must indeed be set 'within a framework of shared values and goals' (BIS, 2009, pg 1). But for values and goals to be shared, what is valued must surely firstly be defined for it to be understood: Highlighted is the importance of clarity of definition of the framework of so-called shared values, and the possible significance for interpretation, engagement and implementation of this policy of alternative discursive domains across the heterogeneous HE sector in terms of conceptualising the nature of innovation, knowledge and valid knowledge and by extension the role of the university and the academics therein.

UK KT policy analysed from the standpoint of evolving developments in the understanding and exploration of different conceptions of the nature of knowledge, of types of knowledge, of academic knowledge, and of valid knowledge in the new economy, reveals in policy discourse a patchwork of different conceptions which appear to offer little coherency in conceptual underpinning. Two strands, not mutually exclusive, regarding conceptualisation of types of knowledge emerge: one which appears to suggest 'academic discipline' remains embedded in much conceptualisation of valid knowledge in the sector, and the other which reflects the increasing opening up of the innovation equation to other, non academic, multiple sources of knowledge. Thus we see heterogeneity potentially confounding clarity: whilst discourse and terminology used therein suggests policy-makers are trying to acknowledge the value for example of both 'hard' and 'soft' knowledge, 'pure' and 'applied' and similarly 'cutting edge research', as well as 'technologies and skills development', some stakeholders (for example certain academic disciplines and certain HEIs) appear to find themselves alienated from policy both conceptually and in practice, arguing for example that certain processes or measures do not capture the value they can contribute. Equally, conceptualisation of the knowledge transfer process in policy discourse appears to be characterised by heterogeneity: the interchangeable use of 'transfer', 'exchange' and 'partnership' and related terminology such as 'supply' and 'demand', 'collaboration' and 'integration', does not provide a consistently clear conceptual statement about the role of the university and what it is envisaged as bringing to the knowledge innovation process. Further, the possibility of knowledge transfer having an objective of greater good and social impact beyond or even instead of a commercial transaction adds to the patchwork of conceptions. Breadth and heterogeneity in policy-making regarding the nature, value and contribution of academic knowledge reveal and promote potentially very different messages about the KT agenda.

Whether these mixed metaphors dilute understanding and clarity for all potential stakeholders of what is valued remains to be explored. What has been acknowledged is that what counts as valuable knowledge is subject to different discursive domains. Identifying, understanding and defining the different interpretations of different stakeholders to the alternative conceptions, metaphors and related terminology portrayed above could be important for all of those involved in designing, implementing and participating in KT in HE. Context is important for understanding how knowledge transfer works (Ozga, 2004) and some greater focus in this respect would be helpful. A key issue that arises out of the attempt by some to put forward suggestions about different meso and micro level conceptions and inflections in KT, is the lack of meaningful comparable data with a KT policy which encompasses so many different interpretations of knowledge and hence so many different permutations of knowledge transfer. Lack of data regarding the level of engagement in third stream activities (Geuna and Muscio, 2009), but also the nature of engagement, but more importantly the lack of definition and hence comparative data in this area hinders meaningful discussion and hence policy-making, as illustrated by the interesting research but wide-ranging assumptions and interpretations underpinning the dedicated studies of a few scholars referred to earlier (Pilbeam, 2006; Francis-Smythe, 2008; Ozga and Jones, 2006).

We need to better understand the heterogeneity of KT but also through research at meaningful and definable discursive meso and micro levels of for example: policy instrument; local policy agent; institution; faculty/school; academic department; individual. Agrawal's (2001) review of KT research and even the conclusions in that paper, exposes the perhaps understandable tendency in KT research to focus on the perspective of the academic stakeholders. We also need to take this sort of research into the discursive domains of the so-called demand-side of those involved in the KT agenda: customer/industry stakeholders. The conceptual framework drawn up in this paper which captures the different views of types of valid knowledge as academic/disciplinary or/and beyond/non-academic, and the resultant framework and underpinning assumptions of the three/four metaphors of the knowledge transfer process is a useful tool in this respect for further research. Applying this conceptual framework to analysis of the discourse of the various domains involved in the knowledge transfer agenda will start to produce understanding of the extent to which current KT policy discourse is meaningful in the interpretation and engagement of various stakeholders in the KT agenda, and ultimately will help inform consideration of the design and implementation of policy initiatives.

References

Agrawal, A.: University to industry Knowledge Transfer: literature review and unanswered questions. International Journal of Management Reviews 3(4), 285–302 (2001)

Becher, T.: Academic Tribes and Territories. Society for Research into Higher Education and Open University Press, Buckingham (1989)

Biglan, A.: The characteristics of subject matter in different scientific areas. Journal of Applied Psychology 57, 195–203 (1973a)

Biglan, A.: Relationship between subject matter characteristics and the structure and output of university departments. Journal of Applied Psychology 57, 204–213 (1973b)

BIS, The Future of Higher Education, Department for Business Innovation and Skills (2003), http://www.desf.gov.uk/hetgateway/strategy/hestrategy/foreward.shtml (accessed, 01/04/2010)

BIS, Higher Ambitions: The future of Universities in a Knowledge Economy, Executive Summary, Department for Business Innovation and Skills (2009), http://www.bis.gov.uk (accessed, 01/04/2010)

Delanty, G.: Challenging Knowledge, The University in the Knowledge Society. Society for Research into Higher Education and Open University Press, Buckingham (2001)

DIUS, Implementing the Race to the top: Lord Sainsbury's Review of Government's Science and Innovation Policies, Department for Innovation, Universities and Skills, TSO (2008a), http://www.tsoshop.co.uk (accessed, 17/03/2010)

DIUS, Innovation Nation Executive Summary, Department for Innovation, Universities and Skills (2008b),
http://www.dius.gov.uk/policies/innovation/white-paper (accessed, 17/03/2010)

Etzkowitz, H.: The Triple Helix. University-Industry-Government Innovation in Action. Routledge, New York (2008)

Fanghanel, J.: Local Responses to institutional policy: a discursive approach to positioning. Studies in Higher Education 32(2) (2007)

Fowler, C., Lee, A.: Knowing how to know: questioning 'knowledge transfer' as a model for knowing and learning in health. Studies in Continuing Education 29(2), 181–193 (2007)

Francis-Smythe, J.: Enhancing academic engagement in knowledge transfer activity in the UK. Perspectives: Policy and Practice in Higher Education 12(3), 68–72 (2008)

Furlong, J., Oancea, A.: Assessing Quality in Applied and Practice-Based Educational Research: A framework for discussion. Oxford University Department of Educational Studies, Oxford (2005),
http://www.essc.ac.uk/assessing_quality_shortreport-tcm6-8232.pdf (accessed, April 2010)

Geuna, A., Muscio, A.: The Governance of University Knowledge Transfer: A Critical Review of the Literature. Minerva 47, 93–114 (2009)

Gibbons, M., Limoges, C., Nowotony, H., Schwartzman, S., Scott, P., Trow, M. (eds.): The New Production of Knowledge. The dynamics of science and research in contemporary societies, Buckingham (1994)

Godemann, J.: Knowledge integration: a key challenge for transdisciplinary cooperation. Environmental Education Research 14(6), 625–641 (2008)

Hammersley, M.: Troubling Criteria: a critical commentary on Furlong and Oancea's framework for assessing educational research. British Educational Research Journal 34(6), 747–762 (2008)

HEFCE, Strategic Plan 2006-11, updated May 2008, Higher Education Funding Council for England, Bristol (2008)

HMSO, Lambert Review of Business-University Collaboration, Her Majesty's Stationary Office, London (2003)

Huff, A.S., Huff, J.O.: Re-Focusing the Business School Agenda. British Journal of Management 12, Special issues, S49–S54 (2001)

Lazzeretti, L., Tavoletti, E.: Higher Education Excellence and Local Economic Development: The Case of the Entrepreneurial University of Twente. European Planning Studies 13(3), 475–493 (2005)

Mayo, P.: Competitiveness, diversification and the international higher education cash flow: the EU's higher education discourse amidst the challenges of globalisation. International Studies in Sociology of Education 19(2), 87–103 (2009)

Neumann, R., Parry, S., Becher, T.: Teaching and Learning in their Disciplinary Contexts: a conceptual analysis. Studies in Higher Education 27(4), 405–417 (2002)

Ozga, J.: Policy Research in Educational Settings. Open University Press, Buckingham (2000)

Ozga, J.: From Research to Policy and Practice: Some Issues in Knowledge Transfer, CES Briefing no. 31, CES, University of Edinburgh (2004)

Ozga, J., Jones, R.: Travelling and embedded policy: the case of knowledge transfer. Journal of Education Policy 21(1), 1–17 (2006)

Peters, M.: National education policy constructions of the 'knowledge economy': towards a critique. Journal of Educational Enquiry 2(1), 1–21 (2001)

Pilbeam, C.: Generating additional revenue stream in UK universities: An analysis of variation between disciplines and institutions. Journal of Higher Education Policy and Management 28(3), 297–311 (2006)

Prior, L.: Following in Foucault's Footsteps – text and context in Qualitative Research. In: Silverman, D. (ed.) Qualitative Research, Theory, Method and Practice, pp. 63–79. Sage, London (1997)

Ray, T., Little, S.: Communication and Context: Collective Tacit Knowledge and practice in Japan's WorkplaceF. Creativity and Innovation Management 10(3) (2001)

Rhoades, G.: Technology-enhanced Courses and a Mode III Organisation of Instructional Work. Tertiary Education Management 13(1), 1–17 (2007)

Rhoades, G.: Capitalism, Academic Style and Shared Governance. Academe 91(3), 38–42 (2005)

Smith, D., Taylor, C.: Knowledge Transfer in the arts and humanities: policy images and institutional realities. Journal for Continuing Liberal Adult Education (38), 13–17 (2009)

Sylva, K., Taggart, B., Melhuish, E., Sammons, P., Siraj-Blatchford, I.: Changing models of research to inform educational policy. Research Papers in Education 22(2), 155–168 (2007)

Trowler, P.: Academics Responding to Change: New Higher Education Frameworks and Academic Cultures. Open University Press, SRHE, Buckingham (1998)

TSB, Connect and Catalyse: A strategy for business innovation 2008-2011, Technology Strategy Board, Swindon (2008)

TSB, Knowledge Transfer Partnerships, Technology Strategy Board (2010), http://www.ktponline.org.uk/strategy/background.aspx (accessed, 08/04/2010)

Urwin, P.: Research, teaching and knowledge transfer: separate and distinct? Conference paper: British Educational Research Association Annual Conference, Herriot-Watt University, Edinburgh (September 11-13, 2003), http://leeds.ac.uk/educol/documents/00003357.htm (accessed, April 2010)

Van Vught, F.: The EU Innovation Agenda: Challenges for European Higher Education and Research. Higher Education Management and Policy 21(2), 13–34 (2009)

Williams, P.: Valid Knowledge: the economy and the academy. Higher Education 54, 511–523 (2007)

Managing Knowledge in the Framework of the Organizational Evolution of SMEs

Philippe Bouché, Nathalie Gartiser, and Cecilia Zanni-Merk*

LGECO – INSA de Strasbourg - 24 bd de la Victoire – 67084 Strasbourg Cedex – France
{philippe.bouche,nathalie.gartiser,
cecilia.zanni-merk}@insa-strasbourg.fr

Abstract. This article describes the research project MAEOS. MAEOS is a project about the modelling of the support to the organizational and strategic development of SMEs. The main objective of MAEOS is to improve the efficiency and performance of business advice to SMEs. To achieve this objective, a multidisciplinary team was created. Two main research areas are represented: artificial intelligence and management sciences. This work aims at establishing a set of methods and software tools for analysis and diagnosis of SMEs. We address three main questions: how to extract knowledge from experts but also practical knowledge from consultants, how to formalize it and how to use it to help a consultant or an entrepreneur.

1 Introduction

One of the major difficulties encountered by the smallest companies, especially today, in a crisis context, is how to manage their evolution. This issue needs the capacity to, not only, perform a global analysis of the whole of its aspects (economical, production, organization, human resources, sales ...) but also to have a sufficient stand back to see this analysis in the perspective of that evolution. Mastery of change becomes a key to success for many firms facing to strong competition. For us, a Small and Medium Enterprise (SME) can then be characterized by its size, but also by its activity sector and its governance.

The SMEs getting involved in this approach look for the help of consultancy services when they do not have internal resources to do this. The general approach of an internal or an external consultant is to diagnose the company situation according to his (her) own resources, knowledge and methods. However, a very big amount of knowledge is now available in the domain of business and management sciences. In this context, there is a recurrent question that arises: how to access

---

* The authors would like to thank Mr. Henri-Pierre Michaud from AEM Conseil for his support and funding.

R.J. Howlett (Ed.): Innovation through Knowledge Transfer 2010, SIST 9, pp. 63–72.
springerlink.com

existing knowledge to, on the one hand, allow diagnosis of the SME and, on the other hand, think about its evolution.

The MAEOS project aims at the development of a software tool, which uses two kinds of knowledge bases[1]. Firstly, expert knowledge issued from different domains such as management sciences or production, which is formalized under the form of ontologies with associated rule-bases. Secondly, there is consultant/practical knowledge, which is formalized under the form of case bases. With the MAEOS system (Fig. 1), a consultant or an entrepreneur can describe his situation or his problem. The system will use all the knowledge bases to propose, based on rules of reasoning and with the help of a Multi Agent System (MAS), interesting characteristics for describing the statement or identifying evolution tendencies to help the consultant make a diagnosis or find solutions for the specific problem of the company.

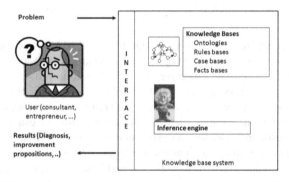

Fig. 1. The MAEOS system

In this way, the users of the MAEOS system will have access to a large quantity of information, which will continue increasing with the use:

- Each new case will be saved in the case base,
- In addition, new ontologies given by experts can be added anytime.

Large and diversified quantity of knowledge can be considered according to three aspects: the manipulation of that knowledge, the use of heterogeneous knowledge and the transfer of that knowledge.

Here, there are several important aspects regarding knowledge transfer. The first and main problem is that the volume of knowledge, both theoretical and "expert", is huge and sometimes, much more detailed than needed. We face a double issue about knowledge capitalization: structuring for managing the large quantities of knowledge and organizing into a hierarchy for permitting different access levels.

Secondly, the knowledge transfer led by MAEOS and the automatic reasoning that is integrated in it might allow the consultant to propose original solutions, generating in this way high level innovations, because they are radically new.

---

[1] A knowledge base (KB) is a special kind of database for knowledge management, providing the means for the computerized collection, organization, and retrieval of knowledge.

This article is structured as follows: In first place, we present knowledge transfer and more precisely the two main sources of knowledge of the system and their representations. Afterwards, we show the way the system can manipulate huge and various knowledge sources. In a third part, we describe the system from the point of view of the user. This last part explains the knowledge transfer from the system to the user. Finally, we present our conclusions and perspectives.

2   Knowledge Transfer

The interest and the difficulty of the project are the combination of a large variety of sources and origins of knowledge around SMEs topics. As it has been introduced previously, a very big amount of heterogeneous knowledge has to be involved in the MAEOS system knowledge bases (Fig. 2):

- Experts' knowledge: It comes from diverse domains linked with SME evolution and change management. This knowledge is the most often available in the shape of written texts, such as scientific publications and books. There is also an evolution of this knowledge in time, with new methods, new models proposed by the researchers. The main problem is not to find this knowledge but to formalize it. Indeed, knowledge, in its current shape, is:

  o   Difficult to be used by a human being because it is needed to collect, read, assimilate and organize a lot of fragmented and scattered knowledge. Moreover, there are too many information sources. Moreover, it is difficult to be continuously aware of the evolution of this knowledge.
  o   Impossible to be directly used by a computer, especially because of the written shape of the information sources.

  We propose then to formalize this knowledge under the form of ontologies with associated rules bases.

- Practical knowledge: It comes from the consultant practice. It is the result of his (her) experience and own ability. Nevertheless, this practical knowledge is very difficult to identify and to formalize because it is embedded in his (her) mind as routines.

We propose then to formalize this knowledge under the form of a case based system.

The interaction between the system and the users will be done through a base of facts (a fact is an instantiation of a concept, as we will see afterwards).

Therefore, we need methods to represent all the necessary knowledge in the same format. In this way, we will be able to manipulate it with a computer system and make associations or comparisons among different sources (to find contradictories points of view, for instance). Two representation models are then used: ontologies and a case based system.

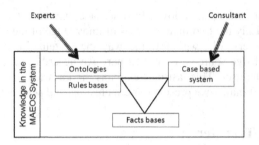

Fig. 2. Knowledge transfer flows

3   Ontologies: The Formalisation of Expert Knowledge

An ontology is a formal explicit description of concepts in a domain of discourse and includes classes (sometimes called concepts), properties of each concept describing various features and attributes of the concept or slots (sometimes called roles or properties), and restrictions on slots, called facets (sometimes called role restrictions). An ontology together with a set of individual instances (base of facts) of classes constitutes a knowledge base (Noy, 2001).

To build de inputs for the MAEOS system, our main sources were the ontology MASON (Lemaignan, 2006), TOVE (Fox, 1992, 1998) and ENTERPRISE (Uschold, 1998). Some parts of specific ontologies have been also considered.

These ontologies cover different areas, such as Professional Learning and Competencies with the ontology of FZI-Karlsruhe (Schmidt, 2007), organization modelling with UEML-1 (Berio, 2005), or Service Oriented Architectures with the SOA Open Group ontology, among others.

Beyond the use of existing ontologies, we have developed our own ones about certain relevant fields for our SME context (organization, quality, production, innovation ...). In first place, we have developed an ontology on the organization models based on the main works of (Mintzberg, 1979). This ontology integrates the concepts that describe the structure and models of companies, the relationships among concepts and the restrictions to those concepts according to the company characteristics (its size, for example, that is relevant for this project) (Renaud, 2009). We have made the choice of using this source because it is a clear reference in the organization theory field, at least at the concept level. The works of this author have been widely quoted, commented and refined. Other works of this author (Mintzberg, 1989) will complete this ontology.

In second place, we have developed an ontology on production systems based on (Courtois, 1989). This reference is a choice of our industrial partner.

Nowadays, a new set of ontologies about management science are being developed on subjects such as leadership, strategy, very small enterprises...

Nevertheless, we risk to be confronted to a double issue about contradictions at knowledge level: contradictions among knowledge sources and contradictions within the same knowledge source. It is because of these issues that we have decided to develop separate ontologies. This choice permits the continuous feeding of the MAEOS system with new ontologies.

As each ontology has the same formal structure, all this knowledge can be easily manipulated with a computer (Renaud, 2009). To finish we can remark three important points:

- We can identify two levels of knowledge in the ontologies:
  - o An "abstract" level, corresponding to high-level knowledge, which is not directly useful for the user. This knowledge will never be presented to him. They are concepts that do not have real existence by themselves, but that are useful to analyse and understand the company situation. They are only used in the MAEOS system for reasoning purposes. For example, at this level in the Mintzberg ontology we will find the different types of organisation which are used to structure knowledge but which will have no physical representation for a user that is not an expert.
  - o A "concrete" level corresponding to low-level knowledge. It contains concepts that a user of the system can manipulate (because he knows them or because he can easily understand what they mean) to describe the situation of its company or its problem. At this level, in the Mintzberg ontology, we find concepts such as age of the company or size of the unit.

- The unification of knowledge in a set of ontologies, which have the same format, provides the means for the computerized collection, organization, and comparison among different sources (to find contradictory points of view, for instance). We can also use complementary sources to solve a certain problem, propose different solutions using different points of view, or other possibilities (Renaud, 2009)

- To combine the ontologies, we define bridges among them with the help of domain experts. These bridges are the identification of the concepts that have the same semantic sense. It is important to note that it is not because we find a concept with the same name in two ontologies that it is the same concept. Therefore, it is imperative to search semantic equivalences with the experts. In this way, the system will be able to use new ontologies from a first one specified by the user.

4   A Case Based System: The Formalisation of Practical Knowledge

The second important source of knowledge is the experience of the consultant, which will become practical knowledge in our system. Each consultant develops his own knowledge base based on all the projects he has participated. In front of a new problem, he uses this knowledge and he searches if he has already faced such a problem or a similar one or if he has partial solutions to explore.

Therefore, it is important to be able to acquire this knowledge. The best solution is a case based system (CBS). The principle of the operation of such a system (Fig. 3) is the following:

- **Retrieve:** Given a target problem, retrieve cases from memory that are relevant to solving it. A case consists in a problem, its solution, and, typically, annotations about how the solution was derived.

- **Re-use:** Map the solution from the previous case to the target problem. This may involve adapting the solution as needed to fit the new situation.
- **Adapt:** After having mapped the found solution to the target situation, test the new solution in the real world (or a simulation) and, if necessary, revise.
- **Retain:** After the solution has been successfully adapted to the target problem, store the resulting experience as a new case in memory.

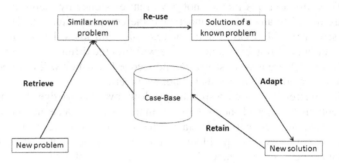

Fig. 3. Case based system

This knowledge base is the result of the experience of the consultant and experimented users of the MAEOS System. It is the result of practice. It is a continuous source of information (because each new case will be memorized to be reused in the future), and the more the system will be used, the more possibilities will be explored.

We have seen the different sources of information of MAEOS system. The use of this knowledge is made with the help of a Multi Agent System (MAS).

5   Using Heterogeneous Knowledge with a Multi-agent System

The operation process that has been chosen is similar to that of a panel of experts. Each expert has an area of knowledge and a set of skills. He examines aspects of the business related to his area of expertise. Once the study is completed, his conclusions are shared with other experts. Finally, an analysis and diagnostic report is created.

In its implementation, the system is a Multi-Agent system (Wooldridge, 2009). The agents use the two kinds of knowledge bases (KB). The first one is a rule-based KB containing academic knowledge. The second one contains cases related to the socioeconomic context of SMEs and some specific cases already studied.

No direct communication among agents exists. All exchanges are made through a blackboard or common bag.

An agent is associated with a particular knowledge base. Therefore, all agents are characterized by a knowledge domain, a collection of facts and/or rules and a set of meta-data (Fig. 4). Each agent picks information up in the blackboard. It accomplishes its deduction tasks. At the end, it adds the results to the bag. The triggering of an agent is made by a set of data corresponding to the characteristics of its knowledge base. The process is considered as finished when the agents have nothing new to add to the blackboard. At the end, a series of post-processing operations aggregate the entire contents of the bag.

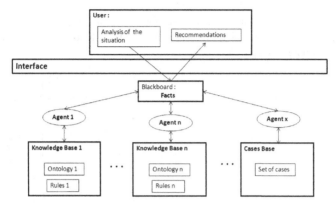

Fig. 4. The multi-agents system

The goal is to provide concise results that are close to the context of the subject of study. A more complete presentation of our MAS can be found in (Bouché, 2010). Now we will see the use of the MAEOS system from the users' point of view.

6   Example of the Use of the MAEOS System

As seen in the previous sections, the user role of the MAEOS system is to give the set of facts (Fig. 5) corresponding to the situation or the problem to the system.

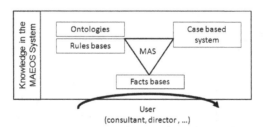

Fig. 5. The use of MAEOS

There are interactions between the user and the system. The main mechanisms are the following:

- We have defined a "company form", it must be completed by the user for each new case,
- Afterwards, the user is invited to give the facts corresponding to the situation, the system can propose specific concepts that the user can instantiate or not,

- These facts launch the activation of the MAS which may find new facts that are proposed to the user,
- From this step on, the user can give new facts and the process iterates until the system has no more rules to execute,
- Conclusions are then presented to user.

To facilitate the acquisition and visualization of results, we have developed the front-end DISKO (Development Interface for SME's Knowledge Organization) that provides a user-friendly interface.

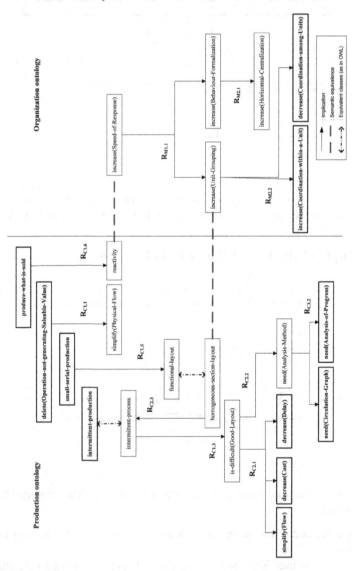

Fig. 6. Working of MAEOS system

Next figure (Fig. 6) shows an example of the working of MAEOS system on an industrial case (for confidentiality reasons we cannot describe the situation thoroughly).

Thick boxes at the top of the figure represent concepts instantiated by the user to describe the problem in the company White boxes are concepts instantiated by the MAS when it executes the rules; arrows between ontologies are semantic equivalences between concepts. Finally, thick boxes at the bottom are conclusions and solutions proposed by the system to the user. This application has been validated by our industrial partner (the interested reader can find the full details of this example in (Bouché, 2010)).

Next tasks include the improvement of the system with the help of more real cases developed with our industrial partner.

7 Conclusions

In this article, we have presented the MAEOS system. The system uses two sources of knowledge: expert and consultants' knowledge. To manage this knowledge, the choice was made to formalize it under the form of ontologies and case bases. A multi-agent system then uses this knowledge to help solve a problem stated by the user. The system can manipulate huge and diverse knowledge sources. We have focused here on three main dimensions: acquiring expert and practical knowledge, formalizing it and using it to help the consultant.

The MAEOS project fundamentally differentiates from other approaches because of the different nature of the pieces of knowledge to manipulate and of the goal of the software tool. The idea is not to build a knowledge server to manage a corporate memory (Van Heijst, 1996) (Nagendra, 1996), but to develop a framework for theoretical and knowledge structuring about strategic analysis of evolution in SMEs. The aim of the software developed in the framework of the project is clearly to help the consultant in his thinking and reasoning; it is a system to help him to manage a huge amount of knowledge by the formalization of it.

As for the perspectives of future work after these first results, there is the problem of introducing the notion of "time" in the ontologies. In fact the current version of our software is able to link facts. Indeed, it helps the consultant to identify the hidden characteristics, i.e. the characteristics he has not observed in the company. The next step will be the introduction of the difference between facts which describe something (they will help the consultant to characterize the company), of facts which will certainly appear in the future (they will help the consultant to identify and to understand the risks linked with the company evolution) and of facts which can appear in the future if the company decides to act in a certain direction (they will help the consultant to formulate recommendations for the future evolution of the company; indeed, new facts can be deduced from the ontologies, they describe interesting evolutions from a theoretical point of view).

Tightly related with these situation is, therefore, the characterization of the company at different moments in time: where it comes from (what the situation of the company in the past was like); what the company is like today; and which aspects people from the company (the manager for instance) would like to develop for the future.

References

Berio, G., Jaekel, F.-W., Mertins, K.: Common Enterprise Modelling Framework for Distributed Organisations. In: 16th IFAC World Congress, Prague, Czech Republic (2005)

Bouché, P., Zanni-Merk, C., Gartiser, N., Renaud, D., Rousselot, F.: Reasoning with Multiple Points of View: A Case Study. In: Setchi, R., Jordanov, I., Howlett, R.J., Jain, L.C. (eds.) KES 2010. LNCS, vol. 6279, pp. 32–40. Springer, Heidelberg (2010)

Courtois, A., Martin-Bonnefous, C., Pillet, M.: Gestion de Production. Editions d'organisation (1989) ISBN 2708136488

Fox, M., Grüninger, M.: Enterprise Modeling. AI Magazine, 109–121 (1998)

Fox, M.: The TOVE project. LNCS (LNAI), vol. 604, pp. 25–34 (1992)

Lemaignan, S., Siadat, A., Dantan, J.-Y., Semenenko, A.: MASON: A proposal for an Ontology of Manufacturing Domain. In: IEEE Workshop on Distributed Intelligent Systems, Collective Intelligence and Its Applications, DIS 2006, vol. 15-16, pp. 195–200 (2006)

Mintzberg, H.: The Structuring of Organizations. Prentice-hall, Englewood Cliffs (1979) ISBN 0-13-8555270-3

Mintzberg, H.: Mintzberg on Management: Inside Our Strange World of Organizations. Simon & Schuster Inc., United States of America (1989)

Negrenda Prasad, M.V.N., Plaza, E.: Corporate Memories as Distributed Case Libraries. In: Gaines, B., Musen, M. (eds.) Proceedings of the 10th Banff Knowledge Acquisition for Knowledge-Based Systems Workshop (KAW 1996), pp. 40-1/40-19 (1996)

Noy, N.F., McGuinness, D.L.: Ontology Development 101: A Guide to Creating your First Ontology (2001), http://www.ksl.stanford.edu/people/dlm/papers/ontology101/ontology101-noy-mcguinness.html

Renaud, D., Zanni-Merk, C., Rousselot, F.: A Compound strategy for ontologies combining. In: KEOD 2009 International Conference on Knowledge Engineering and Ontology Development, Madeira, Portugal (2009)

Renaud, D., Bouché, P., Gartiser, N., Zanni-Merk, C., Michaud, H.P.: Knowledge Transfer for Supporting the Organizational Evolution of SMEs: A Case Study. In: International Conference on Innovation through Knowledge Transfer 2009 (InnovationKT), Hampton Court Palace, London (2009)

Schmidt, A., Kunzmann, K.: Sustainable Competency-Oriented Human Resource Development with Ontology-Based. In: Cunningham, M., Cunningham, P. (eds.) Competency Catalogs eChallenges (2007)

Uschold, M., King, M., Moralee, S., Zorgios, Y.: The Enterprise Ontology. The Knowledge Engineering Review 13, 31–89 (1998)

Van Heijst, G., Van der Spek, R., Kruizinga, E.: Organizing Corporate Memories. In: Gaines, B., Musen, M. (eds.) Proceedings of the 10th Banff Knowledge Acquisition for Knowledge-Based Systems Workshop (KAW 1996), pp. 42-1/42-17 (1996)

Wooldridge, M.: An introduction to Multi Agent Systems, 2nd edn. John Wiley & Sons, Chichester (2009)

Assessing Changes in University Knowledge Transfer Capability to Support Innovation: A Knowledge Intensive Business Service Perspective

John Sparrow

Birmingham City Business School
Birmingham City University, Franchise St
Perry Barr, Birmingham
B42 2SU
Tel.: +44(0) 121 331 5217
john.sparrow@bcu.ac.uk

Abstract. As universities increasingly engage with industry, the need for the management of knowledge transfer to draw upon appropriate measurement of activities is growing. There is little understanding of the relationship between strategy, infrastructure and capacity development, and alternative knowledge transfer activities. Much of the measurement of university knowledge transfer activity, emphasises basic 'output' assessment (e.g. number of patents, licenses, engagements, financial value etc.). This limitation is exacerbated when one seeks to support innovation above and beyond high technology-, science- and research-led initiatives, since innovation processes in service innovation spheres are more complex and diverse.

There are many ways in which knowledge transfer can be categorised. Whilst these frameworks provide some insights into activities, they are essentially (supply-led) 'product' categories and do not reveal the ways in knowledge transfer activities meet the demands of users. Viewing knowledge transfer activities as knowledge intensive business services (KIBS), is one way to more fully understand the ways in which universities are supporting innovation in its broader sense.

Understanding the competence of a university in terms of its service capability allows a university to develop strategies, tactics and initiatives to develop infrastructure and capacity.

The current study examines developments in a case study university over a four year period in a structured assessment of knowledge intensive business services for regional innovation. A number of statistically significant changes in capability are identified which align to the strategic endeavour.

The study demonstrates value in assessing and managing KT activities for innovation in KIBS terms.

Keywords: university knowledge transfer, innovation support, knowledge intensive business service (KIBS), capability, measurement, change.

R.J. Howlett (Ed.): Innovation through Knowledge Transfer 2010, SIST 9, pp. 73–81.
springerlink.com   © Springer-Verlag Berlin Heidelberg 2011

1 Introduction

As universities increasingly engage with industry, the need for the management of knowledge transfer to draw upon appropriate measurement of activities is growing. There is little understanding of the relationship between strategy, infrastructure and capacity development, and alternative knowledge transfer activities. Sharifi et al (2008) in discussing university technology transfer offices note that 'Universities are part of the changing circumstances and, as well as being participants in the process of innovation, perhaps they should apply the same principles that are proving to be the basis for successful innovation in industry. This potential does not seem to have been fully understood nor addressed by universities and policy makers. Issues such as the definition, role, impact, position, practice, management, evaluation and classification of these entities within the context of open innovation and an integrated approach in terms of value chain management." (p 337).

1.1 Measuring KT Activity

Research and evaluation studies of business support initiatives (e.g. Bennett and Robson, 1999a) frequently measure impact in terms of basic measures of business outcome, and shed little light upon the processes through which business development is occurring (Sparrow and Patel, 2007). The categories of support 'product' and agencies (e.g. Bennett and Robson, 1999b) tend to be at such a high level of abstraction that it is hard to understand the dynamics.

Similarly, much of the measurement of university knowledge transfer activity, emphasises basic 'output' assessment (e.g. number of patents, licenses, engagements, financial value etc.). In reviewing the metrics in use, and working with key stakeholder groups, Holi and Wickramasinghe (2008) categorised knowledge transfer activities as: Networks (specifically social, for example, between academics and the business community); Continuing Professional Development (CPD); Consultancy; Collaborative Research; Contract Research; Licensing; Spin-Outs; Teaching; and Other (e.g. access of academics to high technology equipment), and sought to measure volumes of such activities. Similar frameworks have been developed by others (e.g. Bekkers and Freitas, 2008). The UK HE-BCI survey categorises university KT income in terms of: Collaborative research; Contract research; Consultancy contracts; Facilities and equipment related services; Continuous professional development; Continuing education; Regeneration and development programmes; and IP income. It also records outputs in terms of Patent applications; Patents granted; Formal spin-offs established; and, Formal spin-offs still active after three years. The only activities recorded are whether the university provides: Enquiry point for SMEs; Short bespoke courses on client's premises; Distance learning for businesses; and contracting systems for all consultancy.

1.2 Measuring Innovation Support

This limitations of current measures of KT activity are exacerbated when one seeks to measure support innovation above and beyond high technology-, science- and research-led initiatives, since innovation processes in service innovation spheres are more complex and diverse. Abreu et al (2008) note "until recently there has been a prescriptive view of university-business interactions with a narrow focus on technology transfer. Although technology transfer may be important, it is also necessary to focus on the more diverse and varied impacts of business-university knowledge exchange relations" (p 45).

The role of services innovation in knowledge economies has been summarised by Kuusisto and Meyer (2002) and Vang and Zellner (2005). Maffei et al (2005) highlight features of service organisations that challenge innovation support such as difficulties in valuing and financing intangible service assets (like design and marketing), government initiatives not supporting organisational change, and a lack of configuration of innovation support initiatives to service innovation needs. But service innovation is a feature of all organisations, not just service organisations. Services create more wealth than manufactured goods in rich countries. In the UK, three-quarters of wealth comes from the service sectors and even the manufactured goods sectors require a 'service wrap' to differentiate themselves and compete. The boundary between manufacturing and services is blurring for example, cars are sold with profitable financing packages, white goods are sold with post purchase insurance. Although companies can acquire a leading position through exclusive access to scientific know-how through owning rights to patents, they can also achieve it through the associated service design and innovation. To deliver effective service consistently, and in a way that attracts and retains customers is difficult to deliver and replicate as it is has to be embedded throughout the organisation delivering the service. Yet UK government's investment focus for innovation (and measurement systems) continue to be in science and technology.

There is however, some recognition of the need for a broader acknowledgement of innovation contexts and enhancement of understanding of innovation support processes in a wider range of contexts. The Cox Review of Creativity in Business (HM Treasury, 2005) looked across a much wider range of less research intensive sectors and therefore saw innovation in a broader context. The Work Foundation's projects under the knowledge economy theme, has included several studies of innovation in knowledge-based services because the innovation process and its wider economic significance in service terms is not well understood.

1.3 Knowledge Intensive Business Services

There are many ways in which knowledge transfer can be categorised. Whilst these frameworks provide some insights into activities, they are essentially (supply-led) 'product' categories and do not reveal the ways in knowledge transfer activities meet the demands of users. Viewing knowledge transfer activities as knowledge intensive business services (KIBS) (e.g. consultancy, design, accountants etc.), is one way to more fully understand the ways in which universities are supporting innovation in its broader sense.

The specifics of knowledge transfer across the KIBS-client boundary have been studied (Webb, 2002). Muller and Doloreux (2007) note how KIBS were initially mainly seen as providing a "transfer of specialised information" to their clients. KIBS are now acknowledged to allow a change of state of their clients in knowledge terms. "The services they perform can ultimately be seen as leading to a kind of "fusion" of the respective knowledge bases of KIBS and the clients" (p 18).

The role of knowledge intensive services within innovativeness has been considered (Wood, 2002). Providing knowledge intensive business services (Miles et al, 1995) to other organisations as carriers, shapers, creators and co-producers (Hertog, 2000) of innovation is a key facet of innovation. Muller and Zenker (2001) investigated empirically the innovation activities of French and German KIBS and SMEs (small end medium-sized manufacturing firms). As a result they put forward the hypothesis of a virtuous innovation circle linking SMEs and KIBS, to be understood as: "… a circle made virtuous through the knowledge generating, process-ing and diffusing function KIBS fulfill within innovation systems" (p 1514).

1.4 Enhancing Universities' Capacities as KIBS Providers

Sparrow et al (2006) developed a questionnaire to assess university capabilities to support service innovation of a case study university in a case study region. The questionnaire drew upon a number of key frameworks of knowledge intensive business service. Sparrow et al (2009) considered ways in which universities can come to view themselves as KIBS.

The broader university KT evaluation literature has highlighted clear growth in the extent to which universities engage in KT activity. The UK HE-BCI survey data for 2008-09 shows continuation of a trend of increases in income measures since 2004-05 (with the exception of regeneration activities). The value of activities increasing between 25% (for collaborative research) up to a 96% (for IP value) in the period. The changes need to be seen within quite fluid and complex contexts however. National and international KT policies have many ambiguities (Molas-Gallart and Castro-Martinez, 2007). The movement towards enhanced KT has raised issues for academics' boundaries (Henkel, 2007). Barriers to university-industry interaction continue to be addressed (Bruneel et al, 2010). The discourses of KT differ between universities and industry (Werson, 2008). There can be distinct competency requirements (Francis-Smythe et al, 2006; Prince, 2007) and the trajectories (and associated challenges) can be context-specific (see, for example, Geoghegan and Pontikakis (2008) re technology transfer, and Wright et al (2009) re mid-range university linkages with industry). Sparrow (2010a) demonstrated distinct academic staff KIBS capabilities associated with research-led, collaborative project, expertise-transfer and networking KT activities.

Sparrow et al (2010b) reported a study analysing how members of staff being trained and engaging in a role of innovation mentor in a case study university developed personal KIBS support capabilities. What is not clear however, is how universities as a whole can change in terms of KIBS capabilities. This is the focus of the current paper.

2 Methodology

Sparrow et al (2006) developed a questionnaire to assess the ways in which universities utilise knowledge intensive business services to enhance regional innovation, and gathered data within a case study university. The current research administered the same questionnaire within the same university four years later. There have been a number of significant developments that might be associated with a shift in the university's capability.

A re-branding of the university as had been used to signal significant changes in the university's mission and priorities. The university's 2007-2012 Corporate Plan noted that as part of the university's mission it will be "an exemplar for engagement with business, the professions and the community", and that the university will work towards ensuring "our business and industry engagement will encompass considerable knowledge transfer activity as well as support for development of higher-level skills and continuing professional updating".

Furthermore, the university secured a significant increase in its research capability in the 2006-2010 period. The HEFCE research assessment (of activities up to the end of 2007) was based very largely upon the research submissions from the university's centres of excellence. A strategic decision was made to build upon this success with increased funding and support for the existing (and additional) centres. The brief for each centre was for them to engage in internationally significant research but with clear engagement/impact outcomes. There has been significant expansion of the centres within the period under review here.

In addition, a third stream income activity initiative (launched in October 2006) had supported 48 members of staff to be trained and operate as Innovation Mentors. The initiative has been promoted quite extensively and the impact of the approach disseminated widely throughout the university.

The questionnaire drew upon established frameworks of innovation support. Tether (2005) distinguishes between innovation in Outputs, Internal organisation and External organisation. Output innovation (i.e. what is provided to whom) occurs in terms of product innovation (e.g. financial services product, new clinical service etc.) and market innovation(e.g. opening up or breaking into new markets). Innovation in terms of internal organisation (i.e. how organisations 'organise' their own activities for the provision of outputs) occurs through process innovation (defined and repeated processes associated with the production of a service product, such as e-business) and organisational innovation (changes in the way in which provision is organised within the organisation such as 'de-layering', restructuring, introduction of teamwork etc.). Innovation in terms of external organisation of provision (i.e. sources of supply, changes in relations with suppliers, customers, partners, competitors, universities etc. within organisational innovation networks) occurs through changes to the relationships between the organisation and its 'customers', other organisations (e.g. collaborative arrangements, initiating change in other organisations)). University activity in each of these regards was assessed.

The questionnaire drew upon Leiponen's (2005) identification of several alternative organisational processes in achieving innovation. These were: internal cooperation of employees; vertical and horizontal information (from suppliers, customers, competitors, partners etc.); technology adoption; incremental learning (learning by doing); and, utilisation of scientific knowledge. Respondents were asked to assess the universities KT activities in terms of Hertog's (2000) distinction between: universities serving as a *facilitator* of innovation – i.e. supports organizations in their innovation processes, but the innovation at hand does not originate from the university, nor is it being transferred (from other organisations by the university); a *carrier* of innovation – i.e. transfers existing innovations from one organisation or industry to organisations even though the innovation does not originate from the university; and, a *source* of innovation – i.e. developing innovations and initiating the innovation in organizations.

Bercovitz and Feldmann (2006) identified several different basic approaches that universities can adopt to support innovation. These were: specialised research units; joint co-operative ventures; and, interdisciplinary projects that are receptive to industry needs. These were each assessed. Gunasekara (2006) contrasted the impact of a generative role for universities (e.g. science parks, incubators and cluster initiatives) with those associated with a developmental role (e.g. the supply of graduates, regional networking, the provision of information and analysis to support decision making and animateur roles). Respondents were akewd to indicate the extent to which they felt the university fulfilled each of these roles. Koch and Stahlecker (2006) highlighted how the contribution of any specific knowledge-intensive business service organisation within a region depended upon its relationship in the region. i.e. the extent to which it aligns with techno-economic conditions of the region; is embedded in established cluster and other regional networks; and, plays a major role given the number of other KIBS in the region. Respondents gave their assessments of the university's relationships in the region.

Hertog (2000) identified a number of different processes through which innovation can be supported. These include: expert consulting; experience-sharing – 'bees cross pollinating'; brokering – putting sources and users in contact – 'marriage broker'; diagnosis and problem clarification; benchmarking – identifying comparitor/good practice; and, change agency – organisational development from a neutral outside perspective. A further process through which innovation can be supported is evaluation research. University academic staff can adopt a number of different knowledge transfer roles: Educator/lecturer; Trainer; Expert/technical consultant; Coach/Mentor; Formal quality assessor/assurance role; and, Facilitator roles. The extent to which university staff are seen to undertake each of these roles was assessed.

The assessments of university provision of innovation-oriented KIBS were sought from members of academic staff. Participants were asked to assess the extent to which they considered particular aspects KIBS were being practised by the university.. All items were scaled (0 – None/Not at all, 1 – Nominal, 2 – Low, 3 – Moderate, 4 – Considerable, 5 – Very substantial).

3 Analysis and Results

Sparrow et al (2006) obtained responses from 32 members of staff. In the current study, responses were obtained from 39 members of staff. The data were entered into SPSS and analysis of variance tests conducted to explore the statistical significance of differences in mean assessments of the university in KIBS terms, over the four year period. Table 1 details the aspects of activity where significant ($p<0.05$) change was identified.

Table 1. Statistically significant changes in the university in Knowledge Intensive Business Service (KIBS) terms: 2006-2010

KIBS facet	UCE 2006	BCU 2010	Difference
University as a 'source' of innovation	2.20	2.80	F= 4.500, df(1,58) p<0.05
University supporting regional innovation through specialist research units	2.46	3.23	F=6.050, df(1,57), p<0.05
University engaging in joint/co-operative ventures	2.45	3.21	F=5.231, df(1,55),p<0.05
University staff working upon interdisciplinary projects	2.33	3.10	F=6.000, df(1,59), p<0.05
University playing a 'generative' approach towards regional innovation	1.83	2.50	F=5.305, df(1,57), p<0.05
University playing a 'developmental' approach towards regional innovation	2.32	3.10	F=5.551, df(1,57), p<0.05

4 Discussion and Conclusions

The current study demonstrates that it is possible to measure KT activity in universities by means other than output assessment. It is also clear that assessing university capability in KIBS terms provides a more detailed view of the ways in which strategy, tactics and initiatives to enhance KT activity, impact upon capability.

There are some limitations to the study. The sample sizes for the two periods are quite low. The study design can only indicate aligned changes and not direct causal links between the university's management efforts and assessments of capability. The assessments themselves are complex individual judgements. It is possible that the judgements are unduly influenced by any 'internal marketing' efforts within the university and merely confirm the sought 'image' of the university and not its objective capability. The nature of the changes identified in the case study university seem meaningful against the backcloth of the specific suite of changes in its context. There is a need for further study using this sort of approach in universities following different KT missions and trajectories.

There are also other considerations in university change that relate to the study. For example the need to configure innovation support initiatives to meet innovation support needs, issues around academic's boundaries, how re-branding can significantly signal changes in mission and priorities, and the importance of increasing research capacity. Each of these issues warrants study.

Sparrow et al (2010) showed that it is possible to configure training and KT experiences for academic staff that can enhance specific facets of KIBS personal capability. Sparrow (2010) highlighted distinct profiles of KIBS competencies associated with different KT channels. Taken together with the current study, these studies suggest that it is possible to construct strategy, tactics and development endeavours to secure particular forms of KT capability within a university.

References

Abreu, M., Grinevich, V., Hughes, A., Kitson, M., Ternouth, P.: Universities, Busi-ness and Knowledge Exchange, Council for Industry and Higher Education, UK (2008)

Bekkers, R., Freitas, I.M.B.: Analysing knowledge transfer channels between uni-versities and industry: To what degree do sectors also matter? Research Policy 37, 1837–1853 (2008)

Bennett, R.J., Robson, P.J.A.: Intensity of Interaction in Supply of Business Ad-vice and Client Impact: A Comparison of Consultancy, Business Associations and Gov-ernment Support Initiatives for SMEs. British Journal of Management 10(4), 351 (1999a)

Bennett, R.J., Robson, P.J.A.: The use of business advice by SMEs in Britain, Entrepreneurship and Regional Development, pp. 155–180 (1999b)

Bruneel, J., D'Este, P., Salter, A.: Investigating the factors that diminish the barri-ers to university–industry collaboration. Research Policy 39.7, 858–868 (2010)

Francis-Smythe, J., Haase, S., Steele, C., Jellis, M.: Competencies and Continuing Professional Development (CPD) for Academics in Knowledge Exchange (KE) Activity, Report to Contact Knowledge Exchange, University of Worcester (2006)

Geoghegan, W., Pontikakis, D.: From ivory tower to factory floor? How univer-sities are changing to meet the needs of industry. Science and Public Policy 35(7), 462–474 (2008)

Henkel, M.: Shifting Boundaries and the Academic Profession. In: Kogan, M., Teichler, U. (eds.) Key Challenges to the Academic Profession, Report for UNESCO, Kasel, International Centre for Higher Education Research, pp. 191–204 (2007)

den Hertog, P.: Knowledge intensive business services as co-producers of innova-tion. International Journal of Innovation Management 4(4), 491–528 (2000)

HM Treasury, The Cox Review of Creativity in Business: building on the UK's strengths, HM Treasury (2005)

Holi, M.T., Wickramasinghe, R.: Metrics for the evaluation of knowledge transfer at universities. Library House, Cambridge (2008)

Kuusisto, J., Meyer, M.: Insights into services and innovation in the knowledge-intensive economy. Finnish National Technology Agency 134/2003 (2002)

Maffei, S., Mager, B., Sangiorgi, D.: Innovation through Service Design: From research to a network of practice. Joining Forces, Helsinki (2005)

Miles, I., Kastrinos, N., Flanagan, K., Bilderbeek, R., Hertog, P., den Huntink, W., Bouman, M.: Knowledge-intensive business services: users, carriers and sources of innovation. Luxembourg: EC (DG13 SPRINT-EIMS publication no 15) (1995)

Molas-Gallart, J., Castro-Martinez, E.: Ambiguity and conflict in the develop-ment of 'Third Mission' indicators. Research Evaluation 16(4), 321–330 (2007)

Muller, E., Doloreux, D.: The key dimensions of knowledge-intensive business services (KIBS) analysis: a decade of evolution, Fraunhofer Institute of Systems and Innovation Research, Working Paper No. U1/2007 (2007)

Muller, E., Zenker, A.: Business services as actors of knowledge transformation: the role of KIBS in regional and national innovation systems. Research Policy 30, 1501–1516 (2001)

Prince, C.: Strategies for developing third stream activity in new university busi-ness schools. Journal of European Industrial Training 31. 9, 742–757 (2007)

Sharifi, H., Liu, W., McCaul, B., Kehoe, D.: Enhancing the flow of knowledge to innovation: Challenges for university-based knowledge transfer systems. In: Bessant, J., Venables, T. (eds.) Creating Wealth from Knowledge: Meeting the Innovation Challenge, vol. ch. 15, pp. 335–358. Edward Elgar Publishing, Cheltenham (2008)

Sparrow, J.: Development of knowledge intensive business services and knowledge transfer channels to support SME service innovation, Paper to be presented at Institute for Small Business and Entrepreneurship conference, Grand Connaught Rooms, London (November 3-4, 2010)

Sparrow, J., Mooney, M., Lancaster, N.: Perceptions of a UK university as a knowledge intensive business service enhancing organizational and regional service innovation. International Journal of Business Innovation and Research 1.1/2, 191–203 (2006)

Sparrow, J., Patel, S.: The failure of business support to facilitate entrepreneurial learning. In: Proceedings of the 30th Institute of Small Business and Entrepreneurship Conference, Glasgow (November 7-9, 2007)

Sparrow, J., Tarkowski, K., Lancaster, N., Mooney, M.: Evolving knowledge inte-gration and absorptive capacity perspectives on university-industry interaction within a university. Education and Training 51.8/9, 648–664 (2009)

Sparrow, J., Tarkowski, K., Mooney, M.: Assessing changes in university aca-demic staff knowledge intensive business service skills, Paper presented at the Engage-HEI conference, University of Bradford (May 20-21, 2010)

Vang, J., Zellner, C.: Introduction to Special Issue on Service Innovation. Industry and Innovation 12.2, 147–152 (2005)

Webb, I.: Knowledge Management in the KIBS-Client Environment: A Case Study Approach, University of Manchester, PREST Discussion Paper 02-12 (2002)

Werson, A.: Lost in Translation: Academic and managerial discourses of knowl-edge transfer, Unpublished EdD thesis, University of Edinburgh (2008),
http://www.era.lib.ed.ac.uk/bitstream/1842/3229/
1/A%20Wersun%20EdD%20thesis%2008.pdf

Wood, P.: Knowledge-intensive services and urban innovativeness. Urban Stud-ies 39, 993–1002 (2002)

Wright, M., Clarysse, B., Lockett, A., Knockaert, M.: Mid-range universities' link-ages with industry: Knowledge types and the role of intermediaries. Research Policy 37(8), 1205–1223 (2008)

Defining Four Pillars for Successful Applied Knowledge Transfer

David-Huw Owen[1] and Zach Wahl[2]

[1] AEA Knowledge Leader: Knowledge Transfer, AEA Technology plc, Gemini Building,
Fermi Avenue, Harwell, Oxon, OX11 0QR, UK
David-Huw.owen@aeat.co.uk
[2] AEA Group Director of Information Management, 1760 Old Meadow Rd., 1st Floor,
McLean, Virginia 22102, USA
ZWahl@ppc.com

Abstract. Knowledge Transfer (KT) is a broad field with a myriad of applications, from academic to business-focused initiatives intended to harness the knowledge individuals and organisations possess. However, without a true business focus, KT initiatives often suffer from a lack of direction, resulting in expended resources without measurable returns and benefits.

With proper business goals, project management, and supporting use of technology however, KT programs can be managed to yield quantifiable investment returns. As a result, individuals and organisations can effectively benefit from, capture and share their knowledge, connecting the right people within an organisation to the knowledge and experts they need in order to be more effective.

This paper outlines the key features of AEA's proven KT methodology and details AEA's *Four Pillars* to effective KT. The value of this approach is highlighted through a selection of brief case study examples.

1 Introduction

There exist many accepted definitions and working variations to the theme of Knowledge Transfer (KT). Most broadly, KT is accepted as the process through which knowledge held by one entity (often referred to as the knowledge base) is passed to a-another entity (often referred to as the recipient). Effective KT being deemed to have taken place only once knowledge has been embedded within the recipient and whereby the recipient is sufficiently enabled to be able to make use of and employ the knowledge for themselves.

Whilst Argote & Ingram[1] defined KT as the process through which one unit (e.g., group, department, or division) is affected by the experience of another, the UK Government has, (since the House of Commons report in June 2003[2] on The

R.J. Howlett (Ed.): Innovation through Knowledge Transfer 2010, SIST 9, pp. 83–93.
springerlink.com   © Springer-Verlag Berlin Heidelberg 2011

Future of Higher Education, and the associated, and often cited, Lambert Review of Business-University Collaboration[3]), adopted the terminology almost exclusively to refer to university-business[1] relationships. Specifically, this definition identifies the transfer of academic knowhow, research and innovations to businesses. Since the mid 2000s, KT in the UK has therefore increasingly been used as an umbrella term for all university *third stream* or *third mission* activity and an increasing drive for commercial interactions with the business community.

The DTI, DIUS, BIS and TSB have successively promoted a UK-wide definition of KT as an essential and inexorably linked component of innovation. KT therefore being commonly defined in government circles as '*the exchange of information through networks where knowledge transfer is about transferring good ideas, research results and skills between universities, other research organisations, business and the wider community to enable innovative new products and services to be developed."* [4].

The Government's wish throughout the intervening period having been to '*promote the transfer of knowledge generated and held in Higher Education Institutions (HEIs) and Public Sector Research Establishments (PSREs) to the wider economy to enhance economic growth*'[5].

2   Beyond Academic KT

'University-business' interaction, however, is but a portion of the overall KT activity being regularly provided, promoted and promulgated throughout the UK.

Often undertaken, for example, as collaborative engagements, networking, business support, innovation stimuli and or high-growth programmes, KT methods are increasingly being commercially used, often without direct appreciation of their actual use, to frame the interactions between two or more organisations where collaborative knowledge sharing and enablement is a key.

Governments, (e.g. local, regional, domestic, European and international) their departments, executive agencies and NGOs increasingly employ elements of KT approaches to undertake government-to-business, -to-stakeholders, -to-community engagements; in the delivery of support programmes for example, in the delivery of behaviour change strategies, in the delivery of fiscal incentives and legislative measures, and in policy implementation and impact determination.

Companies and independent organisations of all forms are also increasingly employing KT mechanisms both with other external actors (e.g. with other business and organisations, with government, academia and the general public) as well as within their own organisations (e.g. between departments, across siloed or disparate operations). In the latter, for example, the methods are typically being used to identify and share good practise and to implement programmes of issues awareness and behaviour change.

---

[1] NB the term 'university' being employed collectively to describe all UK Higher Education Establishments (HEIs) & Further Education Establishments (FEs) involved in KT activities.

3 Understanding the Definition of Knowledge

Like the term *Knowledge Transfer*, the concept of *knowledge* is itself often mis-construe. Understanding the difference between data, information and knowledge is key to being able to deliver effective KT. The associated mechanisms which may enable data and information transfer will often not, in themselves, translate directly to deliver effective transfer of knowledge. Whilst knowledge is a dominant feature of our post-industrial society, gaining an appreciation of the types of knowledge that it to be transferred (especially that which exists within organisations) allows an understanding of the appropriateness and potential likely impacts of the KT tools and delivery methods being employed and to adjust them accordingly. Clearly understanding the form that the knowledge takes therefore and the potential mechanisms for enablement from that knowledge are important challenges in successfully delivering effective and applied KT.

Though Locke[6] is often cited as giving us our first hint of what knowledge is, countless others have continued to try to refine a definition. Davenport and Prusak[7], for example, usefully defined knowledge as, '*a fluid mix of framed experience, contextual information, values and expert insight that provides a framework for evaluating and incorporating new experiences and information*'.

However, one of the most appropriate definitions for KT purposes is Drucker's[8] in which he states that "*Knowledge is information <and or experience> that changes something or somebody - either by becoming grounds for actions, or by making an individual (or an institution) capable of different or more effective action*". Blackler[9] has expanded on a categorisation of knowledge types first suggested by Collins[10] and codified five knowledge types as: embrained, embodied, encultured, embedded and encoded.

4 Common Challenges and Issues with KT

Our experience shows us that KT programmes, irrespective of size, scope or stakeholder context to which they are being applied, traditionally fall short of delivering effective impacts due to a combination of one of four key issues:

1. A lack of understanding and / or appreciation of the knowledge's form and of the constituent mix of KT actors involved (i.e. the direct knowledge holders, the recipients and all other influencing stakeholders - negatively or positively)
2. A scarcity of enablement, facilitation and or behavioural change support required to effect the productive use of the knowledge being transferred
3. An absence of clear, measurable and realistic impact objectives that can be monitored to evaluate the tangible (or lack of) benefit being realised
4. An inability to adjust the programme and / or adapt the specific KT mechanisms being implemented to match the evolving landscape into which the programme is delivering

Consequently, and often due to an array of differing KT appreciations and ter-minologies employed, simply realising a project or delivery element involves KT (of some form or other) can be habitually frustrated from the outset. Effective, measurable, realistic and achievable KT objectives are often not clearly defined and agreed upon from the programme's inception as a consequence. The time required to complete and effect successful KT is therefore often misguided. And the approaches and activities implemented may not be appropriately aligned to deliver the required outcomes.

Similarly, a programme's analysis and appreciation of the constituent mix of KT actors (i.e. knowledge holders and recipients within the wider stakeholder landscape) can often be (erroneously) overly simplified or omitted entirely. An early and complete appreciation of the breadth of stakeholders involved, their differing needs, likely contributions and demands, the level of involvement required and the degree of interaction necessary to effect KT are critical when planning and delivering KT programmes. The analysis of the stakeholder landscape must also be an ongoing activity. External and / or unanticipated constraints placed on KT actors can often change over the duration of the programme which may in itself limit their ability to respond to the KT activities being delivered and which, therefore, will have a direct bearing on the programmes likelihood of success. The landscape of KT actors and influencers may also evolve as a result of the programme own impacts on that landscape and new actors and stakeholders may emerge which are crucial if the desired benefits are to be realised. There is therefore a need to continually and iteratively review, redefine and contextualise all of the KT actors within any programme, and to, as appropriate, realign the approaches being employed to the evolving stakeholder landscape.

5   AEA's Definition of Applied Knowledge Transfer

AEA commercially employs a broad and mature definition to KT in the delivery of its activities. For AEA, knowledge transfer is the fundamental process through which knowledge, held by an entity (the *'knowledge holder'*), is embedded within another (the *'recipient'*) to enable that recipient to undertake a specific action differently and / or more effectively.

- The knowledge itself being appreciated to contain an enabling blend of experience, contextual information, protocols, processes, values and / or insights as are appropriate.
- The transfer process being the combination of knowledge provision, embedding, enablement and empowerment. Successful and effective KT therefore only having been achieved when the recipient is able to productively employ the knowledge for themselves.
- The *'knowledge holder'* and *'recipient'* able to be individuals, groups and or organisations as appropriate, which can represent a mix of any number of stakeholder contexts from across the commercial, academic, third sectors, government and public sector bodies to the general public at large.

- The KT actors can also, depending on the nature of the knowledge and context of the KT programme itself, be both the same active recipient and holder of part of the 'total' knowledge that is being transferred; i.e. a collaborative exchange of knowledge rather than just the linear transfer from holder to a recipient. Indeed, there can also exist more than a single, primary knowledge holder as there can often exist primary, secondary and tertiary tiers of knowledge recipients each with varying levels of interest, participatory needs, and contributions to make.

6 The Four Pillars of AEA's Approach to KT

Our experience therefore shows us that, in order for a KT program to be successful (irrespective of size, scale or scope), a number of core considerations must be incorporated. We define these as the *Four Pillars* to applied KT.

If successfully incorporated, actualised, and intertwined, any KT programme will possess complementary proportions of these key building blocks necessary to understand the knowledge, to help individuals collaborate and transfer the knowledge successfully. And, in applying the knowledge gained, to realise the outcomes of their participation.

Equally, such an approach will, in building on the knowledge gained, also afford the recipient with a framework to continue to develop and capture knowledge further empowering, supporting the KT achieved and enabling sustainable growth and developmental improvement.

Fig. 1 AEA's Four Pillars to successfully delivering applied KT

- **KT analysis** - this pillar requires the continuous, fit-for-purpose examination of both the knowledge being transferred and of the KT actors involved (i.e. the holder, recipient and wider stakeholder landscape) both at inception and as they evolve throughout the programme. This analysis affords an understanding of the knowledge type, its likely inherent value to the recipient and its potential level of accessibility. Adequately mapping and analysing the stakeholder landscape should similarly provides clarity, for example, of the various likely actors, their degree of influence over the successes of the transfer processes, their motivations for engaging, and the relative importance of the knowledge to them and their organisation. Underpinning these analyses, this pillar also requires the continual and detailed quantitative and qualitative review of the KT programme's outcomes against the central aims and objectives as originally idealised.

- **Programme management** - Providing the strategic coordination, development, planning, management, technical support, and analytics which lie at the heart of any successful KT programme, this pillar makes possible the controlled, measurable, and evolutionary KT required. This pillar necessitates that all actions, activities and mechanism provide value-for-money whilst also allowing the programme to flex and respond, as needed, to any changes in the evolving landscape. This pillar also ensures that the key stakeholders (including the client) are able to clearly appreciate, through analytics monitoring, the tangible successes being realised by the program; focussing on the business metrics of benefits and impacts rather than just a programme's traditional output delivery indicators.

- **Enablement through engagement** - Successful KT cannot be delivered through mass dissemination and communication of the knowledge alone. Adequate enablement in the form of facilitation, collaborative, behavioural and change management support is also necessary to effect the productive use of the knowledge being transferred. This pillar requires that throughout both the programme's development and subsequent delivery, the most fitting complementary mix of engagement and enablement mechanisms are identified, checked, validated and employed. That the choice of designing new and / or the use of exiting routs, channels and enablers is motivated based on the actual KT actors involved and that they are continually challenged to support the participants to share, embed and adopt the knowledge for the long-term sustainability of the KT programme's overarching goals.

- **Knowledge management** - Any thriving KT program will identify and generate 'new' knowledge. This pillar therefore requires that appropriate mechanisms to capture and manage both the existing known and emerging new knowledge need to be put in place. That knowledge management considerations are therefore required to provide the supporting infrastructure, tools, platforms and processes that allow all knowledge within the transfer programme to be suitably identified, captured, codified, and retained, and which can enable the ongoing development, growth and continual exploitation of the knowledge moving forward.

7 Case Study Examples

Examples of our KT activities in these areas which underpin our successful methodologies and are Four Pillars approach are included under four core headings of technology support, market stimuli, behaviour change and partnership brokerage.

Technology Support: Since 1999, AEA has managed the DTI/Technology Strategy Board's R&D Programme for emerging and low carbon energy technologies. Such emerging technologies include wave, tidal stream, offshore and onshore wind, carbon abatement technologies, hydrogen and fuel cells, micro-generation,

PV, bio-energy and grid integration. Today the programme is primarily a grant application scheme for collaborative R&D projects. However, since 1999 we have undertaken detailed technology status reviews to understand the global development status of wave, tidal and wind power. And have undertaken technology route mapping to establish R&D strategy and priorities to enable the programmes resource to be optimally focussed on the critical development issues where the UK could gain a competitive advantage. AEA have also managed and administered each funding competition, including the provision of dedicated applicant support to maximise the quality of projects submitted. And have managed the evaluation of all proposals, including the coordination of the peer-review panel of Independent Assessors. Importantly, AEA also directly monitor the projects throughout their funding periods to ensure that they achieve their objectives and deliver real benefit to the UK. Since 2004 AEA has, through this £8 million programme, supported over 600 project applications, managed over 2,000 assessments by over 60 independent assessors, and successfully supported over 150 multimillion pound technical emerging energy related projects through from proof of concept to commercialisation.

Similarly, AEA has, since 2007, also managed the international NZEC project (Near Zero Emissions Coal for China Phase 1) for the Department for Environment, Food and Rural Affairs (DEFRA) and Department for Business, Innovation and Skills (BIS, formally the Department for Business, Enterprise and Regulatory Reform - BERR) which involves coordinating 23 UK and Chinese academic and industrial partners to examine the issues surrounding Carbon Capture and Storage as a 'medium-term' option for mitigating climate change. AEA is leading on capacity building and facilitating KT between Chinese and UK parties including identifying, coordinating and supporting numerous academic and industrial placements throughout the UK. This first phase of the programme aims to examine the various Carbon Capture and Storage technologies available and to identify potential storage sites in China culminating in a 'roadmap' identifying the technical and policy issues that need to be addressed in taking the technology forward to a demonstration phase.

Market Stimulus: AEA currently delivers the 'Linking Innovation in NERC' (NERC-LIN) programme on behalf of the Natural Environment Research Council (NERC). Its objective is to explore and promote a range of mechanisms to ensure that NERC-funded research addresses the needs of a wide range of market sectors, thus maximising its economic impact for the benefit of UK Plc. The programme focuses on recruiting and developing a network of LIN-funded Knowledge Exchange Fellows in universities and other higher education establishments, whose role is to engage businesses regionally with NERC research and establish links with potential end-users through events, clubs, one-to-one brokering and a restricted-access web portal. Since its inception in 2008, the programme has appointed more than 20 high-calibre KE Fellows across the UK who are actively pursuing opportunities for closer collaboration between researchers and businesses. The NERC-LIN programme has also carried out scoping studies into ways of developing better knowledge exchange with key priority areas for the UK economy, including the marine industries and the energy sector; in particular geo-thermal, nuclear, wind, wave and tidal power generation.

Since 2008, AEA has developed and managed both the RipplEffect and Big Splash campaigns under Envirowise for the DEFRA to provide water efficiency KT and support to business of all sizes across and in all sectors throughout England. Providing structured KT support the programme delivered on-site support to help the companies understand and map their water use, to identify opportunities to start reducing their water consumption and save money, to implement behaviour change mechanisms, to embed this knowledge within the company and to help monitor and measure the water and cost savings made.

In total more than 750 businesses have, to date, participated in this ongoing initiative. Just under 0.5 million cubic metres water savings have already been achieved, with £1.9 million cost savings identified through independent on-site audits, and £415,000 actual cost savings. The wider Envirowise programme, deliver since 1999 by AEA for the Department for Environment, Food and Rural Affairs (Defra) to help UK companies improve resource efficiency, reduce waste and save money has, through various KT instruments, having stimulated to date saving totalling over £1.3 Billion at a total Government cost of £50 Million, and annual savings to UK businesses of over £220 million per year. Since 1994, the programme has dealt with over 172,000 business enquires, and the programme's 140 plus Waste Minimisation Clubs involved over 3,000 active member companies throughout the UK.

Behaviour Change: AEA has an extensive track record of working with NHS Trusts (both Acute and Foundation) with ongoing delivery relationships throughout the UK. Currently providing a mixture of technical consultancy and KT services, these programmes primarily include energy, carbon and waste management technical consultancy (from feasibility studies through to physical implementation programmes) as well as legislative and compliance support services. A major element of these programmes is their KT initiatives for direct enablement and behaviour change with front-line staff to improve, for example, their day-to-day carbon, energy and waste practises. Working, for example, with Birmingham Shared Services (a consortium of three Birmingham primary Care Trusts) AEA recently undertook activities to support them in both reducing their primary energy consumption and to meet their challenging CO_2 emissions reduction targets.

Through complementary programmes of awareness raising, and dedicated KT workshops, enablement and behaviour change initiatives AEA has support the Trusts in making significant improvements in their overall sustainability i.e. reducing energy and water consumption, in reducing their waste output, and in improving their resource efficiency and recycling rates as well as their transport and procurement processes. This has led to an identified potential annual carbon dioxide savings of over 5,475 tonnes as well as some £300,000 per year in cost savings via low-cost interventions and changes in individual activations. Similar programmes undertaking two separate waste reduction specific KT projects for two Welsh NHS Trusts delivered savings of almost 30% in waste arising (including hazardous wastes) amounting to savings of over £70,000 per annum per Trust.

AEA is also working across UK government departments, executive agencies and local authorities on similar carbon and waste reduction programmes, and has recently been appointed to work with the US Government to set Departmental carbon and energy reduction targets, to help implement these through a combination of technical consultancy and KT led behavioural change, and to verify their achievements through impact evaluation and benefit realisation analytics.

Partnership brokerage: In the 'Closed Nuclear Cities Programme' AEA is working with our partners HTSPE to help commercialise non-military products and services from Russian nuclear weapons research centres. The programme involves significant change management within the institutions involved in the programme and, in particular, developing the marketing and business management skills of former weapons experts. The specialist activities delivered by the programme include training and coaching of product/service development and marketing teams, the management of a 4 million Euro per year innovation grant disbursement programme, and the development of commercial (sales and investment) partnerships in the high tech sphere between British companies and closed city organisations. AEA have also provided support for the mobilisation of existing in–country legal and financial advice to help establish new technology-based ventures and for the specialist commercial protection and exploitation of IP in conditions of military and political sensitivity.

As of May 2010, in Russia and other former Soviet Union nations, the programme has secured investments of over £37 million, has brokered nearly 150 in-country grant supported projects, created over 4,700 sustainable jobs, and delivered over 275 specific in-country collaborative projects for training, mentoring, commercial partnerships and economic development.

8 Conclusions

In this paper we have outlined the key features of AEA's proven KT methodology and our *Four Pillars* to effective KT. The success of this approach has been highlighted through a number of specific brief examples which have highlighted the impacts and benefits delivered by those programmes.

By incorporating the Four Pillars to KT effectively, and by employing the continual iterative cycle to KT design, delivery and recalibration, our approach provides organisations the opportunity to realise a number of key benefits:

-   An engaging KT programme that responds with, speaks to and delivers for the needs of all of its constituent stakeholders.
-   A process which equally supports enablement, embedding and engagement, and which recognises that effective KT is only successfully achieved when the recipient is able to make independent, productive use of that knowledge for themselves.
-   A suite of Tools, Design Factors and analytics that can be quickly and effectively deployed, and that are able to be easily refined, as necessary, to match the evolving landscape which they are influencing and informing.

– An underpinning methodology that evaluates the successes of the KT through impact and benefit realisation, and that is subject to continual iterative review of the outcomes that are being achieved rather than just delivery output alone.
– An overarching approach that is sufficiently flexible to deliver a highly tailored KT methodology, per programme, and that is specific, focused and bespoke to each client's particular needs.

9 About AEA

AEA Technology plc (AEA) is a global consultancy firm that has, for nearly 40 years, helped hundreds of public and private sector organisations respond to environmental challenges and opportunities. Delivering solutions predominantly across areas of climate change, energy and environment, AEA provides consultancy for a wide range of clients, from major UK, EC and US Government departments, to FTSE 350 firms and other global businesses.

AEA currently supports some of the world's most complex national and international Knowledge Transfer efforts which account for almost 20% of the company's annual turnover. Our successful track record and the methods employed are founded on more than two decades of active KT engagements and experience. Our approach, in more than 200 major KT dedicated programmes and countless other supporting actions has been deployed across public, private, academic and third sector interactions alike. Our tools and methods have been delivered and refined through application across a host of industry specialisms including health, education, transport, energy and climate change, waste and resource efficiency, air and water, and sustainable production and manufacturing. Our KT expertise has supported programmes for market transformation, high growth, sustainability (socio-economic and environmental), innovation (both technology and processes), as well as specialist research and development, demonstrational and deployment activities. Our KT approaches also leverage extensive complementary expertise in specialist world-leading technical consultancy, in knowledge management, project/programme management, and technology and innovation brokerage services. In so doing, our KT services have successfully stimulated proactive collaboration, the effective sharing of ideas as well as the enablement and behavioural changes necessary. Our KT programmes have therefore not just transferred knowledge, but have stimulated further innovations, have encouraged improved sustainability and economic competitiveness, whilst also delivering social, cultural and (both organisational and) individual enrichment.

References

[1] Argote, L., Ingram, P.: Knowledge transfer: A Basis for Competitive Advantage in Firms. Organizational Behaviour and Human Decision Processes 82(1), 150–169 (2000)

[2] The Future of Higher Education, House of Commons Education and Skills Select Committee (June 2003), Report & formal minutes for the Fifth Report of Session 2002–03 Volume 1 (ref: HC 425-I)

[3] Lambert Review of Business-University Collaboration, HM Treasury (December 2003) ISBN: 0-947819-76-2

[4] This was based on and adapted from - Fagerberg, J., Mowery, D.C., Nelson, R.: The Oxford handbook of innovation. Oxford University Press, NY (2005)

[5] Research Council Support for Knowledge Transfer', House of Commons Science and Technology Committee (June 2006), Report, together with formal minutes for the Third Report of Session 2005–06 (ref: HC 995-I)

[6] Locke, J.: An Essay: Concerning Human Understanding, BOOK IV. of Knowledge and Probability (1689)

[7] Davenport, T., Prusak, L.: Working Knowledge. Harvard Business School Press, Boston (1998) ISBN-10: 1578513014

[8] Drucker, P.F.: The New Realities. Transaction Publishers (1989) ISBN-10: 0765805332

[9] Blackler, F.: Knowledge, Knowledge Work and Organizations: An Overview and Interpretation. Organization Studies 6, 1021–1046 (1995)

[10] Collins, H.: The Structure of Knowledge. Social Research 60, 95–116 (1993)

Organisational Identification of Academic Staff and Its Relationship to the Third Stream

Trevor A. Brown

Business Development Manager Manchester Metropolitan
University Cheshire, College House,
Crewe Campus, Crewe Green Road, Crewe, Cheshire, CW1 5DU
t.a.brown@mmu.ac.uk

Introduction

A university is a organisation where academics study, research and teach students. The archetypal "academic" has an image and identity that is as clear as a doctor or fireman. However the nature of a university is changing, the university is now required to seek out new relationships with businesses and non traditional "customers", delivering learning and knowledge in new ways, frequently driven by commercial demands. University senior management teams are motivated by government and funding to meet these demands and steer the university towards these new goals. These new areas of activity are often referred to as the "Third Stream" TS (teaching and research being streams 1 and 2). The new mission, strategies and definitions of third stream initiatives form a changing organisational identity for a university which may challenge widely held notions of a universities identity by its member staff, the academics. Dutton et. al. (1994, p1) state; *"Strong organisational identification may translate into desirable outcomes"*. If the university wants its members (the academics) to embrace the changing mission of a university and undertake actions in support of the new mission, university managers must understand the organisational members (the academics) relationship to the new identity and aim to engender a strong organisational identity.

Unpicking the academics definitions of aspects such as TS and academic identity, and how individuals engage with, relate to, or define the new organisations identity being formed out of the changing nature of universities, could give indications as to what constitutes their organisational identity and the level of *"identity dissonance"* (Elsbach et. al., 1996, p1). A clear understanding of this dissonance would support university managers understanding of members engagement with the strategy and mission in support of TS. Once modelled this then could lead to recommended actions which generate *"intraorganisational cooperation or citizenship behaviours"* (Dutton et. al., 1994), thereby supporting the strategic direction of the organisation.

Via the literature review the study will research and then establish a set of factors of academic identity and utilise these to investigate the organisational members identity and their perceptions of how this identity is valued by the

R.J. Howlett (Ed.): Innovation through Knowledge Transfer 2010, SIST 9, pp. 95–112.
springerlink.com   © Springer-Verlag Berlin Heidelberg 2011

organisation, described as organisational identity. A set of factors for third stream will be developed and utilised to establish the academics understanding of third stream and its importance to them. This will establish a level of organisational identity and how third stream impacts upon and relates to this identity.

1 Literature Review

1.1 Organisational Identification and Strategy

Current thinking, according to Rughase (2006), is that strategic management practice focuses on logical aspects and gives examples such as the favoured economic resolution and states that other aspects such as values and emotions of organisational members are dismissed. Leibl (2001) and Mezia et. al. (2001), (cited in Elsback, 1996) back these notions up commenting that *"Strategies often fail as they do not join the prevalent concepts and desires of organisational members"*. Andrews (1987, p. 19) sets an early baseline in this thinking, arguing that the values, ideals and aspirations of individuals influence purpose and need to be brought into strategic decision making and that problems within strategy implementation were because, during strategy formulation, the members past thinking, personal values, cultural loyalties, rules and restraints which formed beliefs about their organisation, where not incorporated. Dutton et. al. (1994) introduces a further aspect of identity with strategy describing how the individual organisational member will interpret the various strategic issues, this interpretation will then influence which strategies are noticed and which are not.

Dutton et. al. (1994) models an individual's identity and self concept as a relationship to the organisations identity and how this "organisational identity" can in turn shape an individual's identity. It specifically focuses on the individual's image of their organisation. The degree to which the members concept of their personal identity is perceived (by the member) as having the same attributes as the organisation is described by Dutton et. al. (1994) as "organisational identification" The author produces a strong argument that members of organisations will change their behaviours by thinking differently about their organisation. It is argued that a positive organisational identification may convert into desired outcomes, examples include; intraorganisational cooperation or citizenship behaviours. More recent studies have supported this argument and found interactions between organisational identification, motivation and well-being. (Wegge et. al. 2006). The process of identification is described by Ashforth et. al. (1989) as one of self-categorisation formed through ritual, ceremony and stories which support the communication of the identity to members. Negative relationships between members and the organisational image are also found to produce negative business outcomes, as was found by Dutton et. al. (1994) with the Port Authority New York and Exon executives. This can result in undesirable outcomes such as constrained positive actions towards responsibilities or tasks. An aspect not explored, but which may occur, is a null response, the organisational member may not experience sufficiently strong or negative organisational identity to produce any response of significance.

Mael and Ashforth (1992) point out that a professional and or occupational identity are not automatically specific to the members organisation and that values within a profession may conflict with those of the employing organisation. Here Mael and Ashforth (1992) separates the identity a member has with the organisation (I work for Manchester Metropolitan University) from the identity the member has with a profession (I am an academic). This is referred to as "Professional and occupational identity", the member defines him/herself in terms of what they do rather than who they do it for. It is argued (Vanmaanen and Barley, 1984) that members embrace the archetypal character attributed to individuals within that work.

1.2 A "Conceptual Framework" for Organisational Identity

Following Yin (1994) (cited in Saunders et. al., 1997 p, 348), the study will make use of existing theory to devise a framework within which to conduct the research. The model derived from the literature review is expressed here as a "Conceptual Framework" (Robson,1993 p 63) and is shown in Figure 1.

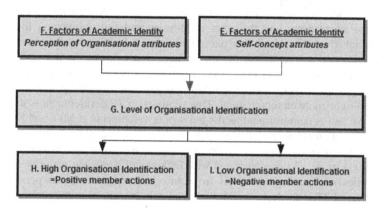

Fig. 1 Organisational Identification (Modified with academic titles, Dutton et. al., 1994)

The original model of member identification developed by Dutton et. al. (1994) can be demonstrated in Figure 1. Please note the researcher has adapted the model to utilise academic member titles, however the general model remains, i.e. that the members perception of self (E) interacts with members perception of the organisation (F), this results in a level of organisational identity (G). Dutton et. al. (1994) establishes that a High Organisational Identification will equate to positive member actions in support of the organisation (H) and that a low Organisational Identification will result in negative member actions in support of the organisation (I). Please note; the arrows denote a causal relationship and influence in the direction of the arrow.

1.3   Third Stream within the Literature and the Entrepreneurial University

This new strategic development area for universities can often be referred to as, or encapsulated within the term "Third Stream" (TS). Academic studies on university TS activity is a relatively new field with literature remaining "*rather fragmented*" (Rothaermel et. al. 2007, p. 1). Rothaermel et. al. (2007) conducted a review of the TS literature and found that reference to academics outside of technology transfer roles are distinctly absent; the majority of papers study professional entrepreneurial staff within the university structure, not academics operating in the mainstream (teaching and research).

This focus on technology transfer roles highlights the entrepreneurial perspective found in the majority of literature on TS. A leading study on the entrepreneurial university undertaken within the Triple Helix University-Government-Industry model (Etzkowitz, 2008) develops the notion of industry, government and university interlinked for the purpose of innovation and entrepreneurship. The focus is on the entrepreneurial expressed as the development of the quazi firm, technology transfer offices and research groups. In regard to the individual academic Etzkowitz (2008) describes the ideal for supporting the Triple Helix as an academic with a foot in both camps, one in academe and the other in industry and firm (company) formation within or linked to universities. The participating individual is a distinct entrepreneurial academic and separate from the mainstream operations of the university. This review of an entrepreneurial university is developed from an entrepreneurial academic perspective, either developing new firms, organisations and patents or developing staff to be more entrepreneurial. This approach to TS development is shown to be effective and is demonstrated as the basis of developments at MIT and Stanford but does not satisfactorily address his findings that : "*Many academics believe that a university best fulfils its mission by limiting itself to education and research*" (Etzkowitz, 2008, p. 4) . As demonstrated with Etzkowitz (2008), the majority of work on TS relies on a definition of TS from an academic entrepreneur perspective. A definition of TS is unclear, yet TS within an entrepreneurial framework is placed within university strategic aims, mission and vision e.g "*An enterprising organisation with enterprising staff and students;*" (MMU 2008) with income targets, within the strategy, defining engagement in entrepreneurial and financial terms. The scope of TS activities is limited, definitions of TS falling almost exclusively within entrepreneurial and commercial/financial activities.

2   Methodological Approach

The research consisted of a small deductive research project, interviewing seven Academics, within an example organisation (MMU Cheshire Faculty), to understand the context and processes within the organisation (Morris and Wood, 1991). The main driver of the study is to explore organisational identification and the organisational members responses to impacts on their identity, derived from Dutton et. al. (1994), in particular the emergence of TS and its impact on members (the academics) identity. This is approached via a qualitative interpretation of a methodology constructed for a study of alumni (Mael and Ashforth, 1992). The original model incorporates generic measures of organisational identity but requires adapting to the specific organisational context. This study will develop new relevant antecedents via the literature review, which will establish a baseline of significant antecedents expressed here as Possible Factors of Organisational Identification / Academic Identification.

2.1 Correlates of Academic Identity

For this study new specific Correlates of Organisational Identification are required which will apply to and contextualise the model of organisational identification, for clarity I will refer to these as "Factors of Identity". Henkel (2005) undertakes a study of academic identity within policy changes, utilising *"communitarian moral philosophy and symbolic interactionism"* as the basis for a review of academic identity. The author unpicks aspects of academic identity prior to the changing environment (of academia) and then discusses the change factors impacting on academics. I have utilised the Henkel (2005) paper to establish the key Possible Factors for academic identity as follows;

2.2 Possible Key Factors of Identity

- Academic as a living tradition, the history and role
- Academic autonomy- (pattern working life / quality of life)
- Academic control of teaching and research
- Academic freedom (research agenda and priorities)
- Bounded academic space, The strong
- Classification and boundaries between groups and disciplines, The strength of
- Community of scholars, The defining
- Community other, The defining
- Department, The
- Disciplinary culture
- Discipline, The
- Epistemology, The
- Institution, The
- Integration into the community, The level of
- Managerial culture, The
- Multiple and contradictory identities (avoiding fixation on a single identity)
- Narrative account of self and changing of identity over time, The
- Obligations, fulfilment and respect of the community, The

- Power of the group/community, The
- Status in the nation "definers, producers, transmitters and arbiters of advanced knowledge"
- Unit, The
- Values and beliefs of the community, The

Although Henkel (2005) outlines these differing parts of the academic identity, this work is undertaken within a presumption that the academic identity is entirely formed within academic related activity or employment. The academic will have a broader range of inputs to their identity than implied by their current role.

2.3 Academic Enterprise and Third Stream

According to Molas-Gallart et. al. (2002)
"There are no magic bullets in indicators of Third Stream activities. A variety of indicators need to be collected. Each of them will, by itself, be incomplete and its interpretation will be open to questioning. Yet when taken together, the result can be a powerful measurement system." This is supported by Alice Frost Head of Business and Community Policy at HEFCE (2008). *"What I have found in discussing different terminologies, is that when any individual or organisation tries to define terms, they become reductionist of the agenda. And while third stream funding has been around for many years now, one person's or HEI's definition of the terms can be very far from another's"*. In reference to the above 2 views the researcher is defining the case study's interpretation of TS by the organisations own measures. These measures will to some extent, represent the strategic direction and drivers for the university. The researcher has utilised the case study organisations internal HEFCE reporting document titled "HEIF 4 Pro Forma". This is a document used to capture TS in faculties for central reporting. This form is supplemented by the Academic Enterprise Strategic plan developed in 2007 and the MMU Cheshire Strategic Plan 2007-2008. Evaluation of these documents revel the following;

2.4 Factors of Academic Enterprise and Third Stream

- Academic Enrichment
- Collaboration development
- Community Engagement
- Conferences
- Consultancy
- Contract research (Business funded or Applied research)
- Curriculum Development Mainstreaming of innovative products
- Development of Knowledge
- Employer led accredited courses
- Engagement with business

- Engagement with regional forums
- Facilities and equipment services
- Formal understanding of business need in region.
- Formal understanding of community need in region
- Funded Projects
- "High Interest" activity development
- Income generation/ commercial income
- Increase graduate recruitment
- Increased Student Numbers
- International links with Universities and HE Colleges
- IP Intellectual Property income
- KTPs
- Outreach and networking
- Partnership opportunities
- Partnerships Brokering relationships/networking:
- Partnerships business assists
- Partnerships joint funding applications.
- Professional Body Links
- Raised awareness amongst businesses.
- Reputation for Knowledge
- Short courses (non accredited)
- Student enterprise
- Raised profile of staff within the business sector.
- Recognition as a Knowledge Centre
- Staff development
- Student enterprise training
- Student Social Enterprise schemes improving employability.
- Utilisation of a wider staff skills base

These forms and the definitions of TS, are the case study organisations main methods for driving and capturing TS activities and as such represent the organisations summary or definition of TS. The categorisation and member interpretation of TS activities will directly influence the member's perception of the organisation and impact on their self concept. This can be illustrated with a revision of figure 1, adding the Third Stream Change Factors to the Organisational Identification model. The original model (figure 1) has been modified here (figure 2) to evaluate the impact of a change factor (A), the Third Stream, on the member academic's Organisational Identity (G), the arrows within the model representing causal links. The level of impact (C and D) of Third Stream will be determined by the member academics interpretation of Third Stream (B). This interpretation is influenced by how the organisation has categorised or implemented this change (A). The extent to which the impacts (C and D) interact with the members organisational identity (G) may then have positive or negative effect on member actions (H and I) in support (or not) of the perceived Organisational Identity (G). The researcher is particularly interested in whether these member actions (H and I) will be actions defined within the organisations Third Stream definitions (A) or not. This may reveal whether; 1. The TS initiatives have become characteristics of the organisation and 2. They have become a part of their self concept. (Dutton et. al., 1994)

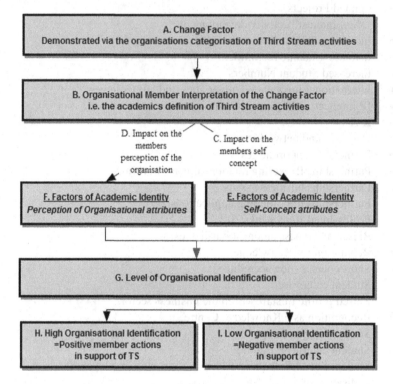

Fig. 2 Organisational Identification with the Third Stream Change Factor. (Conceptual Schema)

3 Interview Schedule

Seven academics where selected for the study. Individual member data was be established, including; Length of service in MMU, Length of service as an academic, Role, Title, Subject area, Prior career, Age and an approximation of the time spent on Teaching, Research, Administration and Other activities was established. An open semi-structured interview on Academic Identity / Organisational Identity followed, utilising the factors for academic identity, issued to the interviewee as cards to prioritise in response to a set of questions.

Academic Enterprise and Third Stream was explored using open semi-structured interview about the individual's definition and how they perceive it. Utilising key factors for Third Stream and Academic Enterprise (separately) issued to the interviewee as cards to prioritise in response to a set of questions.

Note; please contact the author for details on data sets, quotes and processes.

4 Findings

4.1 Qualitative Level of Organisational Identification

The combination of interviews and chosen factors give a qualitative and initial indication of the level of Organisational Identification perceived by the organisational members, the interviewees.

4.1.1 Managerial Culture vs. Academic Autonomy/Freedom/Control, (Administration)

The joint highest priority of all negative factors of OI was that of a Managerial Culture, and this is viewed as a priority for the organisation. Local management was less of an issue, but imposed decisions, from higher in the institution, are the main concern. A key enactment of this Managerial Culture is found within administrative duties. The level of these activities encountered by the organisational member is in direct conflict with the notion of Academic Autonomy/Freedom/Control, the highest positive factor of OI. Although the organisation is viewed as supporting Academic Autonomy/Freedom/Control this is felt to be a surface level of support and the administrative burden imposed by the Organisations Managerial Culture is perceived as impacting on this freedom. Further aspects of the Organisations Managerial Culture appear to impinge on developments wanted by the Organisation and there is no perceived management or operational support for these organisational goals. The Strength of Differentiation between the disciplines was also viewed as a management issue not valued by the academic members, resulting in further perceptions of the organisations management not supporting the members, when cross faculty or department action are required.

4.1.2 Institutional Hierarchy

In most cases the levels of Institutional hierarchy appear to be directly related to the level of IO. There does appear to be OI with the individual unit, a significant level OI with the Department, less but some with the Faculty and little or non with the Institution. However this was not the case with one interviewee with the shortest length of service, 1.3 years, who did state a positive level of OI with the institution.

The Henkel (2005) derived factors of identity where found to be lacking in 2 areas. Comments from interviewees led to 2 new factors emerging during the interviews;
1. Obligations to the Learner and 2. The External Community.

4.1.3 Obligations to the Learner

This factor is joint highest value to the organisational members. Overall the respondents felt that the institution does support a value of Obligations to the Learner, however the actions and disappointments concerned with operations and administration, counter this, leading to the interviewees feeling that it is not valued in reality.

4.1.4 Others

Community is a high positive factor of member's identity but believed to be undervalued by the institution. The level of OI within the Disciplinary Culture (Academic not Managerial) is low, with a feeling that student numbers drive the organisation not the quality of the academic disciplines.

4.1.5 A Summary of the Qualitative level of Organisational Identity

The combination of factors of Academic Identity and an analysis of the interviews leads to an evaluation of the "perceived organisational identity" (Dutton et al. 1994, p.1). From the information gathered it is clear that the small evaluative sample of Organisational Members interviewed, is experiencing a Low Level of Organisational Identification with the University. This can also referred to as "organisational dissonance" (Elsbach and Kramer 1996). There is a an emergent possibility that this relationship is time dependent, with the newest member of staff having the most positive OI with the institution, further study would be required to evaluate this. This does seem to contradict the literature as Mael and Ashforth (1992) found that the length of time a person is associated with an organisation has a positive impact on their level of organisational identity.

4.2 The Academics Definition of Third Stream and Academic Enterprise

The organisational member definitions and understanding of what AE and TS are, is varied and contradictory. Some of the members feel it is the same thing, others view it as entirely different and for those that view TS and AE as the same, they have very different interpretations of this. Money features highly within the interviews and there is a mixed view as to how this defines TS and AE. Overall there is an understanding that AE or TS (dependant on the individual) will include some aspect of income generation, however all the interviewees would tend to contextualise the money aspect, defining it as clean or "*dirty*" money, separating out relationships between community and business and highlighting its links to research and teaching. It is notable that only one organisational member stated that AE was defined by where the funding originated, i.e. not HEFCE funding, however this was stated for AE not TS. From the organisations standpoint this is the TS definition. For those members who embed AE within Teaching and Research the relationship to money is viewed as a negative attribute. Overall the money aspect of the TS and AE definitions is a part of the members definitions, which to a greater or lesser extent was recognised. However during discussions on money, each member preferred to define TS and AE by other characteristics which complemented, or was an intrinsic part of, their teaching, research, knowledge exchange ("*ideas out*"), community obligations and their career choice in becoming an academic.

A further observation regarding this contextualisation is reflected within the subject areas. Those members who rate AE as a high importance are within a Business and Management Department and those with the low importance are academics within Literature subject area, obligations to the community and knowledge exchange being drivers for this differentiation.

Although TS activities are measured by the organisation as separate activities, this is not necessarily the operational experience of the members. The "Other Activities" question raised the definition of how the member perceives AE and TS. Overall AE and TS are perceived as an embedded part of the member's core roles of teaching and research (of which administration, in this context, is a part). This resulted in many of the members being unable to separate the four activities (Research, Teaching, Administration and Other Activities) into distinct sections, as the researcher had envisaged.

4.2.1 Summary of the definition of TS and AE

These results combined with the AE and TS factor choices, gives an understanding of the members definitions of AE and TS being formed almost entirely from an individual and academic perspective. Teaching, research, knowledge exchange ("*ideas out*"), community/learner obligations and their career choice in becoming an academic are the main defining drivers for definitions of AE and TS. Aspects of income generation are viewed as organisational drivers for TS and AE and its importance in defining TS and AE is secondary to individual and academic drivers.

5 Conclusion

5.1 The Effect of the Members AE and TS Definitions on the Level of Organisational Identity

The organisational members understanding of AE and TS is founded within the context of the factors of Academic Identity detailed earlier. The core activities of Teaching and Research enacted through Academic Autonomy, the Discipline, Obligations to the Learner and the Community are reflected in the definitions of TS and AE. Although the factors for AE and TS where all recognised by the organisational members, they are perceived within and/or as a compliment to their core identity. The organisation has developed a set of measures (the factors) formed from funding demands and the changing nature of universities, not based on this core identity. This results in a set of factors and definitions which are the same yet with perspectives, priorities and drivers for engagement which are quite different. The organisational members enact TS and AE activities because of the links to their core identity, not for income generation. As the (AE and TS) definitions are based on the notion of Academic Identity then this change factor is subsumed into the definition of the Academic Identity. TS and AE are a part of the Academics Identity and therefore the qualitative evaluation of members OI remains unchanged. However the financial aspects of AE and TS are negatively viewed by the organisational members and this association may result in an increased level of Negative Organisational Identification as these financial aspects are highlighted by the organisation.

5.2 Does a Low Organisational Identification = Negative Member actions in support of Third Stream? (Please Refer to Figure 1)

The model developed in Figure 3 anticipates that a low level of OI will result in negative member actions. Given the limited scope of this research a tentative initial finding would be that this part of the model stands correct. The impact of the negative OI is either inconsequential or results in negative actions. Where TS activities exist they occur in spite of, or regardless of the organisation, they occur because of the nature of Academic Identity. Where they do not occur, associations with managerial culture (the key negative organisational factor) and its impact on academic freedom, appear as the main aspects of the "organisational dissonance" (Elsbach and Kramer 1996). As there is no data to support a "High Organisational Identification =Positive member actions in support of TS" (please refer to figure 3, section H). No assumptions can be made as to the validity of the section of the model.

5.3 Engagement with the TS Strategy

As the definition and value of TS and AE has been found to be an intrinsic part of; the organisational members identity, their Academic Identity and interpreted through this identity, there seems little evidence (given the limited scope of this small evaluative and qualitative study) that organisational members are engaging with TS as a result of the Organisations Strategy. There is some pragmatism from one member expressing a view that if her job depended upon it (TS) her priorities would change, but this was a single comment. Priorities for all members where focussed on teaching followed by research and the TS activities that emerge, are the result of their complementing and supporting the members core identity and values, not because of any strategic initiatives. In one case the interviewee believed that institutional issues such as the managerial culture, worked against the freedom required to deliver TS initiatives and progress towards organisational goals and strategy was hampered.

5.4 Organisational Identification and Strategy

In answer to the research objective "To investigate to what degree does strong organisational identification in mainstream staff, result in significant engagement with the TS strategy." There is no strong OI, a weak or negative OI exists within the selected members. This negative OI or "organisational dissonance" (Elsbach and Kramer 1996), as a minimum, may have no effect on their engagement with the TS, as the engagement is dependant on the Academic Identity not on Organisational Identity. However this negative OI may also impact on engagement with TS as aspects of the organisations identity, linked to the management of TS (e.g. income generation), are in conflict with the organisational members identity, and may result in a negative impact in engagement with the organisations TS strategy.

One unexpected observation is the shock experienced by the members at the number of AE/TS factors issued during the interview. As is stated within the Enquiry Design section, the selection of these 38 factors was based on various organisational documentation, the individual factors come directly from these documents. It was clear that the range of AE and TS activities had not been presented to the members previously. This raises questions as to how engaged the organisational members are in the strategy making process and the how engaged they are with strategies in operation. Arguably the range and diversity of the factors for TS would be more familiar to the organisational members if a greater degree of interaction had taken place.

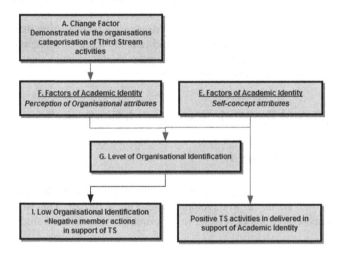

Fig. 3 A model of Academic organisational identification in relation to change and actions

Figure 3 proposes an adapted summary model of Academic OI in relation to change and actions, observed in this study. Factors of self concept or Academic Identity are the drivers for actions in support of TS, the level of OI is negative and equates to negative actions in regard to TS.

5.5 Changing Mission

The literature review established that a changing mission for universities was a key element of the development of TS activities. This study gives some indication that this issue of the changing mission seems less prevalent in the development of TS. The issue for individual academics is how the changing mission manifests within managerial culture. TS appears to be an embedded core value that exists and need not be grafted on, academics need not change their core identity to accommodate these demands. It would appear that it is the management of the engagement of academics in the strategy and processes for TS that require re-evaluation. These strategies currently formed into income targets, are the core strategic measures operated by the organisation and it is these that need to be contextualised against individual academics drive and obligations to deliver TS activities in support and in compliment to their teaching and research.

5.6   Entrepreneurial Academic and Mainstream Academic

The researchers first thoughts on the separation between an entrepreneurial academic and mainstream academic, formed from authors such as Etzkowitz (2008). Etzkowitz (2008) recognises that *"Many academics believe that a university best fulfils its mission by limiting itself to education and research"* (Etzkowitz, 2008, p. 4). The assumption is that this would not be entrepreneurial and by his definitions not include TS. The findings for mainstream academics show that TS is embedded within the teaching and research and is at the core of academics identity, so the proposed opposition between a mainstream academic and an entrepreneurial academic is not as clear as imagined.

6   Strategic and Research Recommendations

1. Expansion of the Study

A further study of a case study organisation that links its strategy to academic identity drivers would be useful in developing these ideas further. This would need to longitudinal and considerably broader in the numbers of academics studied.

2. Broader Innovation and Policy Research

Further studies could evaluate the relevance of organisational identity with business/university relationships and the development of TS activities, in relation to Furman et. al. (2002) and Etzkowitz (2008). Exploring how organisational identity supports the strong relationships required for TS.

3. Strategic Recommendations

A final conclusion of this study is to make an early presumption on the approach managers could take to develop TS activity within the university. The recommendation would be to utilise the factors for academic identity as a key driver of engagement, thereby developing an engagement strategy that complements and is formed from the embedded identity demonstrated by the academics. Engagement of the academic teams in TS strategy development and realigning the strategic goals to be based on academic drivers rather than monetary drivers would support engagement and increase activity, the final outcome being increased TS activity.

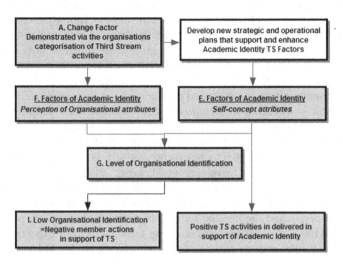

Fig. 4 Strategic recommendations

In speculating an approach for this strategic recommendation university managers could undertake a subject group based approach to understanding the academics perspectives, priorities and drivers. Types of TS activity could be matched to these groups and the complementary benefits of these TS activities highlighted to the group, on their terms. This will involve a redefinition of some types of TS activity and work to remove the separate status TS activity has from the core roles of teaching and research.

A speculative example of this could be proposed with KTP project development. Currently these KTPs are "sold" internally to academic staff as a way of generating income and complementing the TS strategy. A new approach would be to emphasise aspects of the project which complement the individuals "Academic Identity self concept attributes". For example this could include an emphasis on the;

- Academic freedom intrinsic within a KTP (time, budget and subject area).
- Benefits to community organisations (improvements to services, cost savings).
- Benefits to the learners from up to date and direct research through KTP.
- KTP as a different teaching environment (an organisation rather than a class).
- Demonstrated examples of similar academic "types" working on projects which support these" Academic Identity self concept attributes".

7 Boundaries and Limitations

A limitation of my research is my interest with a type of worker within the organisation, who I refer to as the "mainstream academic". An academic may have a
significantly large teaching load during the sample, or opportunity to engage has
not arisen. The current job descriptions of staff do not explicitly include TS as a
duty, so although specified strategically, locally the academic may not see this as
their role or task, the academic may have high organisational identity but not engage in TS, whichever definition is utilised. Assumptions where also made that the
academic is aware of the universities third stream strategy.. Mael and Ashforth
(1992) find limitations in the proposed methodology as they state that the *"causal
sequence from antecedents to identification to consequences" is untestable and
recommend a "within-subjects longitudinal approach to capture the dynamics of
identification over time"*. The study is cross-sectional and this will need to be
accounted for in the findings. The researcher, myself, is working within the case
study faculty as a Business Development Manager, although this has led to my
interest in the research subject, I am a key staff member in TS development directly reporting to the Dean. The risk of unseen researcher bias due to this position
is high; objectivity and the need to remain an external observer are problematic.
There is also the risk of respondents adapting answers to meet expectations associated with my role, and my links to senior management, this link could also be
used to send a message to senior management. The sample group of "mainstream"
academics selected from a random group of teaching academics, is subject to differing operations and cultures within each department. For example the core
delivery hours for teaching vary considerably, the financial reward system is interpreted differently and departmental cultures differ. The chosen sample size of 7
has limitations. Across 2 departments this would averages out at 3-4 Academics
for each department. With the variations between departments this may dilute the
sample.

2 of the interviewees commented on the number of factors for TS, presented as
cards in the interviews. The number of factors reflects the range of factors visible
in documentation used within and forming the TS strategies for the university. As
such this is representative of the scope of activities covered by the TS definition.
Comments are also found in the subtle differentiations between factors for TS,
interviewees commenting that some of the factors are too close in their meaning.
Although the researcher has removed distinct doubles in the development of the
factors, this level of sophistication is required to establish the interviewee's real
understanding of each factor.

References

Andrews, K.R.: The Concept of Corporate Strategy. Irwin, USA (1987)
Ashforth, B.E., Mael, F.: Social identity theory and the organization. Academy of Man-
    agement Review 14(1), 20–39 (1989)
Dutton, J.E., Dukerich, J.M., Harquail, C.V.: Organizational images and member identifica-
    tion. Administrative Science Quarterly 39(2), 239–263 (1994)

Eisenhardt, K.M.: Agency theory- an assessment and review. Academy of Management Review 14(1), 57–74 (1989)

Elsbach, K.D., Kramer, R.M.: Members' responses to organizational identity threats: Encountering and countering the Business Week rankings. Administrative Science Quarterly 41(3), 442–476 (1996)

Etzkowitz, H.: The Triple Helix University-Industry-Government Innovation in Action. Routledge, New York (2008)

Frost, A.: Third stream, taxonomy and the value of ambiguity? Conversations the IKT Blog, London (2008)

Furman, J.L., Porter, M.E., Stern, S.: The determinants of national innovative capacity. Research Policy 31(6), 899–933 (2002)

Gagliardi, P.: The creation and change of organizational cultures - a conceptual-framework. Organization Studies 7(2), 117–134 (1986)

Gioia, D.A., Schultz, M., Corley, K.G.: Organizational identity, image, and adaptive instability. Academy of Management Review 25(1), 63–81 (2000)

Gioia, D.A., Thomas, J.B.: Identity, image, and issue interpretation: Sensemaking during strategic change in academia. Administrative Science Quarterly 41(3), 370–403 (1996)

Henkel, M.: Academic identity and autonomy in a changing policy environment. Higher Education 49(1-2), 155–176 (2005)

Johnson, J., Scholes, K., Whittington, R.: Exploring Corporate Strategy, 7th edn. Pearson Education Limited, Essex (2006)

Mael, F., Ashforth, B.E.: Alumni and their alma-mater - a partial test of the reformulated model of Organizational Identification. Journal of Organizational Behavior 13(2), 103–123 (1992)

Markman, G.: Entrepreneurship from the Ivory Tower: Do Incentive Systems Matter? Journal of Technology Transfer 29, 353–364 (2004)

Martinelli, A., Meyer, M., von Tunzelmann, N.: Becoming an entrepreneurial university? A case study of knowledge exchange relationships and faculty attitudes in a medium-sized, research-oriented university. Journal of Technology Transfer 33(3), 259–283 (2008)

Mintzberg, H.: The Design School - reconsidering the basic premises of strategic management. Strategic Management Journal 11(3), 171–195 (1990)

Molas-Gallart, J., Salter, A., Patel, P., Scott, A., Xavier, D.: Measuring Third Stream Activities, Final Report to the Russell Group of Universities. SPRU Science and Technology Policy Research, Sussex (2002)

Morris, T., Wood, S.: Testing the survey method - continuity and change in British industrial-relations. Work Employment and Society 5(2), 259–282 (1991)

Manchester Metropolitan University, HEBCIS – Higher Education Business and Community Interactions Survey (2009a)
http://www.red.mmu.ac.uk/?pageparent=3&
(Retrieved November 27, 2009)

Manchester Metropolitan University, Vision of the Future (2009),
http://www.mmu.ac.uk/about/vision/

Robson, C.: Real World Research, 2nd edn. Blackwell Publishing, UK (1993, 2002)

Rothaermel, F.T., Agung, S.D., Jiang, L.: University entrepreneurship: a taxonomy of the literature. Industrial and Corporate Change 16(4), 691–791 (2007)

Rughase, O.G.: Identity and Strategy: How Individual Visions Enable the Design of a Market Strategy That Works. Edward Elgar Publishing Ltd., UK (2006)

Sainsbury, D.: The Race to the Top A Review of Government's Science and Innovation Policies (Retrieved Accessed April 15, 2008)

Saunders, M., Lewis, P., Thornhill, A.: Research Methods for Business Students, 1st edn. Financial Times Management, London (1997)

Vanmaanen, J., Barley, S.R.: Occupational Communities - culture and control in organizations. Research in Organizational Behavior 6, 287–365 (1984)

Wedgwood, M.: Mainstreaming the Third Stream. In: Mcnay, I. (ed.) Beyond Mass Higher Education, pp. 134–157. Open University Press, New York (2006)

Wegge, J., Van Dick, R., Fisher, G.K., Wecking, C., Moltzen, K.: Work motivation, organizational identification, and well-being in call centre work. Work and Stress 20(1), 60–83 (2006)

Whittington, R.: What is Strategy - and does it matter?, 2nd edn. Thompson Learning, London

The 4 'C's of Knowledge Transfer and Knowledge Based Working- Emerging Themes in Successful Knowledge Working and KTPs

Steve Ellis

University of Chichester
s.ellis@chi.ac.uk

Abstract. This paper relates evidence obtained from an in-depth study involving detailed questioning of over 250 executives, during the period from 2000-2006, into the successes and challenges that those involved in KTP were experiencing. The analysis subsequently uncovered four fundamental themes (the 4 'C's of KTP) that will provide effective guidance to any future KTP, and more importantly, those responsible for them. The four themes outlined were Confusion, (what is KTP all about?), Convergence (how does the KTP fit with organizational or business strategy?), Commitment (how much time, effort and resources do we put to the KTP?) and finally Culture, how does KTP activity fit with and ultimately change the culture of the organisation?

1 Introduction

This research has used evidence from a number of managers including a major case organizations as evidence providers. One of the outcomes of the research was the formulation of a model which has become known as the 4 'C's, as a shorthand way of describing some emerging success factors or themes that lie behind successful KTP activity.

R.J. Howlett (Ed.): Innovation through Knowledge Transfer 2010, SIST 9, pp. 113–123.
springerlink.com   © Springer-Verlag Berlin Heidelberg 2011

These themes are depicted below in Figure 1 below.

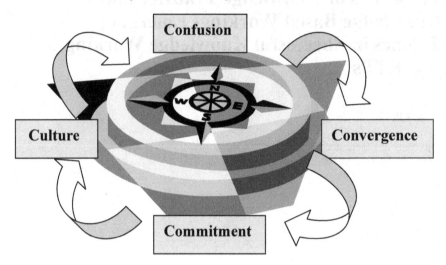

Fig. 1 Emerging themes, the 4 'C's of KTP;

No fewer than 15 different distinct versions of what knowledge transfer is were found in the replies to an open survey question, (65 respondents). Some of the definitions offered however did have some commonality and these are depicted in Figure 2 below.

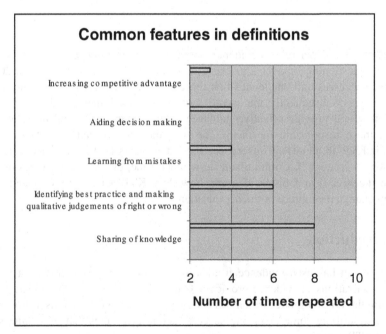

Fig. 2 What do people typically understand KT to be about?

1.1   Thus the First of the Themes That Make Up the Four 'C's Is Confusion

Because everyone, even KT experts, have widely differing views about what KT is, clarity has to be established up-front for the KTP initiative to even get off the ground. O only 2 out of 3 of the people asked could give a valid (ie consistent with a recognized KTP definition) response to the question 'What do you understand by the term Knowledge Transfer?' The range of responses showed that KT is still seen very differently by different people, which perhaps limits meaningful debate until a comprehensive definition is accepted. The most common answers included reference to 'knowledge sharing' and sharing best practice.

1.2   The Second Theme to Emerge as Key Was 'Convergence'

Specifically convergence of the KTP with organizational strategy and overall business imperatives. Organisation strategies that 'bolt on' KTP activities as an afterthought simply will not be as effectively as situations where strategy develops with a keen knowledge based working perspective. Organisations devise strategies for most aspects of activity, but our research showed they were still leaving the key organizational asset- knowledge- out in the cold. Even in my main case organisation, it was reported that even though Knowledge *featured significantly* in strategy, at the same time the level of analysis of knowledge requirements was reported as being *'low'*, leading one to doubt the depth of penetration of knowledge issues into strategy development.

1.3   The Third Theme Identified Was 'Commitment'

Much knowledge based working activity is 'voluntary', it would be difficult to imagine a way to impose behaviours such as 'knowledge sharing'. As a consequence, effective knowledge working relies on people being committed to sharing and developing knowledge both for themselves and others. Gaining commitment to KTP activity is therefore a key success factor. Most people will openly agree that a KTP is a good thing, both for them and the organisation, but getting them to actually do it is another story. My qualitative findings indicated that securing commitment was often hindered by internal procedural or cultural barriers. In addition to this, reward and recognition strategies are very rarely tied to performance in KTP activities and we therefore underuse a major potential lever to encourage the right approaches to KTPs.

1.4   The Final Theme Was the Catch All Requirement for Truly Effective KTP Activity- 'Culture'

My research showed that in most cases KT is fundamentally about working differently. Done well it will empower people, enhance productivity, strip out costs and speed up processes. But effective KT needs above all a 'knowledge-enabling culture'. Without which efforts to change to more effective working practices will fail. I identified two key features to look for in developing, (for it is a journey not a destination), this culture. These features were found to be;

- **Trust**, if I share my knowledge what will happen to it and to me? And
- **Flexibility**, do I or others have the opportunity to use my knowledge or do we adhere to existing inflexible policies and procedures no matter what?

2 Background and Research Methods

The research combined both quantitative and qualitative. It focused primarily on a large, multi-national financial services organisation where questionnaires and interviews were completed by over 200 executives. This evidence was subsequently supplemented by interviews with KM practitioners from another seven industries. The main impetus for the research came from an original desire to investigate the impact of knowledge based working and the whole KT phenomenon on organisations at strategic and operational levels.

Sidebar - the research process

The research process involved gathering data from;
- Questionnaires (written and electronic)
- Semi-structured interviews
- Participant Observation
- Follow up interviews
- Secondary case study data

Altogether views were obtained from over 200 executives, allowing for multiple triangulation of evidence. The data was obtained by using a large multi-national financial services company as a live case study. This data was then compared with evidence from 8 non-case organisations who agreed to allow in-depth semi structured follow-up interviews. These interviews covered companies from the areas of Telecommunications, Food Retailing, Consultancy, Construction, and Information Technology.

The quantitative and qualitative evidence was subsequently analysed to provide a framework with which to understand the current situation with regard to knowledge based working and the four key themes that emerged.

Independent research such as this, not sponsored by a KTP service provider is desperately needed as those who have to implement KTPs still have very few 'role model' organisations, and those that have failures are understandably reluctant to disclose where it all went wrong.

The evidence in this research comes from the experiences of those responsible for KT programmes **and** those who have been on the receiving end of them. It offers clear reasons why the KT plane has often stalled before take off, and where flying it continues to find turbulence.

2.1 The Emergence of 4 Key Themes

As the research progressed four themes emerged through the questionnaire responses and then through in- depth, follow up conversations with managers. Questioned about what activities and approaches to KT were proving to be a success and what difficulties they were experiencing, it became clear that for all the hype surrounding the potential for positive change through knowledge based working, serious difficulties were being encountered.

The first theme of confusion over what KT is about continues to be detrimental to those working in the field. It is crucial to nail this problem for knowledge based working to progress. Without clearly understood and agreed terminology KT is wide open to the criticism that it is too diverse to be effectively operationalised or, even worse, it is offering 'old wine in new bottles', nothing different to 'good management'.

To help with terminology issues I would therefore suggest taking the following three steps;

- Firstly why not admit that the title KTP is not helping us? It does not even describe what those of us working in the field actually do – you cannot actually physically transfer knowledge (except if you are an exponent of Vulcan mind melding a la Star Trek). Knowledge sharing or enhancement would be an improvement.
- Secondly open up communications at the earliest pre-commissioning stages to ensure that what your sponsor or client is looking for really can be achieved by the application of appropriate, tested knowledge work tools and techniques and if it cannot, admit it and suggest they look for another way out. If this hurts too much and you don't want to turn them away then try reframing the problem, so that the KTP can make a difference.
- The third step is to keep the communications open at all times so that when the KTP initiative kicks in both sides are clear about what is happening what is expected and what still needs to be done.

These quotes show the terminological confusion problem in stark relief. Both are describing KT from their perspective. The first comes from a Mobile Telephone Company;

'What we tend to do is use intranet as our delivery vehicle and from that we provide a range of services to the business which arise from demand, e.g. enable SAP systems to gain access to the intranet and all the information tools that are available. In a different sphere we provide a non-business service for third parties that we call 'C Space', or collaboration space which is a secure extranet where suppliers or other third parties can do stuff and have dialogue with us.

Something else we provide is a way of senior executives communicating with our employees via video and on-line 'q and a' sessions. So what you have is about communication, some of it is about collaboration, and some of it is about information sharing.'

I think the above quote is an example of what a valuable and mature KTP is currently offering. The picture painted is clear, it is about practical guidance and resources and it is right in the faces of those who are using it. Any executive of this company would be missing out if they were not signed up for the help that is being offered.

Eight different executives however from the main research case organization, which admittedly inhabits a less technologically literate world, all saw KT very differently. The views gathered in response to what they thought KT was, included;

'Sharing knowledge, experience, centralisation of knowledge for members of the corporation, sharing best practice amongst colleagues, eliminating repeated mistakes.'

'Utilisation of state of the art strategy, of state of the art infrastructure to focus on the right knowledge. Using knowledge in decision making.'

'Making the organisation more competitive. Recording knowledge so that people can retrieve it easily.'

'Fostering innovation and ideas. A repository of information to be accessed by whoever needs it. A way of fulfilling daily obligations to client and company. Maintaining a repository of key information for appropriate dissemination.'

'Transfering of cross border knowledge. The receipt of useful information and how this is handled, used and communicated. Making available and in a visible format the collective knowledge of an organisation and its market.'

'Growing the intellectual property of an organization. Capturing the components of learning from different areas of the group and allowing all to share in it.'

'Knowing who knows what and what they know. Initiate, build develop, share a collective pool of understanding amongst those who need it. Organising the recording, filing/storing of information in a way that makes it accessible to others.'

'Making sure everyone is aware of the existing knowledge in the company. Helping to facilitate the sharing and development of knowledge. Utilising the experience and competencies of each member of the organization.'

These quotes sound utopian but I think they may really be a cry for help, describing some ideals that if achieved would go way beyond the more practical offerings of the first quote. Any KTP that is expected to deliver against such wide-ranging expectations will fail- why not add in 'cure cancer and ensure lasting world peace' for good measure?!

To deal with this confusion issue those responsible for a KTP will find themselves undergoing an exercise that is part education, part investigation into what sponsors expect to get from the programme. This means an assessment of KTP needs and appetites. I have found that a Knowledge Audit is often a good place to start. But my advice is to be very gentle with your sponsor, as any well-crafted Knowledge Audit will probably uncover far bigger and scarier problems than they may be willing to see in the early stages.

Nonetheless until a good match between what is achievable and what is required can be found, don't even start. This means that you have to be clear about what you know you can achieve through a KTP. Don't guess, talk to some people who have been there already and have the bruises to prove it. This is after all a recognized knowledge management activity known more readily as 'Peer Assist'.

The second theme of 'Convergence' emerged from the research when it became clear that where KTP initiatives were having a hard time in achieving their aims, it was also true that the KTP was rarely appearing as part of strategic considerations. As a result much activity was poorly aligned with other organisational priorities. This is fine for a short time or until large amounts of money start to be spent. But once we get into sizable budgets any hint that the activity or programme lives outside of the mainstream strategic direction of the rest of the organization the project is in deep danger of being cut.

In the main case organization it was found that KT was not considered as a part of overall strategy. This quote from a senior KM executive confirmed this;

'...to be perfectly honest, you know, as new as it is I don't mean KT the discipline I mean KT the function, it hasn't had much of a role in strategy development at a corporate level,......my view of the world is that the knowledge proposition is very significant or should be a very significant component of corporate strategy.'

Other KTP practitioners we found seemed to have accepted the non-strategic role for KM as the best of a bad job and were happily battling along regardless.

The following illustrative quote came from a KM director in a Mobile Telephone company;

'I think if you look at it from a pure business viewpoint KT is not a business strategy. KT is tactical rather than strategic within our organisation because of a variety of things, primarily because we are still at an early stage of development within the organisation and there isn't the maturity yet to really see the value of managing knowledge strategically in the organisation. There are various things that might fit into a KT strategy that might be seen as strategic but I don't think there are enough of them to have a really strategic role. The role that I came to play is very much a tactical role not strategic.'

Dealing with this mismatch is probably the hardest test of any KTP. We *'KTP evangelists'* must bear in mind that effective KT is not the *raison d'etre* of organizations it is only a toolkit to be used to help achieve goals. This means that goals for KTP must be stated in terms that speak clearly to the objectives or desires of the organization.

Here are some ideas for indicators of KTP success;

- Savings in direct labour costs
- Improvements in customer response times
- Fewer mistakes being repeated
- Shared experiences between business units doing similar things

This quote from a large UK food retailer demonstrates the way that KT strategy and business strategy can converge.

'So in terms of your original question of how does the KT strategy support the business strategy, it is supporting our Business Transformation Programme. Which is based on the core elements of our Business Strategy. So last year we launched a Portal which will replace the Company Intranet. We are not having a portal and an intranet just a portal with all the information in one place.'

Turning now to the third theme of commitment, I found much of the existing literature to be contradictory. Some studies, (Nijhof et al., 1998)), (Guest, 1998) claim, as one would expect that committed employees will perform better where their oppotunities to be involved in KTPs was increased. Others, (Mathieu and Zajac, 1990), (Gallie and White, 1993), show relationships between commitment and performance to be tenuous and short term. My qualitative data emphasised the need to establish clear employee commitment to the KTP without which it will be difficult to translation desire into action.

But what does an organisational commitment to KTP really consist of? My data suggested the following minimum requirements at the organizational level;

- Top level sponsorship-someone with credible authority must be the sponsor
- A willingness to change as a result of the KTP
- Recognition that the knowledge age is upon us and has not just changed the rules but the whole game
- Breaking away from the short-termism mindset and really invest for the future

Tackling this issue also means looking at commitment at the individual level by asking the 'WIFM' question- 'What's in it for me?' After all, if doing KT was easy and needed no extra effort, wouldn't we already be doing it?

A good KTP should ideally deliver obvious and quick benefits for the participants- a built-in incentive to gain commitment. For example a KTP often yields time savings, but where does this benefit go? If all benefit goes to the organisation via lower staffing levels, or merely increasing the workloads for those left behind then any commitment from individual employees will understandably soon wane. Sharing the benefits is the obvious way to encourage commitment, but this is all rather 'after the event'. If employees are not committed to the KTP upfront there won't be any benefits to share out. A better strategy is to work out what your people are currently committed to achieving, (in their work or non-work lives), and tailor your KTP so that it makes this easier.

Another problem is where a KTP results in people having to do more work. For example logging all learning from projects or completing drawn out after action reviews. These activities might start off with enthusiasm but they will not survive as the pressures of other work apply. This means KTP activity has to be integrated into existing routines and used to replace and remove any activities that are no longer required.

The final theme of culture came out loud and clear from specific questions on the questionnaires used, in the interviews undertaken and through the use of secondary data gathered on case executives scores for both Myers Briggs Type Indicator (MBTI[1]) profiles and TMS[2] work preference profiles. Respondents were convinced that getting the right culture was paramount. My evidence showed the prevailing culture of the main case organisation was closest to the model described by Deal and Kennedy (1999) as a 'Process Culture'- a culture where rules and process are paramount, instruction manuals predominate and a predominantly stifling hierarchy exists. Under this type of culture Deal and Kennedy postulate that compliance to rules is strictly observed and bureaucratic excellence is seen as a highly worthwhile goal. My observation based on the evidence obtained is that such a culture is not very fertile ground for effective KTPs, which after all are seeking to change and reconfigure activity.

This type of culture, which tend to predominate in many large-scale organisations where hierarchy restricts access to knowledge this will not facilitate knowledge sharing, strict rules and procedures tend to stifle change and at it's heart that is what much of knowledge based working is all about.

The good news is that even in the most knowledge unfriendly cultures there will probably co-exist some areas, departments or even individuals who don't want to work in this way, find them - they are going to become your disciples. If you can use them to demonstrate the value and advantages of knowledge based working which circumvents hierarchy and improves effectiveness you will have started the process of chipping away at the old ways of working.

3 Conclusion

Many early gurus, (Drucker 1993, Nonaka 1995), claimed that knowledge management and knowledge transfer was going to transform the way we work, cause wholesale reconstruction of value chains and make much of what we thought organisations were about redundant.

---

[1] MBTI is a widely used indicator of strengths of preferences for behavioural styles across 4 key dichotomies.

[2] TMS is the Margerrison/McCann Team Management System, which uses types of work preference profiling to determine preferred team operating roles.

In the light of this it is perhaps understandable that the achievements of 'the knowledge revolution' seem so far to have been less than expected.

This article describes a new way to approach KTPs from four key perspectives. The first step in the process is to deal with the likely confusion around any new KTP. Clarification with all stakeholders what in the wide, (and likely to be widening), world of possible projects, priorities or practices is being considered? And equally important, is it the right one?

Having got this far the second phase of pre project preparation is to do a reality check with the organisational strategic focus. As a quick guide here it would be helpful to ask where the outcomes from your KTP would fit in the following matrix;

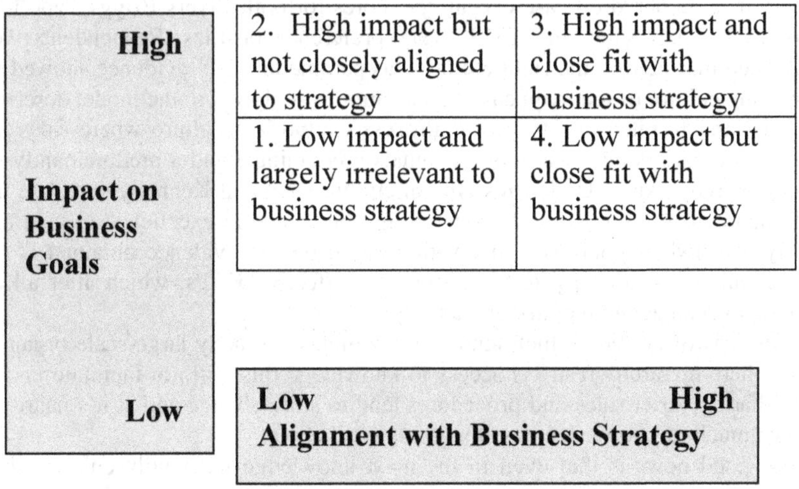

High	2. High impact but not closely aligned to strategy	3. High impact and close fit with business strategy
	1. Low impact and largely irrelevant to business strategy	4. Low impact but close fit with business strategy
Impact on Business Goals		
Low	**Low High** **Alignment with Business Strategy**	

Fig. 3 The KT Alignment and Impact Matrix

Anything that falls in quadrant 1 will be a waste of your time. In quadrant 2 you might find some really neat technology based KTPs (pet products) that will win you no friends- these are often called the 'so what' projects. You will need some initiatives in quadrant 4 as 'quick wins' to establish credibility, but the real meat lies in quadrant 3. Projects that have demonstrable impact and are aligned with business strategy allow you to build superb business cases. Your KTP strategy must live in this quadrant thereby dealing instantly with the convergence issue.

The third 'C', commitment requires that real activity not just acquiescence is the key. This means looking at what drives commitment to other initiatives in your situation. What are your typical employees already committed to and how will your proposal effect this? Incentives and rewards for undertaking KTPs may need to be investigated in order to kick start the process.

Turning finally to the fourth 'C', culture. I found a clear paradox here in that having the 'right' culture for KTPs, (which we concluded must feature trust and flexibility), will make a big difference to your chances of success. But developing such a culture of trust and flexibility, where none exists is not easy or quick. But the very introduction of knowledge based working practices and the whole 'KTP mindset' itself changes culture. What the successful companies have demonstrated is the snowball effect that a good KTP has. Once people start to see jobs and processes from a knowledge perspective, they begin to unleash powerful forces more commonly confined to 'suggestion schemes'. Once this ball gets rolling the rganisation may just have found a way to access the abilities and skills that currently lie dormant.

References

Deal, T., Kennedy, A.: Corporate Culture. Penguin, London (1988)

Deal, T., Kennedy, A.: The New Corporate Cultures. Texere, London (1999)

Drucker, P.: Managing for the Future. Butterworth–Heinemann, London (1993)

Gallie, D., White, M.: Policy Studies Institute, London, pp. 5–9 (1993)

Guest, D.: In: Sparrow, P., Marchington, M. (eds.) Human Resource Management: The New Agenda. Financial Times Publishing, London (1998)

Mathieu, J., Zajac, D.: Psychological Bulletin 108, 171–194 (1990)

Nonaka, I., Takeuchi, K.: The Knowledge Creating Company:How Japanese Companies Create the Dynamics of Innovation. Oxford University Press, Oxford (1995)

Nijhof, W., de Jong, M., Beukhof, G.: Journal of European Industrial Training 22, 243–248 (1998)

Session C
Knowledge Transfer Models and Frameworks

A Proposed Management Framework for Commercialisation of Expertise at Public Universities

Wynand C.J. Grobler and Frikkie van Niekerk

North West University, South Africa

Abstract. Commercialisation at universities, specifically the commercialisation of academic output at universities, has become an economic imperative since the 1990s, forming part of the changing role of universities. Teaching-learning, research and community engagement have traditionally been central to most universities' mission statements. During the 1990s, countries such as Australia, the United States and the United Kingdom developed policies to exploit the collaboration between the higher education sector and industry with regard to technology bases, private sector participation and the exploitation of intellectual/academic output. The need has emerged for the development of a framework for the implementation of expertise and commercialisation at universities so that the academic ethos of the university and scholarship are not undermined. For this reason, it is important that universities develop a suitable framework for implementing expertise and commercialisation – one that is appropriately managed within predetermined guidelines.

1 Introduction

Commercialisation at universities, specifically the commercialisation of academic work at universities, has become an economic imperative since the 1990s and is indicative of the changing role of universities with regard to teaching-learning, research and community engagement, which are undertaken in order to contribute meaningfully to the growth of an economy (Dooris 1989; Etzkowitz and Peters 1991). Teaching-learning, research and community engagement are central to most universities' mission statements. However, until the mid-1980s, commercial activities such as (paid) consultancy would have been seen as part of the community service goals rather than being a specific objective outlined in the mission statement as a goal. During the 1990s, countries such as Australia, the United States and the United Kingdom developed policies to exploit the collaboration between the higher education sector and industry with regard to technology bases, private sector participation and the exploitation of intellectual/academic output (Campbell 2005; Lowe 1993; Tornatzky 2003).

R.J. Howlett (Ed.): Innovation through Knowledge Transfer 2010, SIST 9, pp. 127–137.
springerlink.com   © Springer-Verlag Berlin Heidelberg 2011

For universities, the commercialisation of research and other commercial activities offer the opportunity to generate additional funding but at the same time raised new challenges. These include administrative and ethical issues in terms of accommodating the conflicting values and expectations of the private sector and public sector.

The need has emerged for the development of a framework for the implementation of expertise and commercialisation at universities so that the academic ethos of the university and scholarship are not undermined. Rasmussen (2004) views this as a threefold challenge of increasing the extent of commercialisation, visualising the contribution to economic development and managing the relationship between commercialisation and other core activities at a university. Even though commercialisation may affect both teaching/learning and research, literature remains vague regarding how university management can best manage this commercialisation in order to minimise the potential for conflict, resistance and risk (Etzkowitz 1989; Martin 2003). The purpose of this paper is to provide a theoretical framework for the management of the commercialisation process at a South African university, so that clear policies may be established to manage this process. The authors believe that this framework may be of value to other universities. The changing role of universities, the underlying factors stimulating commercialisation, the domain of commercialisation, the prerequisites for successful commercialisation and the obstacles and potential threats of commercialisation to core academic activities will be discussed. This will be undertaken in order to provide a generic management framework for the commercialisation process at universities.

2 Towards a Definition of Commercialisation within a University Context

No precise definition of the meaning of commercialisation within the university context seems to be generally accepted, concerning in general the commercial exploitation of research. Zhao (2004) defines research commercialisation as a "process of developing new ideas and/or research output into commercial products or services and putting them on the market". Bok (2003) defines commercialisation in higher education as "efforts within the university to make a profit from teaching, research and other campus activities". Commercialisation, according to Harman and Harman (2004), is viewed as the process of turning scientific discoveries and inventions into marketable products and services, and includes licensing patents, creating "spin-out" companies and the movement of expertise or technology from one organisation to another. Sharma et al. (2006) refer to the commercialisation of university technology as the process of a university-industry technology transfer. Upstill and Symington (2002) describe commercialisation as a process that may evolve from a mode of non-commercial transfer (seminars, publications, conferences, informal contacts and symposiums), to a mode of commercial transfer (collaborative research, contractual research, consultation and licensing), to a mode of new company generation (spinout companies). According to Leitch and Harrison (2005), commercialisation entails the emergence of "spin-out" companies and "university founded" companies that may arise from commercial opportunities that may have been identified by the university, which are not necessarily linked to a university research base. In many universities, the term *innovation* includes technology transfer activities, the protection and exploitation of intellectual property (IP) by virtue of licensing agreements and the formation of spin-off companies.

The core activities of universities were traditionally grouped into teaching-learning, research and community engagement. As discussed above, commercialisation activities gradually developed in many universities. In this paper, the third element of university activity is termed *implementation of expertise* and includes non-commercial activities such as professional advice, community service, subsidised developmental engagement, developmental activities as well as activities for income generation, such as short courses, consultation, contract research, internal corporate ventures, the creation of associated subsidiary companies, IP exploitation and the commercial use of university facilities. This broader definition is depicted in Figure 1.

In order to provide a management framework for the effective management of the implementation of expertise at a university, it is important to realise that universities are facing a changing role and changing stakeholder expectations.

3   Changing Role of Universities

Before the 1990s, literature on the commercialisation of universities was limited to a few articles. However, in the 1990s, a debate started on various aspects of university commercialisation. Traditionally, teaching and research have been the core business of universities, as stated in their mission statements. This has now changed with increased globalisation, reduced state funding and the definition of the role of universities (Rasmussen 2006). This change may be seen as a fundamental change in the system of knowledge production at universities. Etzkowitz and Leydesdorff (1997), Etzkowitz and Peters (1991), and Zhao (2004) describe this role change as an "academic revolution" or "academia taking up the role of entrepreneurs".

In the "Triple Helix" model (Etzkowitz and Ledesdorff 1997), it is stated that if universities accept their changing role with regard to commercialisation, the relationship between university, industry and government is seen as a triple helix of evolving networks in which a university has a specific and valuable role to play in terms of innovation. It is expected that universities play a more prominent and significant role in the knowledge economy. In a knowledge economy, the university is not only a provider of human capital but also a seedbed of new firms and entrepreneurial activities. Breznitz et al. (2008) report that the last two decades have seen greater demands being placed on universities to supplement their basic and applied research in order to achieve measurable commercialisation outcomes.

As stated in the OECD report (2000), universities are exposed to more government control and have to increasingly motivate and substantiate the value they add in society (in terms of the traditional role as well as in terms of economic and social impact) in order to obtain public funding.

We believe this imperative partially contributed to universities becoming increasingly independent of public funds. Furthermore, universities embracing the new paradigm have become more relevant and have allowed (and embraced) external stake holders, including the market place, to have a greater influence on their offerings and activities. Universities are ideally positioned to become partners in the triple helix relationship, which allows for the strengthening of the triple bottom line approach in institutions; that is, a strong focus on financial outcomes, social outcomes, as well as on the environment. Even so, it is important that universities do not merely focus on shorter-term sustainability at the expense of their traditional academic role.

By accepting this responsibility, universities contribute to economic development on a regional and national level. Dooris (1989) and Slaughter and Rhoades (1990), also emphasise this responsibility by including technology and knowledge transfer by universities as part of their contribution to the prosperity of a nation. However, there is still a need for well-defined structures on how to integrate this new role of universities into existing activities and structures related to teaching-learning and research in order to ensure effective management of the core business as outlined in the mission statement of universities.

Against this background, it can be argued that universities may experience changing funding structures and changing expectations from society on an external front, which must be synchronised with structures and goal setting on an internal front. The question often raised is what the reasons may be for taking up this new role or responsibility to commercialise.

4 The Rationale Behind Commercialisation

In order to develop a management framework for commercialisation at universities, it is important to identify the rationale for adopting commercialisation activities at universities. The Final Report of the Department of Education, Science and Training in Australia (2002) outlines the reasons why most universities en-gage in commercialisation activities as being to facilitate the commercialisation of research for the public good, promote economic growth, forge closer ties to industry, reward, retain and recruit faculty academic staff and students and to generate income. The work of Charles and Benneworth (2001) indicates that universities engage in collaborative research, business reach out programmes for the purpose of generating income, matching core funding, accessing business networks, consolidating their own research strengths and creating networking opportunities. In developing a management framework for commercialisation at universities, these factors need to be managed as key success factors.

5 Prerequisites for Successful Commercialisation

When examining the prerequisites for successful commercialisation at universities, a number of important aspects have been identified. These include having a proper structure for facilitating university commercial activities (Amidon Rogers 1988), clear and objective policies and management structures (Fuchsberg 1989), a clear decision-making infrastructure to maintain perspective (Chafin 1988), a significant allocation of additional resources (Bureau of Industry Economics report on Commercial Opportunities from Public Sector Research 1990), adequately established measures to guard against potential conflicts of interest and measures to guard against legal and financial problems (Atkinson 1985).

These prerequisites must be incorporated into a management framework for the commercialisation activities at universities. Notwithstanding the above, there are various challenges and risks on the road to successful commercialisation, which need to be considered.

6 Challenges on the Road to Successful Commercialisation

The Australian Department of Education (1989) mentions the poor research infrastructure in higher education and the lack of research groups with a critical mass to undertake competitive industry-orientated research as one of the primary obstacles. Clarke (1986), Chaseling (1989) and Cichy (1990) mention the differences in cultures between the public and private sectors regarding commercialisation as being major obstacles to creating joint industry/higher education ventures. Levinson (1984) indicates that industry and universities may have different cultural climates and organisational values, and even mission statements that may form barriers to effective commercialisation. Universities and industry partners should give due consideration to these factors if sustainable and mutually beneficial partnerships and ventures are to be developed.

If not well managed, the commercialisation activities at a university may pose a threat to the traditional core business of a university and scholarship.. In this regard, the potential negative perception of the public towards the university in terms of the traditional values and culture are highlighted by a number of researchers, such as Clarke (1986), Garett (1985) and Brophy (1988). Garett (1985) argues that due to commercialisation, fewer original research papers are being presented and standards are compromised in terms of academic quality and methodology. Ellison (1988) provides another risk of commercialisation as "the delay in the publication of research because of the secret nature of the research" and the lack of peer review processes, which poses a risk towards a university.

Boyer (1987) and Slaughter and Rhoades (1990) identify mission drift at universities, from their traditional core focus on knowledge generation and dissemination towards the interests of business, as a threat because of the tendency to shift from fundamental research to applied research. The implication is that universities increasingly engage in research led by industry, controlled by industry and probably directed by industry.

Conflict of interest, as a risk, is widely discussed by many researchers (Wheeler 1989; Feller 1990). The potential for conflict of interest increases in the entrepreneurial university. Etzkowitz and Peters (1991) indicate that commercialisation may also have a negative impact on teaching and learning.

In order to manage a university strategically, it is important to consider the potential threats that commercialisation poses to a university so that these may be embedded in strategy formulation.

7 Critical Success Factors for Managing Commercialisation at Universities

According to Kernich (2002), successful commercialisation may depend on a number of management factors. These include the development of a sound business plan, a clear indication of intellectual property ownership, the isolation of any business incubator from the daily activities at the university, acceptable incentive schemes for staff involved, independent financial auditing and the establishment of clear commercialisation outcomes and direction, including legal and financial aspects. Additional factors include being consistent with the university's strategic plan, decisive decision making by management, university-wide management of commercial assets and sound governance within the university.

Zhao (2004) indicates that there are two prerequisites for successful commercialisation, namely adequate financing from government and industry, and effective innovation management. Zhao further indicates that universities need to foster a supportive structure, which requires top management commitment, the inclusion of commercialisation aspects into the university's strategic plan and reforms in reward systems.

8 Prerequisites for a Sound Management Framework for Commercialisation

Given the traditional role of universities and the associated entrenched value systems, processes and policies, it is important to develop an appropriate management framework and support systems, should a university embark on a trajectory of developing into a more entrepreneurial and enterprising organisation. Such a framework would not only have to be enabling in terms of the newly identified opportunities, it would also have to protect the university's core as a knowledge generating and disseminating institution. For a university to be effective in its commercialisation efforts, these activities need to be managed in such a manner that they are aligned with and strengthen the university's traditional core functions and scholarship.

A University needs to position itself in such a way that its core business is strengthened, its scientific and financial sustainability guaranteed and its reputation amongst all its stakeholders reinforced. In extending its activities, within its mandate, beyond the traditional core business, a university should incorporate a number of important prerequisites into developing these commercial activities, including the alignment of commercialisation with its mission and vision.

A lack of alignment with the university's vision and mission may lead to vision drift, a loss of reputation and the ultimate demise of the organisation. The mission statement should then culminate into well-defined goals for commercialisation, in addition to the core functions of the university, which includes the offering of short courses, consultation, internal ventures and technology transfer.

It is proposed that the University's interests are protected by appointing suitably qualified internal and external directors to represent its interests on the boards of companies in which the university has shares, according to the relevant shareholders' agreements. Members of the university's traditional governance structures (Council and Senate) are typically not appointed with commercialisation activities in mind.

Management and hence management systems at the university, are typically developed around the needs associated with its traditional core activities. Appropriate management systems are needed for managing commercial activities, which may include elaborate project management modules, etc.

In South Africa, universities do not pay tax on their first and second income streams and when competing with the private sector in areas falling outside the university's traditional mandate it may be seen as unfair competition. Aligning its investments and commercial activities with the needs of its customer base allows the university to optimise investments, serve its customer base and avoid unfair competition with the private sector, while simultaneously building strong ties with this sector.

The risks associated with commercial activities are different from those risks associated with the traditional core business of the university and because universities are traditionally risk averse, risk management tools like corporate and management structuring may be used.

Besides its own academic and general quality requirements, the university has to adopt appropriate quality systems and practices for its activities related to the implementation of expertise. These wide-ranging activities and associated quality requirements call for a proper and regular assessment of quality and, if needed, the improvement of quality systems, policy and practices at the university.

9 Proposed Management Framework for Commercialisation

For the university and its staff, the advantages offered by the implementation of its expertise and by the flexi-appointment of staff into a number of professions outweigh the disadvantages, if they are properly managed.

Figure 1 depicts the NWU's organisation of traditional core functions, research and teaching/learning with its understanding of the concept *implementation of expertise*. The links between internal departments (schools and research entities) and external structures (companies, trusts, etc) are shown (Van Niekerk 2008).

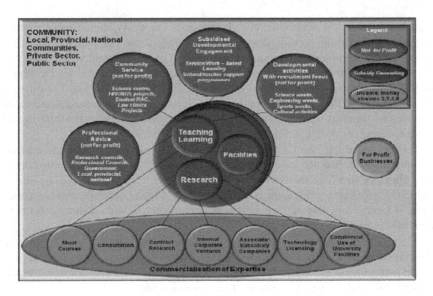

Fig. 1 Nwu Core Activities and Implementation of Expertise

Figure 1 depicts the University's traditional subsidy generating core functions, with teaching-learning and research in the centre. Taking into account the changing role of the university, as discussed in this article, the traditional role of teaching and learning extends in the modern university into for profit commercial activities and not for profit community engagement. This extra set of activities, previously managed as "add-on activities", has become part of the core activities of the university in the knowledge economy, where government, society and other stakeholders increasingly rely on universities to become strategic differentiators, influencing society and the economy at large. Short courses on a level below what is traditionally offered by higher education institutions and the mission of the university should not be presented by a university. In order not to lose the community engagement associated with some of its lower level courses, these courses should be transferred to an appropriate legal entity specifically created for this purpose or perhaps offered in collaboration with suitable partners. Short courses falling within the mandate and mission of the university are presented from within the existing faculties, schools or research entities.

When considering technology transfer, the university uses flexible mechanisms to identify, protect, manage and commercialise its IP. By successfully meeting this objective, the university, its staff and students will be in a stronger position to benefit in terms of attracting industry and government research funding, establishing a national and international reputation for research excellence, establishing linkages with other universities, industry and government, generating financial returns from commercialisation, growing knowledge-based industries through research and the commercialisation thereof, and generating national wealth through new products, services and employment.

A technology transfer office is responsible for identifying, protecting and commercialising the IP arising from research, in accordance with its IP policy.

This is done in close cooperation with the researchers and inventors and their line management, as well as external role players such as IP law firms and private companies. Depending on the nature of the IP, the market barriers to entry, growth potential, availability of relevant local expertise, availability of finance, prevailing regulatory climate, the university may choose to exploit its IP by licensing the IP directly (or less frequently, by assigning it) to an existing company to exploit the IP optimally. The existing company takes the risk for developing and commercialising the IP, and the university receives license fees, as agreed upon between the parties. The university can also create spin-off companies and may choose to transfer the IP rights to the newly formed company, either by licensing or by assignment. The university may take shares in the spin-off company. The spin-off company develops and commercialises the IP.

Corporate structuring is a practice whereby enterprises with different risk profiles are encapsulated into separate legal entities (usually being private companies of the relevant jurisdictions) to protect the various enterprises from the commercial risks not inherent to them. It is based on and utilises the principle of limited liability of shareholders. It also provides for appropriate governance and management of these entities.

Management structuring takes corporate structuring one step further by relinquishing the management control to which a shareholder may be entitled to in an independent organisation. Corporate structuring aims to protect the shareholder against exposure to the commercial risks of its subsidiary's activities and attempts to limit abuse of such buffering. A prerequisite for the success of the university's commercial activities will be the disengagement of decisions influencing those activities from the general considerations affecting the university.

The structure for implementation of expertise and commercialisation seeks to retain those activities closely linked to the university's traditional core activities, while attempting to remove the impediments associated with the academic environment and, simultaneously, maintain the "ownership" needed for sustainable success.

In this model the following commercial activities are retained within the university environment: the technology transfer office; short courses aligned with the university's mission and vision; contract research; consultation by staff and commercial entities managed as internal ventures within faculties.

Commercial activities to be moved to an external entity(ies) include: commercially viable internal ventures to be formed into companies; all spin-off companies; property investment and facilities commercialisation and non-mission related commercial activities.

10 Conclusion

Commercialisation and the commercialisation of the output of academic work at universities have become an economic imperative, which is indicative of the changing role of universities with regard to research, teaching and community engagement. *Implementation of Expertise* has been proposed as a more inclusive concept commensurate with the changing role of universities.

For universities, commercial activities offer the opportunity to generate additional income and thereby make up for shortfalls that are the result of decreasing subsidies from governments. More importantly, triple helix relationships increase sustainability and offer the possibility of more impacting research and innovation at a university.

Commercialisation at universities generates new challenges such as the risks associated with entrepreneurial activities and the administrative and ethical challenges in terms of accommodating the conflicting values and expectations of the private and public sectors. Should a university wish to embrace the new role for universities which includes entrepreneurial activities, the university needs to include the associated focus in its mission statement and in its management structure. In this regard, a framework for commercialisation that resonates with the academic ethos of the university in terms of culture and values is critically important.

The emerging and developing new role for universities calls for management and management structures that resonate with public and private institutions in a manner that allows universities to fulfil their role in triple helix relationships, while strengthening and growing their traditional role.

References

Atkinson, S.: University-industry research agreements: major negotiation issues. Journal of the Society of Research Administrators 17, 67–83 (1985)

Amidon Rogers, D.M.: Meeting the global challenges of a new era. Engineering Education 70, 222–223 (1988)

Boyer, E.: American higher education: the tide & THE UNDERTOW. Journal of Institutional Management in Higher Education 11, 5–11 (1987)

Breznitz, S.M., O'shea, P.R., Allen, T.J.: University Commercialization Strategies in the Development of Regional Bioclusters. The Journal of Product Innovation Management 25, 129–142 (2008)

Brophy, J.: Technology Transfer and Economic Development in Arizona, Arizona (1988)

Bok, D.: Universities in the marketplace: The commercialization of Higher Education. Princeton University Press, Princeton (2003)

Bureau of Industry Economics, Commercial Opportunities from Public Sector Research. Report 32. Canberra. Australian Government Publishing Service (1990)

Campbell, A.F.: The evolving concept of value add in University commercialization. Journal of Commercial Biotechnology 11(4), 337–345 (2005)

Chafin, S.: Of science and virtue: University research and technology transfer. Change 20, 45–47 (1988)

Chaseling, C.: Charting the critical success factors from concept to reality. Directions in Government 3, 18–19 (1989)

Charles, D.R., Harles, D.R., Benneworth, P.S.: The regional mission: Higher Education's Role in The Region, London, UK (2001)

Clarke, A.: Intellectual property-problems and paradoxes. Journal of tertiary Educational Administration 8, 13–26 (1986)

Department of education,science and training, Best practice processes for university research commercialization: Final Report, Australia (2002)

Disraeli, B.: Address to the House of Commons, Quoted in BARTLETT. J. Familiar Quotations (1980)

Dooris, M.J.: Organisational adaptation and the commercialization of re-search universities. Planning for Higher Education 17, 21–31 (1989)

Ellison, D.: Questioning the patent system. Australian Journal of Bio-Technology 1, 31–33 (1988)

Etzkowitz, H.: Entrepreneurial science in the academy: a case of the trans-formation of norms. Social Problems 36, 14–29 (1989)

Etzkowitz, H., Leydesdorff, L.: Introduction to specialissue on science policy dimensions of the trilpe Helix of University-industry-Government relations. Science and Public Policy 24(1), 2–5 (1997)

Etzkowitz, H., Peters, L.: Profiting from knowledge:organizational innovations and the evolution of academic norms. Minerva 29, 133–166 (1991)

Feller, I.: University patent and technology-licensing strategies. Educational Policy 4, 327–340 (1990)

Fuchsberg, G.: Universities said to go too fast in quest of profits from research. Chronicles of Higher Education 35, 28–30 (1989)

Garett, L.: There are problems. Journal of the Society of Research Administrators 17, 91–97 (1985)

Harman, G., Harman, K.: Governments and Universities as the main drivers of Enhanced Australian University Research Commercialization Capability. Journal of Higher Education Policy and Management 26(2), 154–169 (2004)

Kernich, G.: Commercialization. Seminar presentation. RMIT. Melbourne, Australia (2002)

Levinson, N.: Industry –university research arrangements: an action – oriented approach. Journal of the Society of Research Administrators 16, 23–30 (1984)

Leitch, C.M., Harrison, R.T.: Maximising the potential of university spin-outs: the development of second –order commercialization activities. R&D Man-Agement 35(3), 257–272 (2005)

Lowe, J.: Commercialization of University Research: A policy Perspective. Technology Analysis & Strategic Management 5(1), 27–37 (1993)

Martin, B.R.: The changing social contract for science and the evolution of the university. In: Science and Innovation. Rethinking the Rationales for Public Funding, Edward Elgar, Cheltenham (2003)

Rasmussen, E., Moen, O., Gulbrandsen, M.: Initiatives to promote commer-cialization of University Knowledge. Technovation 26(2006), 518–533 (2006)

Sharma, M., Kumar, U., Lalande, L.: Role of University Technology Transfer offices in University Technology Commercialization: case study of the Caleton University Foundry Program. Journal of Services Research 6 Special Issue (2006)

Slaughter, S., Rhoades, G.: Renorming the social relations of academic science: technology transfer. Educational Policy 4, 341–362 (1990)

Tornatzky, L., Waugaman, P., Gray, D.: Innovation U:"New University Roles in a Knowledge Economy. Southern Growth Policies Board (2002)

Upstill, G., Symington, D.: Technology transfer and the creation of companies: the CSIRO experience. R&D Management 32(3), 233–239 (2002)

Van Niekerk, F.: Implementation of Expertise at the North-West University. NWU Internal Policy Framework (2008)

Wheeler, D.L.: Pressure to cash in on research stirs conflict –of-interest issues. Chronicle of Higher Education 35, 29–30 (1989)

Zhao, F.: Commercialization of Research: a case study of Australian universities. Higher Education Research & Development 23(2), 224–236 (2004)

Vademecum for Innovation through Knowledge Transfer: Continuous Training in Universities, Enterprises and Industries

Francisco V. Cipolla Ficarra[1,2], Emma Nicol[3], and Miguel Cipolla Ficarra[2]

HCI Lab. – F&F Multimedia Communic@tions Corp.
[1] ALAIPO: Asociación Latina de Interacción Persona-Ordenador
[2] AINCI: Asociación Internacional de la Comunicación Interactiva
[3] Department of Computer and Information Sciences, University of Strathclyde, UK
Via Pascoli, S. 15 – CP 7, 24121 Bergamo, Italy
ficarra@alaipo.com, emma.nicol@cis.strath.ac.uk,
ficarra@ainci.com

Abstract. In the current work we present a first state of the art in technology transfer from the university –public-private environment to the enterprises and industries in Spain and Italy and vice versa. In it are described the main causes that boost and damage that two-way relationship. Additionally, a first vademecum is established to avoid those environments where technology transfer is either nonexistent or difficult to carry out because of the human factors this process entails. This short guide allows one to detect easily through the Internet whether we are in a real or false technology transfer process.

Keywords: Innovation, Knowledge, Software, Hardware, Telecommunications, Education, Virtual Campus, Enterprises, Industries.

1 Introduction

One of the main problems between the productive and the educational sector is the technology transfer in both directions to ensure that the last advances are quickly spread to the rest of the national and international society, with the purpose of increasing the quality of life of human beings [1-4]. However, we can see how these interrelations are not only nonexistent, but they can also slow down the technological growth of a local community, for instance. The main reason for these distortions are economic factors, where the private educational sector and the businesses have an active role. Chronologically these distortions originate in the era of the Internet's explosion, which in Mediterranean Europe can be dated to around 1995. In other realities, this phenomenon does not cause so many distortions like the cases that will be described in the current work, since it is a natural phenomenon between the educational sector and the entrepreunerial and/or industrial sectors [1] [5].

R.J. Howlett (Ed.): Innovation through Knowledge Transfer 2010, SIST 9, pp. 139–149.
springerlink.com   © Springer-Verlag Berlin Heidelberg 2011

One of the main advantages of private university teaching in Southern Europe is the elasticity of their curricula and easiness to introduce changes, as compared to the public universities, where the modification process is slower. That is, the private universities theoretically adapt more quickly to the contents of the university courses in the B.As, engineering, masters, specialization courses, etc. However, this false reality in some educational environment ends up being a mirage in the desert of the educational trade. At the same time, we can find modern public university institutions, which, in given contexts of a state territory, work as a private university body. It is the case of those teaching centres located in territories where the religious or nationalistic factor regulates the changes have to be introduced or not, deriving from the new technologies, in their communities. Evidently, we are in the face of situations that do not keep the indispensable principles of university teaching, that is, free, universal, egalitarian and secular. To the extent to which the technological breakthroughs are faster in the labs, the mistakes made in the transfer towards the educational and productive sector are bigger, especially due to the human factors [6] [7].

To introduce or generate educational programs it is necessary to count on a professional team who do not only know the technological breakthroughs from the technical point of view, but also from a practical point of view. To this end, some businesses try to present their latest novelties in the educational sector to grasp the attention of the future professionals and eventual consumers of those products or services. The open spaces inside the private educational centres in the late 90s were the continuous training courses and/or the masters. In close analysis, it was easy to detect in those courses how the directors running the masters did not have any knowledge of the subjects that would be taught, but rather it was to aggregate a series of education professionals, of the business world and government institutions, especially those who represent the local and/or regional authorities. This happened in the then fashionable multimedia sector.

The direction of masters in new technologies, in our case multimedia, was assigned to those people who possessed a PhD. title, obtained in a private institution in the USA. If one analyzes the college curricula of these PhDs we can see how in the nineties already a kind of hybrid species of the current Bologna plan had set in, especially of the obtainment of doctorates. For instance, a technical engineer in digital signal processing (it takes three years to get that title), with two more years in private centres of the USA, obtained the title of PhD. However, these meteoric PhDs were those virtually responsible for the transfer of technology in the main cities of the European Mediterranean. The term "virtually" refers to the fact that in reality those neo-professionals lacked the necessary training in the formal and factual sciences to autonomously exercise those functions, especially in the multimedia sector. In spite of these real shortcomings, in certain university environments, the whole machinery of technology transfer from the educational environment towards the productive sector and vice versa turned around the meteoric PhDs through continuous training [8].

The continuous training of some private college centres features great investments of money in the publicity campaigns in the main traditional media such as the printed press, internet, television, billboards, buses, etc. The goal is to sell the main technological breakthroughs under the format of seminars, specialization courses, masters, etc. In the case of the masters, you can even attend those modules of the programme that are interesting to the potential students or clients. The mercantilist factor of education degenerates into many situations in which the students of that continuous training are regarded as simple clients to be pleased in their demands, instead of students who wish to receive a specialized training in a short time.

The problem that arises in the face of educative mercantilism is the heterogeneous composition of the courses, either because of the age of the attendees, the training and/or previous experience in the subjects that are being taught, the real reasons why they have registered in the continuous training courses (widening of knowledge or to be promoted inside a public or private institution, for instance), etc. This heterogeneity may impair the correct transfer of the latest technological breakthroughs, since many theoretical concepts are taken for granted at the moment of structuring the program and their respective modules.

2 The Importance of Empathy in the Educational Structuring for Technology Transfer

We can define the empathy in the interactive design as the interactive systems designer's mental ability to put himself in the shoes of the potential user. It is the result of the triad confirmed by the cultural knowledge, mental ability to occupy the place of the other in the communicative process and the competence in advancing the user's behaviour in front of certain situations [9], [10]. For instance, in multimedia design traditionally we talk about cognitive models, that is to say, the solution would be to frame it in the psychological context. Obviously, it is a valid alternative for the first hypertext and multimedia systems in the late eighties and the decade of the nineties. With the advent of the use of information networks, whether it is Internet or extranet from international entities, since the late nineties it has been a matter of communicability. A communicability that stems from the design process in the interactive systems and is translated to its usability. If we analyze some multimedia products aimed at the education of the nineties, we can find how in the design of their structure one resorts continuously to two quality attributes such as are prediction and self-evidence [10]. A priori, prediction and self-evidence can seem similar, but it is not so. In self-evidence, the navigation of pages with dynamic elements (i.e. audio, video, animation, etc.) and the structure of the system can be anticipated by the user from the first moment, even if the user has scarce experience in the use of hypermedia. On the contrary, in prediction the user must have previous ability in order to navigate efficiently and overpass complex situations, after having previously navigated the hypermedia system. These two attributes are related to the concept of isotopies inside the context of communication. That is to say, those elements that must be maintained continuously in each one of the design categories to favour the interaction of the users with the content of the multimedia system. For instance, the location itself of the navigation keys in the different screens. The same modes of activating and deactivating the dynamic means, the synchronization between the audio and the images in movement, regardless of whether they refer to a video or an animation, etc. The presence of the isotopies in the interactive design indicates a high degree of empathy towards the potential users.

In the face of such varied situations in the confirmation of the participants in the technology transfer courses, it is important to resort to empathy to organize the contents. These are contents that can be explained in the classical training classrooms, in presence lessons or in virtual classrooms or virtual campuses. In the latter cases, it is an interesting work that is made by the virtual agents, especially for the explanation of the functioning of technological components or theoretical knowledge. Now empathy can be applied in the elaboration of contents of the interactive systems aimed at the education, thus dividing the potential users or students into several groups and intentions in the acquisition of knowledge. In the off-line and on-line multimedia systems of the late 90s and early 2000s it was feasible to have a defined profile of the potential users such as are the eventual ones (less than an hour of navigation, for instance, consulting a topic of tourist information, intentional users (between one and two hours, generally, are users interested in the content of a subject and want to go deeper into it) experts (unlimited time, such as a scientific researcher), inexpert and intentional (unlimited time, for instance, students who have no experience in the use of computers but who are keen on learning).

However, in the continuous training courses aimed at the transfer of technology is where some Lombardian public universities usually mix the university students with those who hail from professional training, it is very complicated to reach these goals in the short term. Consequently, the presence of empathy in the process of structuring the contents in the presence classes can do little or nothing to solve the problems that arise at the moment of approaching the issues of technology transfer due to the knowledge differences and/or experiences of the students who attend those courses. The only natural solution is the division of the courses between university students and those who attend professional training.

The instruments deriving from e-education may help to balance those disparities, but they do not solve the problem of the lack of knowledge and/or experiences, nor the diversity of ages among the participants in the courses. Through the chats, ads boards, videoconferences, interactive whiteboards, etc., it is possible to help to cohere the group, but with merely informative purposes instead of didactic.

The empathy in the communicability of those and other interactive systems is very positive when in the design process of the interactive system a communicability analyst or a team of professionals intervenes. In this sense we are not approaching the organization issues of the collaborations independently among themselves, that is, whether it is multidisciplinarity, interdisciplinarity or transdisciplinarity. To the readers interested in the differentiation of these terms you can look up the following bibliography [5]. Obviously the designers must take into account that these contents will have to be adapted to very different users and the strategies used in the videogames are advisable in these cases. In them empathy plays a very important role since the contents may be presented in a progressive way, in relation to the breakthroughs the user makes at the moment of interacting with the system. Now we can also detect new problems when we have false professionals in the sector of the interactive systems, or dynamic persuaders [8] [11], where the human factors seriously damage everything related to the new technologies and their transfer to the educational and/or productive world [11] [12]. In these cases, it is interesting to work with beta versions of the interactive systems and carry out tests with real users of the products made by these false professionals or dynamic persuaders, for instance. In the case of the educational presence courses one-day modules can be allocated to them and evaluation questionnaires to the participants at the end of the lesson to detect the quality of the educational process. These prevention measures are due to the fact that in Southern Europe there are professionals who are alien to the areas of knowledge and/or experiences about which it is necessary to carry out the technology transfer. Besides, although we have banked on the empathy, the communicability analyst in the interactive systems to support the educational process, etc. sometimes they turn out to be difficult to detect before being inserted in the professors body [8] [9]. For instance, in the Balearic Islands we may have a PhD in computing and mathematics who never made an equation or a computer program because in fact he/she works in the fine arts. In contrast, in Catalonia there are industrial doctors who define themselves as artists in computer synthetic images (animated and/or static) without ever having made a drawing, painting or sculpture, whereas we can meet telecommunications PhDs without knowing what a wire is or which are the main components of a parabolic aerial. At the same time, in Lombardy we may come across with a graduate in computing who defines himself as PhD and researcher in multimedia, pedagogy, philosophy, etc., but who in reality has worked in the trade unions and landslides in the mountainsides. These are mere examples of the reality that those who attend as professors can find in the technology transfer courses in the Southern European areas. That is, we may have variegated classes, not only considering the students but also the professors, such as those previously enumerated. In both situations, the empathy to speed up the learning process and the communicability to improve the interrelation of the students with the real or virtual professors, little or nothing can they do in the face of the presented human factors.

3 Technology Transfer: Premises and Theoretical Models

Next a classical model followed for those universities who inside the triad administration (state, regional, local government bodies, etc.) university and enterprise, intend to achieve continuous training in the transfer of innovative technology with private bodies in Southern Europe:

• Denomination: avoid the confusion with other university titles
• Fulfillment of the minimal numbers of established hours
• Course programme: number of the subjects and credits
• Listing of institutions collaborating: practices, sponsors, etc.
• Professor body of the course: specifying titles and institutions to which they belong
• Director/s or coordinators of the course. At least one of the coordinators or directors will be professor of one of the professorship categories stated in the university by-laws.
• Explanation about the individualized evaluation of the students. In no case titles or diplomas will be delivered only because of attendance.
• Definition of the entry profile of every student.
• Justification of the adequacy of the course to the centre areas
• Adjustment to the demand of the market. Justification for the programme of courses that has not been successful in previous editions.
• The programming of the continuous training activities will be free for the centres and the certification will be responsibility of the organizing centre.
• The centre will inform about the activities to the university general secretariat in regard to the number of registered students in every course and on its development.

In this innovation, where groups of interconnected enterprises and related institutions belong to a sector or market sector and they are linked through common and complementary elements [13], they join to create joint projects and thus increase their competitiveness [14], but in keeping with the vademecum presented to the intrinsic goals that they usually manifest in the Internet, such as:

- The creation of a cooperation axis, arriving together where one would not arrive on its own.
- Increase of productivity of the participating enterprises.
- Increase of the innovation ability at a low cost : collaborative innovation, creation of synergies, development of consortia and collaborative projects.

The main agents that intervene are the business associations, service provider enterprises, enterprises demanding services, university, public administrations, professional training institutions, etc. As their main goals we can mention: Development of thematic platforms according to sector, selection of the participating agents, definition of goals and operative organization of the technological innovation core and the relationships among their members. So far we can see the theoretical aspect of these activities. Now in the creation and development of the core of technological innovation appear the first economical aspects that may seriously distort the main challenges to be met in the technology transfer with a high quality level. In other activities it is possible to mention:

- Integral management of the innovation process
- Specialized reports (continuous follow-up of the technology trends and the national and international market).
- Specialized advice in management, strategy and organization in the setting in motion of activities.

- Creation of a common innovation space, that is, through forums and seminars of innovation. In these forums and seminars the participants are supposed to have a science, technology and market expert profile. However, here it is where the dynamic persuaders can destroy the whole innovation and technological transfer project, especially with the points that follow.
- Search of aids and subsidies, public and private
- Access to the community of enterprises and entrepreneurs.
- Collaboration with external agents, that is, interchange activities and promotion of the technological innovation process in national and international events.

Finally, we have a stage that is called consolidation and fostering of the innovation core, through such tasks as:

- Development of new concepts, technologies, collaborative and chartered projects: Promotion of larger projects; R+D national and international projects; New enterprise models; and Management of the generated knowledge (here the agents aim at the return of the investment that was made).
- Valuation programme of the innovation: Tech valuation; and Innovation market, through the access to external platforms of open innovation.
- Reaching a high local, regional, national and international media impact

Evidently these are the theoretical aspects of some models and classical principles in the analogical or paper support, but sometimes we can come across a mirage of this reality in the Internet., especially in the transfer sector of the latest technological breakthroughs [13] [14]. Here is the main reason why it is important to have a vademecum that will allow us to save money in the face of some oasis mirages in the search of excellence of technological innovation

4 Vademecum towards Excellence in the Multidirectional Technological Transfer

The current vademecum represents the synthesis of a ten-year research process carried out in several European universities, with private, public and hybrid by-laws on a "sui generic" basis (we use this notion when they are realities that go against the rest of the rules). The universities and enterprises are located in the following autonomous regions of Europe: Aragon, Catalonia, Balearic Islands and Lombardy. The negative factors are marked with an "N", the positive with a "P" and indifferent with a "I". It is important to consider the order of the contents of the vademecum and the logical connectors of the sentences (and, or, and/or equal, not equal, etc).

• The courses promoted on line generally have the following structure: presentation, goals, contents, direction, duration, schedule of activities, titles, access to a virtual campus, job vacancies, cost of the registration and funding (I).
• The publicity campaign encompasses the main local, national and international media (N).
• The courses are taught in cities where several languages are simultaneously spoken (N).

- The classes are taught in several languages and dialects (N).
- The contents of the programmes are presented in dialects and eventually in languages (N).
- The classes are "presence" ones in a physical classroom (P).
- The classes are virtual (N).
- The university campuses are real (P).
- The university campuses are virtual (N).
- The universities are between 1 and 10 years old (N).
- The universities are between 10 and 30 years after their foundation (N).
- The universities are between 40 and 50 years after their inauguration (I).
- The universities are between 50-100 years old after they were opened (P).
- The universities have been working for centuries (P).
- The classes are combined into presence classes and virtual ones (P).
- The professors are from the university/enterprise/industry from which the courses are taught (I).
- The professors come from other universities and/or enterprises/industries (I).
- The professors have their resumes posted on-line (P).
- The professors have a varied and/or meteoric titling from the universities. For instance, degrees in psychology and history, master in journalism and PhD in telecommunications (N).
- The professors have a titling of technician and engineer obtained in software and hardware enterprises, such as HP, IBM, Xerox, etc. (N).
- Most of the professors have a PhD (P).
- The courses of the professors are in keeping with their training (P).
- The experience of the professors in the subject they present has extended for years (P).
- The professors use outsourcing resources in their lessons (N).
- The number of collaborators in the courses is lower than the number of professors (N).
- The practice labs have a unique technology brand that is presented to the student (N).
- The access to all the similar technologies is guaranteed at the moment of teaching the theoretical and/or practical classes (P).
- The accessibility to the technology that is being studied in the curricula is only for those firms for which there are mutual collaboration agreements (N).
- The courses imply a stay in the enterprises of the professors who teach the lessons (N).
- The students' selection depends on the University (P), the enterprise/industry (N) or both (I).
- The students must submit a real project before ending the technology transfer course (P).
- The access to the possibility of carrying out a final exam requires to overcome several previous tests or exams and the approval of a practical project (P).
- The students must pass several tests to obtain a certification or diploma (P).
- The obtainment of a diploma requires a minimum of hours of attendance in the presence courses (P).
- The evaluation of the knowledge is in the hands of the professors (P) or those responsible for the enterprises/industries (N).
- The exam of the acquired knowledge will be in a real space, with supervisors (P).
- The final exam to have access to the title is made through virtual campuses (N) or classrooms (P).
- The evaluation of the knowledge is done in a progressive way through the internet (P).
- The course is made through the academic rules of the university (P).
- The course is made in relation to the demand in the work market (I).
- The potential students must possess an academic (P) or entrepreneurial profile (N).
- The total amount of the registration fee must be paid before the beginning of the course (N).
- The student is oriented towards scholarships, loans and aid towards the financing possibilities under advantageous conditions (I).
- The reached results have a high media impact in the community where the core of the technological innovation has been generated (I).

Next we sum up graphically the main contents of the vademecum resorting to two *smiles* for the information on-line on these issues:

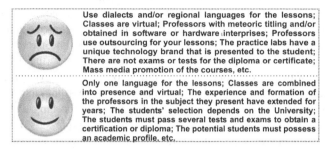

Fig. 1 Set of elements that boost and damage the transfer of technological innovation

This is the first state of the art in Southern Europe where it has been avoided to insert the names of the institutions, people responsible, training programmes, etc., with the purpose of keeping the anonymity and the respect to the privacy of information. Lastly, we insert the following graphic that demonstrates the result of applying the guidelines or vademecum to 50 Spanish and Italian portals chosen at random among university portals, industrial unions, associations and bank foundations, and local, provincial, regional, statewide bodies, etc. which are devoted to the transfer of innovating technology. The results are divided into two big groups: private and public institutions and the elements that boost the transfer of technological innovation.

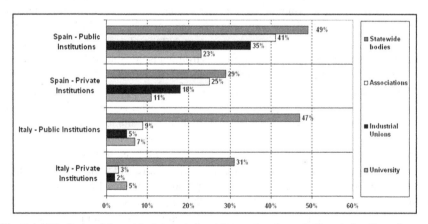

Fig. 2 A little State-of-the-art in Internet about Innovation through Knowledge Transfer

5 Lessons Learned

With the development of the information society and its overall diffusion in the world, on-line training for the transfer of technological innovation will be in an arithmetic increase in quantity but in a geometric decrease of quality due to economic and financial costs. Factors that vary between the north, centre and south of Europe, such as the expected benefit margins and the time to achieve them. For instance, in the long term in the north and with figures that do not go above a dozen numbers in the north and centre of Europe, compared with those who aim at three digits in the short term in the south of Europe, regardless of the core of technological innovation that has been generated. Although these new virtual learning environments, profiting from the functionalities of the ITC (Information and Communication Technologies) offer environments for teaching and learning free from the restrictions set in time and space by presence teaching and capable of ensuring a mutual communication between students and professors, it is necessary to keep on the positive aspects of the transfer of technological innovation. On the other hand, these environments (with a wide implantation in university, work and occupational training) also allow to complement presence teaching with virtual activities and on-line credits that can be developed at home, in the teaching centres or in any place that has a connection to the internet. The latter can be reached if the main premises of the vademecum presented by the organizing agents of the technological innovation core centre are guaranteed in writing.

6 Conclusions

It is in the transfer of the technological innovations that is currently focused a mercantilist factor of the sciences that contradicts one of the main principles of scientific knowledge: communicability, understood as a knowledge that is not private but public. The communication of the obtained results in the R+D labs in the private industries does not only improve and perfects overall education but it multiplies the chances of confirmation or rebuttal. These last two terms sometimes generate uncertainty in some scientific environments in Southern Europe in the continuity in receiving national and international financial subsidies. This is the reason why there is a kind of resistance in the scientific community to organize these courses. On the other extreme we have those who see in the results of the R+D labs a source of financial revenue. Between both situations those agents of the academic sector who need guidelines like those presented in the current work to make decision in a short lapse of time. These guidelines will be widened with the future works including other realities from the centre and north of Europe, like those hailing from Asia and America. Addtionally, with this first version of the guidelines it is possible to know the reliability and seriousness of the agents who take part in the educational process, since in many cases the students must self-finance courses that theoretically and from the standpoint of the main principles of science should be free of charge. Luckily, there are still public and private institutions that ensure that this principle is kept in our days.

References

1. Foley, J.: Technology Transfer from University to Industry. Communications of ACM 39(9), 30–31 (1996)
2. Myers, B.: A Brief History of Human-Computer Interaction Technology. Interactions 2, 44–54 (1998)
3. Tillquist, J., Rodgers, W.: Using Asset Specificity and Asset Scope to Measure the Value of IT. Communications of ACM 48(1), 75–80 (2005)
4. Cipolla-Ficarra, F.: Quality and Communicability for Ineractive Hypermedia Systems: Concepts and Practices for Design. IGI Global, Hershey (2010)
5. Cipolla Ficarra, F.V., Villarreal, M.: Strategies for a Creative Future with Computer Science, Quality Design and Communicability. In: Cipolla Ficarra, F.V., de Castro Lozano, C., Nicol, E., Kratky, A., Cipolla-Ficarra, M. (eds.) HCITOCH 2010. LNCS, vol. 6529, pp. 51–62. Springer, Heidelberg (2011)
6. Dubberly, H.: Toward a Model of Innovation. Interactions 15(1), 28–34 (2008)
7. Vicente, K.: The Human Factor: Revolutionizing the way people live with technology. Taylor & Francis, New York (2003)
8. Cipolla-Ficarra, F.: Persuasion On-Line and Communicability: The Destruction of Credibility in the Virtual Community and Cognitive Models. Nova Science Publishers, New York (2010)
9. Preece, J.: Empathic Communities. Interactions 5, 32–43 (1998)
10. Cipolla-Ficarra, F.: An Evaluation of Meaning and Content Quality in Hypermedia. In: CD-ROM Proc. HCI International, Las Vegas (2005)
11. Tudor, L.: Human Factors: Does Your Management Hear You? Interactions 1(1), 16–24 (1998)
12. Vliet, H.: Reflections on Sofware Engineering Education. IEEE Software 23(3), 55–61 (2006)
13. Cunningham, J., Harney, B.: Strategic Management of Technology Transfer: The New Challenge on Campus. Oak Tree Press, Cork (2006)
14. Speser, J.: The Art and Science of Technology Transfer. John Wiley, New Jersey (2006)

Session D
Knowledge Transfer Insights

Tri-partnerships in Knowledge Transfer: Changing Entrepreneurial Mindsets

Christopher J. Brown and Diane Proudlove

University of Hertfordshire, de Havilland Campus, Hatfield, Herts, AL10 9AB

Abstract. The UK's drive towards a low-carbon economy is an example of the challenges facing Small- to Medium-sized Enterprises (SMEs). There is still a relatively low take-up of this initiative across all sectors. Is this because of the mindsets of the business managers in SMEs? Certainly research into entrepreneurial mindsets surrounding external environmental factors, particularly on the need to create, develop and deliver green values, suggests that adaptation is significantly influenced by their cognition processes. This paper reports on a longitudinal study into six enterprises involved in Knowledge Transfer Partnerships (KTPs) where uncertainty and ambiguity in their marketplace drove the need to change, and solicit outside help. We report on the business entrepreneurs belief systems, and their sensemaking associated with their business models during the medium-term KTP projects. A comparative analysis was performed between the six enterprises studied and a framework was developed from the four major emergent constructs: environmental factors, entrepreneurial sensemaking, strategic orientation and the business model.

1 Introduction

The UK government has created initiatives like the Knowledge Transfer Partnership (KTP), one of Europe's longest and largest running knowledge transfer programmes, to help businesses cope with environmental, social and economic changes (Narayanan and Fahey 2005). These KTPs bring together knowledge portals, like Universities and private research institutions, with Enterprises who need knowledge to help drive strategic change, and lastly associates who will carry out the work. This government driven initiative heralds the potential to enact an important transition for universities, transforming them from mere teaching/research institutions into establishments that combine teaching, research and business focused activities (Etzkowitz and Leydesdorff 2000), and thus benefit both the public and business community alike. Yet, research focusing on the Top 6 research-intensive universities found that only 18% of the engagements with outside enterprises were

R.J. Howlett (Ed.): Innovation through Knowledge Transfer 2010, SIST 9, pp. 153–162.
springerlink.com   © Springer-Verlag Berlin Heidelberg 2011

related to providing work or technical experience (developing new skills and competencies). Over 50% of the knowledge exchange was associated with the university disseminating generic knowledge on the universities facilities or its research outcomes, and not that of the enterprise's (University_Of_Cambridge 2003). Universities that are either research- or business focused must engage with enterprises at a level that delivers strategic value, value that helps transform their business models to become more sustainably competitive.

This study looked at the strategic behaviours and broader belief systems of business entrepreneurs, and the role these played in their decision-making, as evidenced through their strategic orientations.

2 Importance of Knowledge Exchange to Business Models

At the heart of the challenges and issues associated with sustainable development is the drive towards a low-carbon economy, and the impact this has on enterprises' business models (BERR 2010). How will these enterprises and business entrepreneurs gain access to important knowledge and technology to support this transition? The future success of SMEs and entrepreneurs to unlock talent, knowledge and skills held in these knowledge portals, will largely depend on their ability and motivation to engage with them. Previous research on knowledge/technology exchange from these knowledge portals suggested that absorptive capacity of the recipient enterprises was highly important, but equally so was the encouragement and reward given to staff to actively engage with these public/private knowledge portals. Knowledge portals or knowledge hubs are often a combination of government, sector or knowledge institutes that have the responsibility to facilitate and speed-up the diffusion of innovative knowledge to the business/research community (van Baalen, Bloemhof-Ruwaard et al. 2005). Entrepreneurs, intrapreneurs and innovators have always used outside professional and business networks to gain access to important knowledge and skills essential to their respective roles and responsibilities in driving innovation.

Yet, at the heart of enterprises' search for external knowledge, is the approach they take in identifying appropriate knowledge sources and the depth of knowledge available. This identification of external knowledge sources is linked very closely to previous methods of obtaining knowledge/technology: informal networking with public and private entities, R&D collaboration and specific technology acquisition (Kang and Kang 2009). Knowledge exchange is a critical tool or activity in sustaining an enterprise, and just as importantly in developing the business entrepreneur. These business entrepreneurs are unlike ecopreneurs, those that have pre-existing green values and are sometimes referred to as "green-green" entre?preneurs (Kirkwood and Walton 2010). Though substantial numbers of entrepreneurs may indeed have positive environmental attitudes, but these do not manifest themselves into the businesses practices or business model?

These business entrepreneurs can describe their business model as "how a firm will make money and sustain its profit stream over time" (Morris, Schindehutte et al. 2005) pp.727. This business model is often broken down into three constituent parts: industry model – means of creating white space; revenue model – how product and services contribute revenue and profits; process model – how these achieved through the business processes. Fundamentally, these business models are based on an entrepreneur's mindset or sensemaking of his enterprises' environment.

3 Entrepreneurial Sensemaking and Strategic Orientation

3.1 Entrepreneurial Sensemaking

Enterprise research has suggested that individuals, entrepreneurs and intrapreneurs, analyze the action-outcome relationships associated with specific enterprise processes (green values), and then create cognitive maps (Alexander 2004). These individual cognitive maps ascribe specific interpretations to observed collective actions, enterprise events and innovative outcomes, and it is the reinforcement and modification of these that supports, or amends, their future actions (green practices). This sensemaking process has some hierarchy, a taxonomy of sensemaking is presented below (Brown 2006), see figure 1.

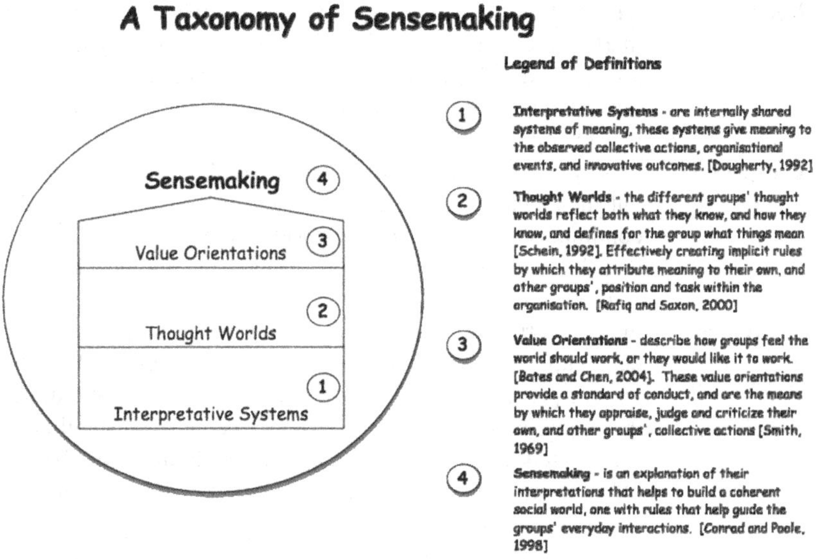

Fig. 1 A Taxonomy of Sensemaking (Source: Brown, 2006)

Sensemaking then is an integrative process of communicative sharing of relevant information pertaining to the challenge (initiating green processes); interpretative act of directing and shaping of that information; and then interpreting it (Smith 1969; Dougherty 1992; Conrad and Poole 1998; Rafiq and Ahmed 2000; Bates and Chen 2004; Schein 2004; Neill, McKee et al. 2007). The business entrepreneurs' sensemaking of the need for change is critical to both their development of green values, and more importantly positive green practices.

3.2 Entrepreneurial Strategic Orientations

Previous research has suggested that problems in enterprises are most often rooted in past decisions rather than any present marketplace dynamics or events (Miles and Snow 1978; Greiner 1998; Aragon-Sanchez and Sanchez-Maron 2005). Research by Miles and Snow presented a typology that linked strategic orientation to those enterprises' evaluation of internal and external environmental factors, and that these triggered changes in their strategic orientations – management style, structural, cultural and process orientated. The research suggested that these strategic orientations result from the business entrepreneurs' and the enterprises' analysis of internal and external environmental factors (competitiveness, marketplace uncertainty and ambiguity, market orientation, economic growth), and reflect their values, attitudes and practices (sesnsemaking) towards ecologically-driven innovations:

Defenders	these enterprises often focused on a narrow or limited product market, creating a niche for themselves where they have subsequently developed a leading position. These enterprises fall into a strategy of trying to protect their market share and revenues/profits.;
Prospectors	these enterprises often start with a single successful product, but then steadily grow their product/service portfolio by their continuous search for new market opportunities by applying their knowledge and know-how to innovate and develop superior customer-valued products and services;
Analyzers	these enterprises can act both, defensively, or prospectively, depending on their analysis of the environmental challenges and the perceived innovation-resources that would be required;
Reactors	these enterprises are characterized by perpetual instability and inconsistency in their strategies, predominantly because of their incapacity to respond effectively to environmental changes.

This approach to reflecting on the entrepreneurs' sensemaking and their enterprises' strategic orientation is very valuable, and an example of reflection-in-action (Adams, Turns et al. 2003). Reflection by the entrepreneurs is used to drive their tacit knowledge associated with green practices and experiences to the surface, helping them to construct meaning and value from these (Raelin 1997). The author's own research into business entrepreneur's mindset changes (Brown and Proudlove 2008) suggested that two of the strongest drivers for business model change are values and attitudes.

4 Research Methodology

Our general approach follows the classic grounded theory methodology of using the analysis of the data collected through interviews, observation, workshops and documentation research to generate links between emergent themes and subthemes (Parker and Roffey 1997). Glaser and Strauss's perspective was that:

> "the grounded theory approach is a general
> methodology of analysis linked with data col-
> lection that uses a systematically applied set of
> methods to generate an inductive theory about
> a substantive area" (Glaser 1992) pp. 16.

The initial analysis involved over thirty Small- to Medium-sized Enterprises (SMEs), we carefully screened these businesses to find atypical problems/challenges that would highlight the broad range of business entrepreneurial mindsets, associated with driving green values and processes, within SMEs facing particular environmental opportunities and threats, and focused on just six SMEs. We used QSR Nud*ist NVivio software to help collate, store and code the data. We coded all transcripts using core and axial coding methods, and analyzed these using constant comparison approach (Glaser 1992). Data reduction in presenting this qualitative research was required, and only small portions of the transcripts are used to illustrate the views of these business entrepreneurs (Silverman 1997).

5 Key Findings

The business entrepreneurs in this study were motivated to engage in KTPs because of the environmental challenges in their marketplace, and therefore to their business model. These challenges were focused around three common drivers:

1. Incorporating green values and processes;
2. Identifying new markets for these green products;
3. Making the business profitable.

As shown in Table 1, these drivers above directly influenced their starting their strategic orientations at the start and finish of the KTP.

Table 1 Enterprise A Entrepreneurial Mindset, Strategic Orientation and Business Model Changes (Started[1] and Finished[2])

Business	Environmental Factors	Business Model	Strategic Orientation	Expected Impact
Enterprise A	Sector uncertainty – falling sales[1] How to engage these new customers – green values[2]	Develop new markets and increase share of existing[1] Marketing best practice to existing markets[2]	Reactive[1] Analyzer[2]	Previous success based on quality of product delivered – not turning into revenue/profit streams?[1] Slowly changing value orientations from being process-driven towards being market-oriented[2]
Enterprise B	Stagnation in European Sales[1] Managing the European Sales Network[2]	Develop new markets and increase share of existing[1] Understanding how to drive sales efficiencies[2]	Defender[1] Defender[2]	Quick solutions to market identification and sales inefficiencies[1] Realization that the European Sales Network was not delivering full value[2]
Enterprise C	Growth into other sectors[1] Training market uncertainty[2]	Develop new markets and increase share of existing[1] Increasing the value of training provision – accreditation with Universities[2]	Prospective[1] Prospective[2]	New market sectors and needs analysis[1] New market sectors and needs[2]
Enterprise D	Concept of Market[1] Concept of Market[2]	Understanding of the market demands and buyer needs[1] Product technology evaluation[2]	Reactor[1] Reactor[2]	Proof of Market and Product Concept[1] Some proof of market[2]
Enterprise E	Market development[1] Market and product development[2]	Understanding of the market demands and buyer needs[1] Product gaps and software development[2]	Analyzer[1] Analyzer[2]	Increase marketing performance[1] Increased marketing performance[2]
Enterprise F	Integrated Product/market development strategy[1] Integrated Market and product development[2]	Understanding of the market demands and buyer needs[1] Market-place/Technology Roadmapping[2]	Prospector[1] Prospector[2]	Integrated Marketing and Product Strategy[1] Integrated Marketing and Product Strategy[2]

For brevity, the cognitive mapping process by which the authors linked the environmental challenges with their sensemaking, and the resulting changes to their strategic orientation, and the anticipated performance impact expected from these, are not shown. However, the resulting business model change framework is shown below, showing how in the six enterprises studied the different challenges caused different levels of change in the business model.

5.1 Development of a Business Model Change Framework

The business model change framework below, see figure 2, captures the level of changes in the business model as a consequence of environmental challenges, such as customer or competitor activities, requiring incremental innovations, or broader marketplace changes requiring product/market innovations, or more dramatic legislative changes at the sector level requiring substantive innovations.

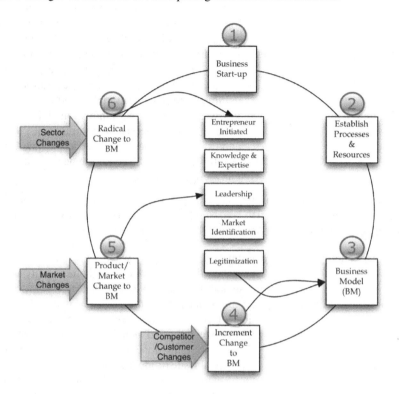

Fig. 2 Business Model Change Framework

The business model change framework, above see figure 2, helped describe the business entrepreneurs cognitive process by which they developed their business model, the key components of this model are:

1 Business Start-up	the entrepreneur would readily describe what the main premise was for his start-up, what the market was, what he would deliver and how they would make money from it – this is the 1st business model?
2 Establish Processes	this helped explain the business processes, set up to deliver on the 1st business model – financial and non-financial;
3 Business Model	either a strategic plan identifying the principal strategic markets targeted, with detailed information about the products and services

	delivered, and the expected revenue and profit streams resulting from this.;
4 Incremental Change	Small changes were often evident in the Business Model as the enterprise reacted to customer/competitor changes, and/or initiated incremental innovations to their products/services;
5 Product/Market Changes	more dramatic changes in the market stimulated some more dramatic changes to the product or market strategy, maybe a next generation product line to help re-position the product;
6 Radical Changes	this was where product innovation alone can not address the external environmental threats and/or opportunities, they needed a new direction for the business, hence the need for external knowledge/expertise.

The inner circle is something that often happened as a consequence of the KTP initiation, the process was fairly consistent, though the order changed depending on the nature of the challenge/issue, and the sector they operated in:

Entrepreneur Initiated	ultimately, this was where the enterprise approached the knowledge hub, the university in this case, looking for help and guidance;
Knowledge & Expertise	an important part of the project was the identification of relevant knowledge and expertise in the knowledge hub, and matching this to the requirements of the entrepreneur's skills and competencies;
Leadership	the critical issue of identifying the key gate-keepers, those managers or individuals who hold sway over the needed resources for the project to work;
Market identification	identifying the important 'white space' for the new products/services originating out of the project, and the projection of sales. All important information relating to the industry & revenue components of the business model;
Legitimization	this is ultimately how the business entrepreneur perceives the new mindset business model, and how they will legitimize it to their other senior managers and work colleagues. The weighting of the benefits and costs, the likely impact on the existing business model, is sufficient to address the business challenges and sustain the business for the medium- to long-term?

The close interrelationship between an entrepreneur's sensemaking of the environmental challenges in their sector, and their reflection on the sustainability of their mindset business model, and the underlying business processes, suggests a push and pull strategy. There are still many unanswered questions on what motivates these entrepreneurs to adopt green values, and more importantly to push these green values into their processes.

6 Conclusions

The outcome of this study is twofold. First, we were able to develop a business model change framework that helped understand the interrelationship between the business entrepreneurs' sensemaking of the environmental challenges to his business model. This framework has its theoretical basis from the work on innovation system frameworks for understanding technological innovation change (Hekkert, Suurs et al. 2007) and that on entrepreneurial mindset business models (Morris, Schindehutte et al. 2005). The resulting empirical findings do provide further new insights into how important entrepreneurial sensemaking is to determining the motivations and commitment to undertake business model changes, especially their overall strategic orientation towards adopting green values and processes, which from the government's perspective is all important to working towards a low-carbon economy. The second outcome, was a useful exploratory study on the value of knowledge portals, and particularly Knowledge Transfer Partnerships (KTPs), to help stimulate changes to entrepreneurs' strategic orientations, moving them away from reactive or defensive strategies and towards prospector and analyser strategies. This study does have limitations, the number of businesses selected, and the diverse sectors they come from. Future research should focus on individual sectors, choosing those that are perhaps the poorest performers on environmental initiatives?

References

Adams, R.S., Turns, J., et al.: Educating Effective Engineering Designers: the role of reflective practice. Design Studies 24(3), 275–295 (2003)

Alexander, D.E.: Cognitive Mapping as an Emergency Management Training Exercise. Journal of Contingencies & Crisis Management 12(4), 150–159 (2004)

Aragon-Sanchez, A., Sanchez-Maron, G.: Strategic Orientation, Management Characteristics, and Performance: A Study of Spanish SMEs. Journal of Small Business Management 43(3), 287–308 (2005)

Bates, R., Chen, H.-C.: Human resource development value orientations: a construct validation study. Human Resource Development International 7(3), 351–370 (2004)

BERR. SMEs in a Low Carbon Economy (2010), http://www.berr.gov.uk/files/file49761.doc (retrieved 04/02/2010)

Brown, C., Proudlove, D.: Business Entrepreneur's Mindsets on their Enterprises' Business Model. In: EDINEB, pp. 1–25. Malaga, Spain (2008)

Brown, C. J. The Acquisition and Dissemination of Ideas: Managing the Innovative Initiative. Human Resource Management, Middlesex University Business School. Doctor of Philosophy, 274, London (2006)

Conrad, C., Poole, M.S.: Strategic Organizational Communication: Into the Twenty-first Century. Harcourt Brace Colleges Publishers, Troy (1998)

Dougherty, D.: Interpretive Barriers to Successful Product Innovation in Large Firms. Organization Science 3(2), 179 (1992)

Etzkowitz, H., Leydesdorff, L.: The dynamics of innovation: from National Systems and "mode 2" to a Triple Helix of university - industry - government relations. Research Policy 29, 109–123 (2000)

Glaser, D.A.: Basics of Grounded Theory Analysis. Sociology Press, Milley Valley (1992)

Greiner, L.E.: Evolution and Revolution as Organizations Grow. Harvard Business Review, 55–65 (May-June 1998)

Hekkert, M.P., Suurs, R.A.A., et al.: Functions of innovation systems: A new approach for analysing technological change. Technological Forecasting and Social Change 74(4), 413–432 (2007)

Kang, K.H., Kang, J.: How do Firms Source External Knowledge for Innovation? Analysing effects of different knowledge sourcing methods. International Journal of Innovation Management 13(1), 1–17 (2009)

Kirkwood, J., Walton, S.: What Motivates Ecopreneurs to Start Businesses? International Journal of Entrepreneurial Behaviour & Research 16(3), 204–228 (2010)

Miles, R.E., Snow, C.C.: Organizational Strategy, Structure and Process. McGraw-Hill, New York (1978)

Morris, M., Schindehutte, M., et al.: The entrepreneur's business model: toward a unified perspective. Journal of Business Research 58(6), 726–735 (2005)

Narayanan, V.K., Fahey, L.: The Relevance of the Institutional Underpinnings of Porter's Five Forces Framework to Emerging Economies: An Epistemological Analysis. Journal of Management Studies 42(1), 207–223 (2005)

Neill, S., McKee, D., et al.: Developing the organization's sensemaking capability: Precursor to an adaptive strategic marketing response. Industrial Marketing Management 36, 731–744 (2007)

Parker, L.D., Roffey, B.H.: Methodological Themes: back to the drawing board. Revising theories grounded theory and the everyday accountants' and managers' reality. Accounting, Auditing and Accountability Journal 10(4), 273–290 (1997)

Raelin, J.A.: A Model of Work-Based Learning. Organization Science 8(6), 563–578 (1997)

Rafiq, M., Ahmed, P.K.: Advances in the internal marketing concept: definition, synthesis and extension. Journal of Services Marketing 14(6/7), 449 (2000)

Schein, E.H.: Organizational culture and leadership. Jossey-Bass, San Francisco (2004)

Silverman, D.: Qualitative research: theory, method and practice. Sage, London (1997)

Smith, D.: Social Psychology and Human Values. Aldine Publishers, Chicago (1969)

University_Of_Cambridge, University of Cambridge: Comunity Engagement Report 2003-4, University of Cambridge (2003)

van Baalen, P., Bloemhof-Ruwaard, J., et al.: Knowledge Sharing in an Emerging Network of Practice: The Role of a Knowledge Portal. European Management Journal 23(3), 300–314 (2005)

The Barriers to Academic Engagement with Enterprise: A Social Scientist's Perspective

Linda Reichenfeld

MPhil AMInstKT, Senior Lecturer and Departmental Academic
Enterprise Coordinator, Department of Interdisciplinary Studies,
Manchester Metropolitan University, Crewe Campus, Crewe Green Road CW1 5DU, UK
0044161 247 5247
L.reichenfeld@mmu.ac.uk

Abstract. This paper explores the barriers to academic engagement with enterprise from a social scientist's perspective and in relation to United Kingdom post-1994 universities in particular, expanding key themes from previous literature to consider both progress and the limiting factors which still face university managers in their attempts to implement their 'change' agendas.

The current strategy of re-orienting and branding universities as professional, managerial and efficient organisations, within which knowledge must be generated in a deliverable and transferable form to external recipients, is unpopular with many social science and humanities academics in particular, owing to the prevailing view among the latter that their identity is under threat. Hence the apparently widespread academic disengagement discussed in this article, which is explored in the context of the reluctant academic pressured to extend their role into often unfamiliar business-speak, commercial enterprise and industrial environments by a university strategy that assumes all academics are capable of incorporating academic enterprise into their day to day activities. The paper concludes that for universities to successfully rebrand as professional and commercially successful institutions they must adopt a more business like approach requiring first that they overcome the fear prevalent among many UK social scientist academics, at least, that their managers, and the higher education sector at large, have shed essential values which, since Humboldt's time, have underpinned the very purpose of higher education institutions. *Key terms: academic identity, branding, engagement, knowledge transfer, innovation, autonomy, multidisciplinary,*

1 Introduction: Knowledge Transfer and the Changing University

The current economic pressures upon the higher education sector and the increasingly important role of knowledge transfer in university strategies have 'transformed small, elite institutions, managed by academic peers in a collegial way,

R.J. Howlett (Ed.): Innovation through Knowledge Transfer 2010, SIST 9, pp. 163–176.
springerlink.com
© Springer-Verlag Berlin Heidelberg 2011

into large, multi-task organisations in need of new governance structures to manage all the tasks and roles of today's institutions' (Geuna and Muscio 2009:94).

Knowledge transfer in higher education institutions is not new, but what is new is the 'institutionalisation' of university-industry linkage referred to by Wedgewood as 'mainstreaming the third stream' which demands, yet is far from achieving, significant engagement from academics (2006). Wedgewood argues that the main business of Universities needs to change; to become more than teaching and research establishments, that Universities should deliver social and economic impact; this requires collaboration with third sector organisations. This may require a second, Humboldtian transformation (the first being the early-nineteenth century synthesis of teaching with research) where academics are expected, as part of their normal roles, to consider a revised identity, expanded from the autonomous researcher and teacher to incorporate what Whitchurch refers to as the 'third space' (2008), or what is commonly referred to as academic entrepreneurship. This latter term is unsatisfactory where it infers that knowledge transfer is only concerned with profit-making and should, rather, be viewed as the academic in collaborative partnerships with not only commercial but social enterprises including non-governmental organisations, charities and other not-for-profit organisations, where such partnership gains may form strategic and efficiency improvements in the partner organisation. The difficulty of arriving at a clear and acceptable definition of academic enterprise; what exactly it constitutes, who exactly it applies to, has further alienated sceptical academics and the broader research community, and this is yet to be satisfactorily resolved. The same applies to the term third stream; indeed the semantics of this entire territory defy clear definition and consequently lack authority. For most academics, their roles revolve around and are defined by research and teaching; for those of us engaged in knowledge transfer, however, there are no conflicts here; rather academic enterprise can provide a spur to research and is of invaluable use to teaching , as well as offering student project and work experience opportunities.

2 Knowledge Transfer and the Social Sciences

Historically, university collaboration with industry has been regarded as "proper" knowledge transfer, and it is commonly delivered from science and technology university departments, where it has become a strategic and 'policy tool for economic development' as well as 'commercialisation of discoveries' (Geuna and Muscio 2009 p95). However, it is much more than this where it reaps, as is increasingly the case, community and social benefit objectives. Social scientists clearly have important roles to play in knowledge exchange where, for example social policy or the voluntary sector are concerned. Enlisting academic research to support strategic development of small and medium enterprises (SMEs); that is to say firms or organisations with fewer than 250 employees), in particular, is sensible where financial gains are to be had but that is not to devalue its worth as a tool to assist public and social organisations deliver local services more effectively. This, though a comparatively recent acknowledgement in the Government's knowledge transfer policy, holds great potential in its ability to engage academics who hitherto have been disinclined to engage with largely technology-based knowledge transfer projects.

Whilst engineers and scientists can see intellectual property and spin-off opportunities in industrial collaborations, social scientists, artists and creative academics have been less inclined to get involved, perceiving knowledge transfer as something that only applies to goggle wearing, lab-coated technicians. This further confuses the issues of academic identity and intellectual territory, since innovation and problem solving are traditionally associated with the "hard" sciences, particularly where economic outputs and impacts are concerned. There is a clear and comparatively recent bias in higher education in favour of commercialisation of research and profit-led investment; arguably contributing to the inexorable drift in higher education away from the value of intellectual and scholarly practice, towards economic outputs and impacts, and this is especially apparent during times of general economic stress (see for example leader articles in the Times Higher Education January 2010). Once again, the imperative that is the "profit-making" academic overrides the value of the academic as scholar and intellectual.

3   KTPs for Social Enterprises

Knowledge transfer partnerships in the UK are operated via the Department of Business and Industry's Technology Strategy Board and new programmes have to pass a Programmes approval Group which selects projects on the basis of potential financial and business-relevant impact for the company, as well as on the basis of the quality of the proposed Knowledge Transfer. MMU Cheshire recently submitted a (successful) bid for a two year KTP with Wulvern Housing Association, the objects of which were not, as is the case with most KTPs to make £100,000 worth of profits, but rather to *save* equivalent monies. There is a new trend emerging – especially in the public sector and allied organisations such as Housing Trusts - where financial savings are even more important than profits. KTP now have a number of partnerships which have efficiency savings at the heart of their objectives and which also target social and cultural benefits in the locality and region.

Housing Associations, like other social enterprises, are under constant scrutiny from auditors and quality assessors to improve efficiency and reduce wastage. The MMU/Wulvern Housing KTP is targeted towards reducing voids and vandalism in the estates and neighbourhoods owned by the Association, which involves customer engagement and intervention: a serious piece of research for the University to be involved with but yielding powerful data for the company to incorporate into its evidence base when planning future spends. MMU social scientists are also talking to third sector partners about addressing Youth Aspirations and both of these projects fit the Economic and Social Research Council (ESRC) objectives to support projects which have social benefits.

For the academic, writing a KTP bid with a partner from a social enterprise organisation is both problematic and requires some patience. This is because both the language and framework of the KTP are clearly targeted at firms from a profit-oriented perspective, requiring detailed management and year-end accounts from the company partner, and making it difficult for not-for-profit organisations to fit the model (Knowledge transfer partnerships/Technology strategy board. Application form, Part A; July 2009). The semantics employed in the application process leave little scope for the justification of social or community benefits, since the target goal for the partnership is to generate additional profits for the company.

However, recent experience at MMU has shown that knowledge transfer partnerships are also valuable to not-for-profit organisations such as housing associations, as for example evidenced by the current partnership between Manchester Metropolitan University and Wulvern Housing. Here, the language, frameworks and selection processes of knowledge transfer appear inappropriate and, if not off-putting, then certainly unhelpful. The language of knowledge transfer not only acts as a potential deterrent to social organisations but also to non-business academics who find the business context and language equally off-putting.

Here, then, is another barrier to academic engagement. If one's identity as an academic includes a strong philosophical and ethical affinity with social organisations, charities, social movements, or indeed any non-profit-making field of expertise, then one is unlikely to seek networking opportunities or collaboration with business executives, with whom one shares little in common. Indeed, some colleagues in my own academic and largely social sciences department strongly resist what they consider "going over to the dark side" such is their identification with an anti-business perspective (albeit an imaginary one).

4 The Changing Role of Universities

Knowledge transfer partnerships clearly require a demand from organisations and companies for the sort of expertise and input that they require and, most important, for that input to be customised to their company needs. It is critical for higher education institutions to 'flatter their wares' in an appropriately produced package and language so that it attracts firms (Winter 2009), and to provide university personnel who can communicate effectively with industrial or business managers and their staff. Here lies a potential barrier for academics whose identities are constructed and then delineated by the traditional view of what academics have been in the past and what many academics want to be in the future. Winter's (2009) study of 'managed academics' acknowledges an apparently growing schism between academics' concepts of their roles and identities with their perceived and expected activities, given the shifting perceptions of what higher education is for and what academics should do beyond traditional roles of teaching and research. This also remains a contention between 'traditional' and 'new' universities in terms of their relative prowess, cudos and positioning in funding authority and national student survey league tables. Consequent to this dualism, academics employed by the former group maintain higher levels of academic standing than their "new" colleagues who are often referred to as teachers rather than researchers and thus not quite the "experts" in their field that "proper academics" are.

It is true that in recent years universities have worked hard to reposition themselves as more relevant to the business community and less distant from the social context. Many have undergone expensive re-branding designed to emphasise their relevance to twenty-first century business culture (see for example Manchester

Metropolitan University's *Vision for the Future* 2009). Concomitantly, university procedures have become more formalised and business-like; employing the language of business far more readily in outlining, for example, mission statements and management hierarchies. This redefinition of higher education as a more business-like, efficient producer of employable graduates is challenging the traditionally held views amongst some academics that 'corporate' values and 'clients' should stay in the boardroom and firmly away from pedagogic and research activities. Interestingly, it is only now, in this new climate of business-efficiency aspirations, that universities are perhaps capable of collaborating with companies effectively; after all, successful partnerships require both parties to share a common language and at least part of their identity.

Thus the acceptance by Universities of this revised definition of their role in society necessitates an equivalent acceptance among academics that their roles have also been expanded to include a more entrepreneurial approach; one that is congruent with higher education strategies in academic enterprise (Whitchurch 2005). That this does not necessarily involve profit-making activities has been lost on some and there is evidence of resistance among many academics who resent what they perceive to be not only a threat to their autonomy and identity, as if that was not enough of a deterrent, but also an implied values change in the pursuit of knowledge. Full academic engagement with this new interpretation of the role demands strategies to reduce the schism in academic identity, coupled with a re-valuing of academic success to included engagement with academic enterprise activities, of which knowledge transfer is one. This is a contemporary theme within the higher education academic community: recent government statements, referring to the need for universities to work more closely with businesses, have added to the general perception that higher education is undergoing a re-branding (Department of business, innovation and skills, November 2009). This necessitates not only the repositioning of universities in relation to the business sector, but also the adoption of a more business-friendly vernacular, referring to corporate values and relying on managerial efficiency structures to deliver such change agendas. The recent rebranding of many universities to promote their Enterprise and Entrepreneurship agendas leaves little doubt in the minds of many academics that their value is attached more singularly to measurable outputs and, preferably, impacts rather than scholarship and education. Somehow, the equally or, some would argue, more important components of this new vision; which relate to social development and widening participation are less well highlighted than the economic aspects so that the overall perspective is of a generally more business-like, output-related rather than purely educational set of institutions.

As reported in Waeraas and Solbakk's (2009) article the branding of higher education is becoming universal practice (their research focuses on Norwegian universities) to redefine what universities 'stand for in terms of values and characteristics, and how they are perceived'. This is occurring, they point out, 'within a context that is characterised by an increasing transfer of business management practices from the private sector'. However, if university governance fails to communicate its redefined purpose to the members of the institution, only confusion can result; leaving academics wondering not only what their once straightforward roles now comprise but what their university is trying to achieve; what kind of values it now holds and whose values these are. There is as yet little research into the impact that rebranding has on university community identities, but Waeraas and Solbakk (2009) suggest this exercise is risking the identity of its academic community. In its efforts to improve competitiveness and enhance its reputation, the branded university risks alienating its staff, particularly in those institutions which have deep-rooted values and traditions revolving around academic integrity and educational value rather than managerial efficiencies.

The organisation of academic resources into subject-based units, required by the research assessment exercise (but maintained under its replacement research excellence framework from 2008-09) further inhibits the interactions often required to collaborate effectively with businesses, whose problems often require multidisciplinary solutions. For this reason small schools or faculties whose academics interact, both informally and on, for example, academic enterprise committees, are more likely to form relationships conducive to knowledge transfer than those who live a more isolated 'ivory tower' existence in large, single subject schools (Prince 2008). Winter (2009) describes the benefits, to successful knowledge transfer, of 'generative conversations' where communities of practice can be formed by academic colleagues across several disciplines.

5 The Perceived Threat to Academic Autonomy

D'Este and Patel's (2007) study of United Kingdom academics found autonomy among researchers highly influential in determining their propensity to collaborate with external agencies. More 'successful' academics with an established publication record, were more confident in collaboration, perhaps perceiving less risk than new or emergent researchers whose reputations were not yet established. Shattock (2007) reiterates this point, arguing that successful academic enterprise needs to be part of an agreed strategy where the individual academic is not liable for partnership or project failure. Where business managers are aware of, and even relatively comfortable with, taking financial risks and take appropriate steps to mitigate failure, this is not a path familiar to many academics. It is not surprising that fear of failure is sufficient to deter many (Shattock 2007 p19). An additional barrier to academic acceptance of a revised role is that, historically, academics have been highly successful at 'resisting initiatives to change' in their roles and in the system at large (for example see van Vught 2008). Consequently, while higher education managers and their funding bodies attempt to **diversify** institutions, academics cling on, in the main, to their 'favoured' identities and reject what they perceive as attempts to reduce their autonomy. Such resistance tends, ironically, towards homogenisation rather than diversity, inhibiting 'processes of differentiation' (van Vught 2008).

As Huisman et al found (2007) even the trend towards large, merged, higher education institutions has not enhanced diversity, but the reverse. As financial pressures have grown, higher education institutions have responded by competing within the sector and in theory this should encourage differentiation towards the search for a unique selling point. While this may be reflected in the range of traditional academic programmes offered it has yet to be translated into a widespread involvement with knowledge transfer. It is interesting to note, however, the disproportionate number of small higher education institutions in the United Kingdom who have used both location and accessibility to enhance their knowledge transfer portfolios; for example see Prince's study which highlights the apparent propensity for small institutions or faculties to gain the most successful access to the surrounding business community (Prince 2008).

MacFarlane's discussion of a 'disengaged academic' maintains that academics in Western cultures are part of a wider and more general civic disengagement with public life and social responsibility (2008). In addition his study found significant differentiation between the pre- and post-1992 higher education institutions in terms of the degree of academic autonomy both allowed and expected. Staff in 'new' universities are more managed, since they are generally uninvolved in policy-making which affects their work patterns, whereas staff in traditional universities are more used to self-governance and the autonomy afforded by "the Senate" (witness also the Oxford 'congregation' and 'the Regent House' at Cambridge). This ought to make post-92 university academics more at home in a 'professional' and 'management' context but does not necessarily mean they are prepared to shift their identities to match the redefinition, as discussed earlier in this article. As we know, academics resist change particularly when it threatens to undermine their autonomy, but also because they are tied into a hierarchical system of performance and reward along a linear path of progression, which all too often becomes a plateau from which the only way is down. If it is the institution which is creating barriers to diversification of roles and activities, perhaps it is time to consider alternatives.

6   The Role of Policy and Funding Councils

The higher education funding council for England points to an additional £1.94 billion income generated from knowledge transfer activities in English higher education institutions, maintaining that universities are successfully integrating third stream policy into their strategies, with the result that 'considerable progress' has been made in engaging academics (Progress in third stream funding report, 2009). However, as several commentators have observed, successful integration of third stream activities requires an 'internalisation of values' before higher education institutions can sustain this policy, demanding a cultural change in respect of academic identity (see for example Hatakenaka 2005; van Vught 2008; Ozga 2004; Berman 2008 amongst others) . Van Vught (2008) agrees that academic resistance to diversification remains a significant barrier to widespread acceptance of a new identity for higher education institutions. In Scotland this has been partially overcome through a broader interpretation of knowledge transfer objectives which includes community and social benefits in addition to commercial gains. Given many academics do not see themselves as entrepreneurs this is a powerful incentive to those who hitherto regarded knowledge transfer as only applicable to engineers and scientists. Ozga (2004) points out the Scottish higher education funding council mission for knowledge transfer to play 'an increasing part in Scotland's economic **and social** wellbeing, delivering the most gains possible for the

Scottish economy **and quality of life'** (my emphasis) (Scottish Executive
2003:40). Scotland provides a specific 'cultural engagement' element to its knowledge transfer funding to support 'the great diversity of cultural knowledge transferred from higher education institutions into their local and national communities' (Universities Scotland briefing, 2008). Similarly, the organisation of
economic co-operation and development's (OECD) view of knowledge transfer is
that it should incorporate social as well as economic impacts (Hatakenaka 2008),
expanding the predominant triple-helix model (as suggested by Etzkowitz H & de
Mello JMC 2003) "universities-industry-government" to incorporate a fourth, social domain, and to which we can also add an opportunity for sustainability to contextualise these relationships, thus:

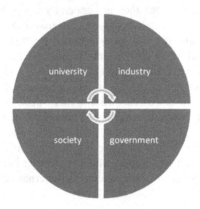

Source: adapted from Etzkowitz H & Leydesdorff L (2000)

Fig. 1 Adapted triple-helix model showing the addition of a society domain

The Organisation of Economic Co-operation and Development has as part of its
remit a preoccupation with deprived societies and communities, which emphasises
the importance of a broader interpretation of knowledge transfer, but that does not
mean wealthier nations should ignore the social and community perspective. Indeed one could argue the moral case for employing academic research for social
benefit applications should always outweigh the private profit motive, however
much the latter group might shout the loudest. In wealthy regions, higher education institutions may remain distant from their local communities whereas those in
more deprived areas work hard to enhance their community-facing status and to
develop out-reach, widen participation and disseminate information in a format
accessible to local people or social organisations. Knowledge transfer in companies is often disseminated only within the firm or the sector, especially where
intellectual property is associated to technological innovation. The aim is predominantly to benefit the company, and, perhaps, researchers, but not the wider
community. Knowledge transfer in the cultural and social spheres, however, is
much more widely disseminated, creating a far broader range of impacts. Glasgow
Caledonian University has demonstrated models of best practice in knowledge

exchange projects within health and social care and community planning (see, for example, case studies in social justice (policing and ethnic youth in Edinburgh and Glasgow) and care of the elderly (maintaining independence and preventing falls) knowledge transfer projects (case studies accessible at www.gcal.ac.uk).

Higher education funding in England, then, has much to learn from the Organisation of Economic Co-operation and Development and Scottish higher education funding council interpretation of knowledge transfer if it wants to offer non-business and non-engineer/scientist academics opportunities to engage in knowledge transfer. The funding body does appear to have accepted the moral perspective. Berman's study of industry-university collaborations found that even having persuaded academics to become involved, it is another matter **sustaining** the relationship (2005). Berman highlights several barriers preventing sustainable partnerships between university and company, even without allowing for the external and macro-economic conditions prevalent at the time. Studies suggest that while small and medium enterprises, in particular, recognise the value of working with a local university to support local students and therefore contribute to the local community, there were numerous obstacles preventing this from coming to fruition in any meaningful way. For example, Berman found that company managers were 'put off' knowledge transfer collaboration by a range of negative factors, including being faced with academics who had no apparent business or project management experience, who did not 'know how to talk to industry', by the 'debilitating slow pace' of university administration and bureaucracy and by the attitude of some academics who appeared rather arrogant, failing to consult the company partner before decisions were taken.

7 Sustaining Collaborative Partnerships

Little has, to date, been written on the subject of how industry-university relationships, once fostered, can be sustained. Certainly there are barriers to successful establishment of such relationships and more so in some institutions than in others. Several authors have pointed to the importance of a specific knowledge transfer office as a bridge between university and industry (see, for example, Geuna and Muscio, 2009; Whitchurch, 2008). Increasing numbers of higher education institutions are operating knowledge transfer offices to coordinate their knowledge transfer and other third stream activities. Are they essential to a university's strategy though? Do they actually facilitate and, more importantly **sustain** partnerships? Ideally, the bridging function of an administrative office dedicated to academic enterprise overcomes the sometimes vastly separate culture and practice of the academic and the company partner. But, if there is already a relationship between the two partners; for example, they have been 'engaged for a while', perhaps the company partner is an alumnus, or the academic is known to the firm through volunteering, contract research or community participation projects, and is thus familiar with the workings of the business sector, then arguably the involvement of a third player in the partnership is less important, and could even, perhaps, be a hindrance to the continued 'marriage' of the two organisations. Universities with well-established knowledge transfer systems recognise the need to smooth the processes of bureaucracy just as companies appreciate working with an HEI which recognises it needs to speed up some of its operations. Where universities do not attempt to do this, there is a real possibility that current university knowledge transfer strategies may actually inhibit partnerships and especially individual academic entrepreneurship through what Geuna and Muscio (2009) describe as 'over-regulation and bureaucracy'. Possibly, these barriers can be overcome, or at least reduced, through the emergence of a new participant in academic enterprise. Aptly described by Whitchurch as 'third space professionals'; these are university employees who have project-management experience, can talk to academics and to business leaders and employees and who understand both the needs of industry and the complexities of the university administrative bureaucracy (2008). There is an important opportunity where such staff operate a knowledge transfer office for academics to offload some of the non-academic aspects of knowledge transfer activity onto support staff who are so much better at it anyway; it seems the issue is not whether we should fund knowledge transfer offices but rather how we staff them. At Manchester Metropolitan University (Cheshire) knowledge transfer is a recent development and the administrative and programme management support is part time and of limited capacity, yet the personnel involved are essential to the smooth operation of each partnership; they are vital links in a complex process yet are often not mentioned as such. It can be the dedication of these knowledge transfer professionals that sustains relationships at times of stress.

Findlow's study reminds us that the need to regularly audit and regulate higher education activities may create tensions between universities and businesses and consequently may discourage innovative practice (2008). Potentially then, institutionalising knowledge transfer will not help if it introduces another layer of bureaucracy and accountability.

8 Concluding Barriers

A Canadian study, conducted in 2006, found the difficulties of transferring academic research to the practice community were nothing new, but that as far back as the 1970s, it was becoming evident that 'as research methods and techniques (become) more sophisticated, they also become increasingly less useful for solving the practical problems that (businesses) face' (Susman and Evered 1978:582). This is the phenomenon of 'sticky knowledge' and is a further potential barrier to knowledge transfer (von Hippel, 1994). Ozga and Jones' study considered the problem of transferring new knowledge between industry and universities (2006). It was not only issues of confidentiality and ownership that needed resolving, but also the nature of writing for academic publication, which has often created a language barrier between the academic community and the public and industrial domains. For far too long the research assessment exercise in the United Kingdom, for example, has rewarded academics for publishing research in 'prestigious' journals rather than engage with practical application of their research; this has not encouraged academics to produce their research in a particularly accessible format, whether oral or written. This 'stickiness' extends to most disciplines and arguably also extends to differences in values between researchers and practitioners;

value of the knowledge and what to do with it can be an intractable barrier requiring careful negotiation between the collaborative partners! The replacement of the research assessment exercise by the research excellence framework offers some long overdue merit for applied research and particularly recognises that which creates new knowledge in non-profit-making spheres.

9 Conclusion

In concluding this discussion it is apparent that knowledge transfer partnerships are a valid application of academic research and do not necessarily require the academic to shift to a new business identity. University-business collaboration does require of the academic, though, an ability to be *entrepreneurial* in terms of subject creativity, to be *managerial* in respect of negotiating roles within the partnership, and above all to be *effective in negotiating* the seemingly endless bureaucracy for which universities are renowned. There are undoubted rewards for academics involved in knowledge transfer, but they are hard-won and this means that the partnership must demonstrate significant impacts. Those of us fortunate to work in partnerships in the world of public and social benefit have no doubt that knowledge transfer is worthwhile, but this does not make the activity or indeed the partnerships easy to sustain. This latter point is worthy of further discussion since it will be key to the success of university strategies as they attempt to meet their 'change' agendas in achieving the vision of "Universities of the Future" (Etzkowitz, H et. al., 2000).

References

1. Berman, J.: Connecting with industry: bridging the divide. J. High. Educ. Pol. Manag. 30, 165–174 (2008)
2. Boden, R., Deem, R., et al. (eds.): Geographies of knowledge, geometries of power: higher education in the twenty-first Century. Routledge, London (2008)
3. Davenport, T.H., Prusak, L.: Working knowledge: how organisations manage what they know, Boston (1998)
4. Delanty, G.: Challenging knowledge: the university in the knowledge society, OUP (2001)
5. Department of business, industry and society. White paper: Higher ambitions: the future of universities in the knowledge economy (November 2009)
6. D'Este, P., Patel, P.: University-industry linkages in the UK: What are the factors underlying the variety of interactions with industry? Res. Pol. 36, 1295–1313 (2008)
7. Etzkowitz, H., Leydesdorff, L.: The dynamics of innovation: from national systems + "mode 2" to a triple helix of university-industry-government relations. Res. Pol. 20, 109–123 (2000)
8. Etzkowitz, H., de Mello, J.M.C.: The rise of a triple helix culture. Int. J. of Tech. Manag. and Sustable Devt. 2(3), 159–171 (2003)
9. Etzkowitz, H., Webster, A., Gebhardt, C., Cantasino Terra, B.R.: The future of the university and the university of the future: Evolution of ivory tower to entrepreneurial paradigm. Res. Pol. 29, 313–330 (2000)

10. Findlow, S.: Accountability and innovation in higher education: a disabling tension? Stds. in High. Educ. 33(3), 313–329 (2008)
11. Francis-Smythe, J.: Enhancing academic engagement in knowledge transfer activity in the UK. Perspectives: Policy and Practice in Higher Education 12(3), 68–72 (2008)
12. Francis-Smythe, J., Haase, S., et al.: Competencies for academics in knowledge transfer activity. In: Proceedings of the British Psychological Society, 2007 Occupational Psychotherapy Conference, Bristol, England (2007)
13. Geuna, A., Muscio, A.: The governance of university knowledge transfer: a critical review of the literature, vol. 47, pp. 93–114. Springer Sciences Business Media, Heidelberg (2009)
14. Grant, R.M.: Contemporary strategy analysis. Blackwell Publishers, Malden (2002)
15. Hall, R.M.: The management of intellectual assets. Journal of General Management 15, 53–68 (2009)
16. Hannon, A., Silver, H.: Innovating in higher education: teaching, learning and institutional cultures. Studies in higher education. Open university press, Buckingham (2000)
17. Hardill, I., Baines, S.: Personal reflections on knowledge transfer and changing UK research priorities. Journal of the Academy of the Social Sciences 4(1), 83–96 (2009)
18. Hatakenaka, S.: Development of third stream activity. Lessons from international experience. Higher education policy institute, Higher education funding council for England (November 2005)
19. Higher education funding council for England 2009, Report on third stream funding (2009), http://www.hefce.co.uk/reports/2009 (accessed)
20. H.M. Government 2003. The future of higher education. Government white paper
21. H.M. Government 2004. Science and innovation investment framework: 2004-2014. H.M. Treasury (2004)
22. H.M Government 2006 Science & innovation investment framework 2004-2014. H.M . Treasury (2006)
23. H.M Government 2009 Department of Business, Innovation and Skills white paper: Higher ambitions: the future of universities in the knowledge economy (November 2009)
24. Jacobson, N.: Social epistemology: theory for the "fourth wave" of knowledge transfer and exchange research. Science Communication 29(1), 116–126 (2007)
25. Knowledge transfer partnerships/Technology strategy board. Application form, Part A (July 2009)
26. Lam, A.: Embedded firms, embedded knowledge: Problems of collaboration and knowledge transfer in global cooperative ventures. Organisation Studies 18(6), 973–996 (1997)
27. Laredo, P.: Revisiting the third mission of universities: towards a renewed categorisation of university activities? Higher Education Policy 20(4), 441–456 (2007)
28. Litan, R.E., Mitchell, L., Reedy, E.J.: Commercialising university innovations: alternative approaches. Social Sciences Research Network (2007), http://ssrn.com/abstract=976005 (accessed)
29. Macfarlane, B.: The disengaged academic: the retreat from citizenship. Higher Education Quarterly 59(4), 296–312 (2005)
30. Manchester Metropolitan University, The vision for the future. Published at Manchester Metropolitan University website (2009), http://www.mmu.ac.uk/about/vision/
31. Marley, L.: Quality and power in higher education. Oxford University Press, Maidenhead (2003)

32. Meyer, M.: Academic entrepreneurs or entrepreneurial academics? Research based ventures and public support mechanisms. Research and Development Management 33, 107–115 (2003)
33. Milne, P., McCormack, C. (2004) Academic autonomy in a climate of change. A Paper Presented at the Society for Research into Higher Education (SRHE), Bristol (December 2004), http://www.heacademy.ac.uk/ (accessed)
34. Nedeva, M.: New tricks and old dogs? The third mission and the reproduction of the university. In: Boden, R., Deem, R., et al. (eds.) Geographies of Knowledge, Geometries of Power: Higher Education in the Twenty-First Century. Routledge, London (2008)
35. Nedeva, M., Boden, R.: Dante's third circle? Shaping the meaning, practice and consequences of the third mission. In: 2007 Conference Proceedings Society for Research into Higher Education, SRHE (2007),
    http://www.srhe.ac.uk/conference2007 (accessed)
36. Osbaldeston, M.: 'Global partnerships' paper. In: ABS Conference, Cranfield School of Management, May 3-4 (2007)
37. Ozga, J.: From research to policy and practice: some issues in knowledge transfer. In: CES Measuring Performance: Challenging Inequalities Conference (2003) Briefing Paper 31 April 2004.
38. Ozga, J., Jones, R.: Travelling and embedded policy: the case of knowledge transfer. Journal of Education Policy 21(1), 1–17 (2006)
39. Parent, R., Roy, M., St Jacques, D.: A systems-based dynamic knowledge transfer capacity model. Journal of Knowledge Management 11(6), 81–93 (2007)
40. Prince, C.: Strategies for delivering third stream activity in new university business schools. Journal of European Industrial Training 31(9), 742–757 (2007)
41. PACEC/ Centre for business research, University of Cambridge report: Evaluation of the effectiveness and role of HEFCE/OSI third stream funding (April 15, 2009)
42. Rherrad, I.: Effect of entrepreneurial behaviour on researchers' knowledge production. Evidence from Canadian universities. Higher Education Quarterl 63(2), 160–176 (2009)
43. Scottish Executive, A framework for higher education in Scotland. Edinburgh, Scottish Executive (2003)
44. Shattock, M.: European universities for entrepreneurship: their role in the Europe of knowledge: The theoretical context. Higher Education Management and Policy 17(3), 13–27 (2005)
45. Tierney, W.: Academic freedom and organisational identity. Australian Universities Review 44, 7–14 (2001)
46. Trowler, P.: Academics responding to change: new higher education frameworks and academic cultures. OUP/SRHE.Universities Scotia, Buckingham (1998); Briefing: Knowledge Transfer and Commercialisation. Public Information Briefing 5 (2005)
47. Van Vught, F.: Mission diversity and reputation in higher education. Higher Education Policy Palgrave Journals 21, 151–174 (2008)
48. Von Hippel, E.: "Sticky information" and the locus of problem-solving: its implications for innovation. Management Science 40, 429–439 (1994)
49. Waeraas, A., Solbakk, M.: Defining the essence of a university: lessons from higher education branding. Higher Education 57(4), 449–462 (2009)
50. Wang, Y., Lu, L.: Knowledge transfer through effective university-industry interactions. Empirical Experiences from China. Journal of Technology Management in China 2(2), 119–133 (2009)

51. Wedgewood, M.: Mainstreaming the third stream. In: McNay, I. (ed.) Beyond Mass Higher Education, Ch. 11, Open University Press, Stony Stratford (2006)
52. Weerts, D.J., Sandmann, L.R.: Building a two-way street: challenges and opportunities for community engagement at research universities. The Review of Higher Education 32(1), 73–106 (2008)
53. Whitchurch, C.: Shifting identities and blurring boundaries: the emergence of third space professionals in UK higher education. Higher Education Quarterly 62(4), 377–396 (2008)
54. Whitchurch, C.: The rise of the blended professional in higher education: a comparison between the UK, Australia and the United States. Higher Education 58(3), 407–418 (2009)
55. Winter, R.: Academic manager or managed academic? Academic identity schisms in higher education. Journal of Higher Education Policy and Management 31(2), 121–131 (2009)
56. Yin, R.K.: Case study research: design and methods. Sage. Thousand Oaks, California (2002), Prince, C. Strategies for delivering third stream activity in new university business schools. Journal of European Industrial Training 31(9), 742–757 (2007)

A Toolbox for ICT Technology Transfer Professionals: A Preview of an Online Toolkit Aiming at the Acceleration of the ICT Technology Transfer Process*

Itxaso Del Palacio Aguirre and Annelies Bobelyn

Imperial College Business School, London, South Kensington Campus, Exhibition Road, London SW7 2AZ, United Kingdom

Structured Abstract

Purpose – This paper describes a technology transfer toolbox for practitioners of which the design currently is being finalized. The toolbox was developed within the scope of a European ICT project entitled FITT – Fostering Interregional exchange in ICT Technology Transfer. The goal of FITT is to centralize and improve the available instruments for transferring ICT research results from science to business. The strategic objective is to maximise the exploitation of research in order to pursue social and economic prosperity. Its main tactical purpose is to make a set of tools readily available with the aim of accelerating the various steps involved in the technology transfer process.

Design/methodology/approach – We propose a practice-based approach using the combined experience of a focused group of ICT technology transfer officers from five different European countries, being Belgium, France, Germany, Luxemburg and United Kingdom. Their insights and practices will be centralized into a toolbox that will be made available through a dedicated website. This website will allow technology transfer staff to: 1) visualize the technology transfer process and its methodologies; 2) get detailed insight into the main steps covered by the technology transfer process; 3) download case studies within highlighted steps of this process; 4) get access to specific tools in order to quickly accomplish certain important tasks; 5) focus on the assessment of research projects geared towards technology transfer and 6) enhance market driven aspects of the technology transfer process.

Originality/value – This methodology attempts to facilitate the complex everyday job responsibilities of technology transfer officers by offering a one-stop-shop

---

* The FITT project web site: http://www.fitt-for-innovation.eu

R.J. Howlett (Ed.): Innovation through Knowledge Transfer 2010, SIST 9, pp. 177–187.
springerlink.com   © Springer-Verlag Berlin Heidelberg 2011

of freely available tools. Through a point-and-click user friendly interface this complete toolset provides unique value as users will be able to select the most appropriate tools for their specific working environment. Besides, an adapted technology transfer training programme will be provided in order to familiarize new employees with the toolbox in an optimal manner. The transnational character of the project allows the integration of various levels of regional approaches.

Practical implications – The practical implications of using this toolbox are numerous. Five core topics largely covering the technology transfer process have been included: 1) Opportunity Identification; 2) IP Management; 3) Human Resources Management; 4) Value Creation and 5) Networking & Clustering.
Within each topic, the online tool will focus on providing :

- A better understanding of the technology transfer process and involved methodologies
- Adoption of a common language through the development of a codebook which provides access to established definitions in the realm of ICT technology transfer
- Better assessment of the economic potential of research projects
- Faster and easier execution of specific tasks thanks to the online usage of specialized tools (e.g. how to calculate the market value of an innovation)
- Access to detailed reference material recommended by the FITT project members

Such outcomes will improve the socio-economic impact of research results and fasten science-to-market turnaround. Through the practical application of these tools, an ad hoc practitioner network will be created. The tool itself will continue to evolve through enhancements from this user network.

Keywords: Technology Transfer, Information and Communication Technologies (ICT), Toolbox, Training, Online.

1 Introduction

Excellent research does not automatically lead to economic and social prosperity: the implementation of research results into marketable products and services is a crucial condition for success. The process of commercialisation through sales, licensing or start-ups is very challenging and many promising ideas consequently fail due to the lack of expertise in this specific area. Commercialisation skills are particularly important in the field of ICT: as a cross-sectional technology, ICT generates economic growth not only for the sector itself but also for other user industries.

The FITT project – Fostering Interregional Exchange in ICT Technology Transfer –, which is jointly funded by the European Union, is the first European project dedicated to technology transfer for Information and Communication

Technologies. It has seven partners in five countries: Germany, Belgium, France, Luxembourg and the United Kingdom. The objective is to support and encourage the exchange of inter-regional technology transfer methods to reduce the gap between research and industrial development. The collaborative efforts are focused on initiatives to improve commercialisation of Europe's best research results and maximise their exploitation for social and economic prosperity.

2 Project Outputs and Benefits

All participating regions in Europe have developed successful strategies and tools for transferring innovations from universities to the market. The FITT project offers a unique opportunity to combine these valuable experiences and bring them together into one practical tool for technology transfer practitioners. The market driven aspects of the technology transfer process will hereby be enhanced.

The FITT project will deliver the following concrete outputs :

- Complete Toolbox as the cornerstone of the project with a set of methods and instruments that offer practical guidance for technology transfer officers

- Transnational Training Program for technology transfer practitioners, based upon the FITT Toolbox, which will allow practitioners to selectively adapt the toolbox to their needs

- Community of Practice for ongoing fruitful cooperation

- Bibliographic material specific for the technology transfer realm

3 FITT Toolbox–Main Features

The FITT website will allow the practitioner to browse through several levels of information for each of the below mentioned topics. Various types of documents are provided in order to allow the user to access information about the covered areas in different shapes and formats. Overall, every topic is described and explained in detail together with its context and its positioning within the technology transfer process. Besides, the toolbox will contain detailed descriptions of selected practices that are considered to be references or 'best practices' within each specific topic. Furthermore, specific real-world examples allow to illustrate the usage of selected practices in a more pragmatic manner. These cases enhance the sense of reality of the toolbox. In those cases where tools are needed, like spreadsheet templates for cost analysis calculations or market projections, the required practice specific tools will be made available. Finally, referenced external sources/experts/articles will be pointed out whenever used throughout the toolbox

The FITT Toolbox covers five main parts which are essential to the process of technology transfer in an ICT environment. The first one is '*Opportunity Identification*'. The detection of opportunities is crucial as it is the starting point of the technology transfer process. Once a research result with commercial potential has been spotted, the technology transfer officers can start to investigate the appropriate ways to protect the related intellectual property and bring it to the market. This leads us to the second part of the toolbox, called *'IP Management'*. In a fast moving sector as ICT it is not always easy to defend proper IP management. It describes the way an organization handles its IP through different processes of protection, valuation and exploitation. The third part of the toolbox covers *'Human Resources Management'*. Within the context of technology transfer, human capital is crucial: the researchers are the key resources of the institutes as they come up with new ideas and develop novel technologies. The fourth part deals with the development of *'Value Creation'*. Defining a value proposition is essential. After researchers have searched for and detected ways to protect their new ideas or technologies, they will have to define their exact unique value proposition and explain how money will be made. Finally, the fifth and last part is centered around *'Networking & Clustering'*. Current trends of innovation emphasize new models where organizations commercialize both external and internal technologies. Following this trend of open innovation, networking and clustering activities which enable innovation to move more easily and efficiently between the external environment and the internal R&D processes, become increasingly important.

The toolbox is designed to be used as a meta-rule or 'a rule to make rules'. The rich experience of the different FITT-partners allowed for thorough internal validation.

The FITT team will also seek endorsement for the toolbox by experts in the field in order to be referenced in academic literature.

The next paragraphs will further elaborate on each of the activities described above and provide deeper insight into the processes that underlie them. In total, well over one hundred original documents have been created. They will be accessible via the toolbox framework. As the underlying FITT logic assumes that humans are central to the process of technology transfer, we will go more deeply into the human resources management activity.

3.1 Opportunity Identification

Detection of opportunities captures the way in which research organizations identify the research results that could be successfully transferred to industry. It should be stressed that early detection of valuable technologies will increase the ability of a technology transfer office (TTO) to decide on an optimal way for transferring the identified results. The main responsibilities of technology transfer officers with respect to opportunity identification appear to be threefold : managing the information flow between the labs and the TTO and monitor activities ; boost the global number of proposals coming to the TTO by creating awareness amongst researchers and finally increase the 'commercial' quality of these ideas by evaluating or assessing them at a very early stage.

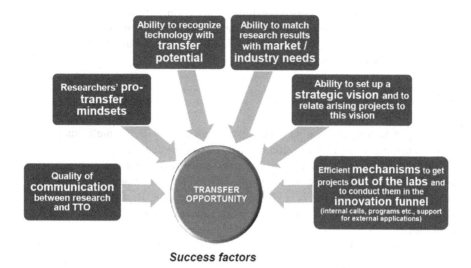

Success factors

A selection of practices that can be installed to perform these tasks will be available in the toolbox. Particular attention is drawn to the following practices:

- Creation of transfer awareness: The performance of a public research organization in generating innovations and transferring technology is largely influenced by the transfer awareness of researchers. The research staff who are interested in transferring results and understand this process are more likely to follow properly the disclosure and patenting procedure, anticipate and communicate with the technology transfer teams in the right moment, be more product and market oriented or participate in choosing the transfer strategy and potential partners. Some examples of actions which the technology transfer office (TTO) can use to create transfer awareness among researchers and engineers are presented in this process.
- Monitoring of activities: This includes different ways of managing information flow between laboratories and the technology transfer offices, which ensure that detection of transfer opportunities is successfully done early in the process, on the basis of continuous deal flow, without missed potential. The pro-active approach allows to have the maximum overview of relevant activities in the labs and to act in the right moment with strategy proposals and necessary guidance. The practices available in the toolbox concern monitoring of such activities as: invention disclosure, collaborative research with industrial partners and satisfaction survey, as well as consultancy provided by researchers of a public research institution.
- Evaluation of transfer projects: When the deal flow from the laboratories to the TTO is optimal, technology transfer officers are susceptible to receive many inventions with commercial interest. A lot of early-stage inventions require substantial human and financial resources to be developed into marketable products. In order to establish whether or not the resources of the TTO

should be spent to seek a commercial exploitation, a first-stage evaluation is often performed shortly after an invention has been identified. The complex decision process that technology transfer officers go through in this regard is presented and detailed within the toolbox.

3.2 IP Management

IP management explains the way an organization deals with its intellectual property through processes of protection, valuation and exploitation.

The process of protection of the IP is an essential stake. In a fast moving sector as ICT, it is not always easy to define the best protection strategy: patenting is a strong protection mean, but is costly and time consuming. Besides secrecy, many other types of protection (e.g. mark, copyright) also exist and offer a wide range of possibilities.

During the process of valuation, the technology can be evaluated by a quantitative or qualitative approach or as a combination of both. How to bring the technology to the market, how to create business etc. are questions that are dealt with during the exploitation process. Exploitation is the process through which outputs from research and development activities can be exploited and technology transfer officers can initiate and support the transfer. We can focus on a practice entitled "Exploitation Scenarios". The objective of this practice is to suggest the drill to conclude a technology transfer agreement by giving a certain number of basic rules related to ordinary, legal and other aspects, and related to a negotiation process. It also regroups necessary initial first steps in order to lead to successful exploitation as follows: the valuation of the technology, then the elaboration of transfer scenarios and finally, management of financial compensation.

For each process, complementary strategies apply, guaranteeing the best choice of commercializing the IP portfolio. This FITT topic contains best practices in all three areas. Furthermore, practices around standardization and certification are shared as well.

A best practice charter is also included in the toolbox relating to intellectual property and knowledge and technology transfer. This charter, which is already in force in many research institutions in France, has also recently received support from the FITT consortium partners.

3.3 HR Management

The following figure reflects the three crucial Human Resources practices in a technology transfer environment. First, researchers need to be motivated before they will actively take part in the technology transfer process. 'Motivation and incentives' as HR systems are therefore key domains of interest. Once researchers are aware and motivated to be involved in the technology transfer process, they generally need some additional 'training', for instance on how to write a business plan for a start-up in order to create a viable business. Furthermore, in the case of a spin-off, the start-up team needs to be constructed. Finally, the key process underlying all three before mentioned HR goals is a clear 'Communication & Collaboration' strategy. To sum up, the crucial HR-tasks of TTO's consist of making researchers aware in taking their ideas to the technology transfer office, motivating them to actively take part in the technology transfer process, providing training in how to create value and assemble balanced teams. The glue that holds everything together is the communication and collaboration process throughout the institute.

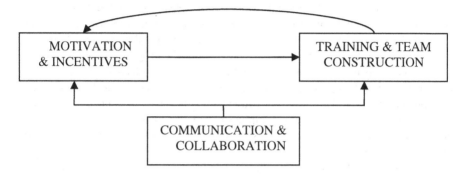

HR Management Process in Detail: *'Training'*

Within a technology transfer context, training opportunities can be provided for both researchers and TTOs. The former are typically well educated in technological courses, but often lack business related knowledge and skills. However, if they should decide to join a start-up, market insights are a sine qua non. Specific courses like financial planning and entrepreneurial marketing seem appropriate to overcome this. In this case, specialized entrepreneurial teaching programmes can be the solution. As there is a nascent need for entrepreneurial engineers, in the private as well as in the public sector, TTOs all over the world are heavily interested in the development of such programmes. At IBBT (Belgium) for instance the Entrepreneurial Bootcamp formula was introduced. The entrepreneurial bootcamp is an intensive personal development program for entrepreneurial researchers.

HR Management Practice in Detail: 'Entrepreneurial Bootcamps'

The Entrepreneurial Bootcamp is a set of focused workshops in which an entrepreneurial multidisciplinary team is created and coached. The workshops are a balanced combination of teaching, coaching and doing with a strong exposure to business executives, industry and financial experts. The final goal of this team is to deliver a presentation at the end of the boot camp that can survive a professional investment board. Everyone connected to IBBT can participate in the program provided that their business idea gets accepted.

The iBootcamp starts with an open enrolment of ideas: Every collaborator from IBBT projects is encouraged to submit his idea. The idea owners need to 'prove' that they understand the primary business drivers motivating and guiding the development of the ideas. Therefore, the promoter has to describe the idea following the Need/Approach/Benefits/Competition method (NABC) in four or five slides. They need to answer four crucial questions corresponding with the four letters of the acronym: 'What is the important customer and market Need?', 'What is our unique Approach for addressing this need?', 'What are the Benefits from this approach?' and 'How are these benefits superior to the Competition and the alternatives?''.

Once the enrolment has officially closed down, all the submitted proposals are pre-screened by a team of experts. They need to be convinced about the value of the idea. Proposals with a technical as well as an economical potential get chosen. The selected ideas are taken on to the next stage of the program. In this phase the start-up team has to be formed. At a dating event, the team leaders of the projects are introduced and can talk to individuals that show an interest to be part of the entrepreneurial team. This way, the iBootcamp is a means to connect committed and talented people with multi-disciplinary/complementary skills and expertise across the virtual boundaries of the institute. Once the teams are formed, the actual iBootcamp can start.

The iBootcamp entails three residential weekends at which the participants are coached to go through the necessary steps of the venture formation process. Each weekend consists of a workshop corresponding with a specific part of the business plan. The workshops are organized along established lines: The participants first learn the theory behind certain crucial aspects of starting up a company. Then they are asked to put the theory into practice and prepare a presentation tailored to their own company with some preliminary results, potential experienced problems and so on. The iBootcamp team of coaches then gives detailed personal feedback.

During the first workshop the participants learn how to develop a successful business plan. They get some teaching on opportunity development and freedom to operate. The second workshop evolves around business models in ICT and entrepreneurial marketing. During the final weekend participants learn about entrepreneurial finance and human resources. After the last workshop, each team needs to bundle all the received and gathered information in a presentation for the IBBT Innovation Board. This final meeting simulates the defence of an opportunity plan to venture capitalists. The iBootcamp teams have to act as though they need funding from these VC's and convince these external financiers to invest in their company. In other words: the entrepreneurial teams should not only present their business but also sell it.

After this final presentation, the IBBT Innovation Board decides whether or not to create a real business around the idea. Some teams are ready for take-off after this intensive two-month training. They can start up an internal or external venture. Others need some extra time to incubate. Still others might experience the business idea not to be valuable after all; they may leave the idea of starting up a company behind entirely.

3.4 Value Creation

A value proposition should convince a potential consumer that a particular product or service will add more value or solve a problem better than other similar offerings. It entails the crystallization of all strategic decisions that need to be taken regarding the customers' demand, the available infrastructure and the financial part. This activity is crucial to start a business and functions as the heart of the organization's strategy. Within the context of technology transfer, defining a value proposition is a key point. After researchers have found out, detected and searched ways to protect their new ideas or technologies, they will have to define the value created for a potential user. Because the ideas of researchers are often very technical in nature, they might experience difficulties to align them with potential commercialization options. Therefore, it can be extremely useful, if Technology Transfer Officers can help these researchers in formulating their value proposition, as a good way to make the business feasible and realistic. Therefore, it is important for Technology Transfer Officers to assess the commercialization opportunities of these technologies. The FITT toolbox contains several practices related to marketing and business modeling that offer support for researchers to get to know the market and to integrate the obtained information in a business model. The toolbox guides transfer practitioners in the assessment of market opportunities, offers templates for technological marketing, provide insights for choosing the business model and provides an Open Source Business Model as an example for exploiting open source software.

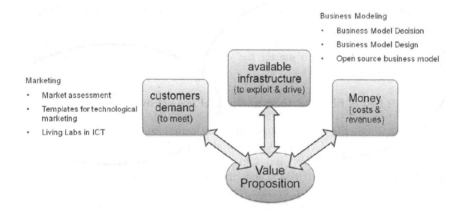

3.5 Networking and Clustering

Technology Transfer is by nature sensitive to openness and environmental interactions, so networking activities are very important. TT-related networking and clustering activities are defined as relationship-based activities that support the sharing and the development of competencies, knowledge and methods involved in the Technology Transfer domain. TT networking is mainly focused on national and international collaboration between public and private companies as a support to the commercialization of innovations. The FITT project emphasizes TT-related networking and clustering practices regarding two main issues: TT collaboration and Network management.

4 Availability/Timeline

The toolbox will be made available through a staged process. A first prototype will be tested internally by the FITT project team during the first part of 2010. A second version allowing an increased level of testing will follow, where the proposed website and associated tools will be used by technology transfer staff within the project member organizations. Subsequent versions will be made available to the general user community and specifically geared towards European technology transfer organizations.

5 Conclusions and Practical Implications

The joint undertaking FITT project involving seven European technology transfer offices from five different countries has allowed to form a framework of tools, reference material and case studies.

The illustrated toolbox acts as a central repository of practices that are used throughout the technology transfer process. Through its online web access, users may browse through the contents in order to learn from others, pick up and use specific tools that are required in their environment and get illustrative examples from the provided use cases.

The pragmatic architecture and modern design of the FITT toolbox will allow technology transfer staff to master the involved processes and techniques in a completely new manner.

Appendix : List of FITT partners and a brief project overview

Since June 2008, FITT has brought together technology transfer professionals from several countries: Germany via the Innovation Agency for ICT and media in Baden-Württemberg (MFG), Belgium via the Interdisciplinary Institute for Broadband Technology (IBBT) and the Walloon Universities' Companies-Universities network (LIEU), France via the National Institute of Research in IT and Automation (INRIA) and the DIGITEO research cluster based in Saclay (south of Paris), Luxembourg via the Henri Tudor Public Research Centre and the UK via Imperial College Business School (ICBS). The objective of the FITT project is to support and encourage the exchange of inter-regional technology transfer methods by pooling tools and sharing results and promotion opportunities. The FITT project is running over a time span of 3 years and has been allocated a total budget of 4.3 M€. It is co-financed by the European Commission (INTERREG IVB NWE).

The project has a dedicated web site accessible at the following address :

http://www.fitt-for-innovation.eu

Further information about the involved partners can be found on their respective web sites:

- Centre de Recherche Public Henri Tudor

    ❖ www.tudor.lu

- INRIA – Institut National de Recherche en Informatique et en Automatique

    ❖ www.inria.fr

- Imperial College Business School

    ❖ www.imperial.ac.uk/business-school

- MFG Baden-Württemberg – Innovation Agency for ICT and Media

    ❖ www.mfg-innovation.eu

- DIGITEO – Fondation de coopération scientifique Digiteo Triangle de la Physique

    ❖ www.digiteo.fr

- LIEU, Liaison Entreprises Universités – University of Liège

    ❖ www.reseaulieu.be

- IBBT- Interdisciplinary Institute for Broadband Technology

    ❖ www.ibbt.be

Knowledge Diffusion in a Specialist Organization: Observational and Data-Driven Case Study

Mounir Kehal

Research Faculty, ESC RENNES SCHOOL OF BUSINESS,
2 rue Robert d'Arbrissel, 35065 Rennes
Brittany, France
mounir.kehal@esc-rennes.fr

Abstract. Management practices and information technologies to handle knowledge of satellite manufacturing organizations may prove to be complex. As such knowledge (with its explicit and tacit constituents) is assumed to be one of the main variables whilst a distinguishing factor of such organizations; amidst those specialist in nature, to survive within a marketplace. Their main asset is the knowledge of certain highly imaginative individuals that appear to share a common vision for the continuity of the organization. Satellites and their related services remain a good example of that. From early pioneers to modern day satellite manufacturing firms, one can see a large amount of risk at every stage in the development of a satellite or a related service, from inception to design phase, from design to delivery, from lessons learnt from failures to those learnt from successes, and from revisions to design and development of successful missions. In their groundbreaking book The Knowledge Creating Company (1995), Nonaka et al laid out a model of how organizational knowledge is created through four conversion processes, being from: tacit to explicit (externalization), explicit to tacit (internalization), tacit to tacit (socialization), and explicit to explicit (combination). Key to this model is the authors' assertion that none are individually sufficient. All must be present to fuel one another. However, such knowledge creation and diffusion was thought to have manifested and only applied within large organizations and conglomerates. Observational and systematic (corpus-based) studies – through analysis of specialist text, can support research in knowledge management. Since text could be assumed to portray a trace of knowledge. In this paper we are to show how knowledge diffuses in a specific environment, and thus could be modeled by specialist text. That is dealing with the satellite manufacturing domain, and having embedded within the knowledge about the business sector and knowledge domain.

Keywords: Knowledge Management, Corpus-based Analysis, Satellite Manufacturing, Knowledge Diffusion, Text Analysis, Small Medium Enterprise, Observational Study.

R.J. Howlett (Ed.): Innovation through Knowledge Transfer 2010, SIST 9, pp. 189–200.
springerlink.com   © Springer-Verlag Berlin Heidelberg 2011

1 Knowledge Diffusion in Specialist Domain

In order to investigate the gap in knowledge diffusion within an organization we did carry an observational study, within an SME (Small to Medium Enterprise) in satellite manufacturing, a specialist domain. Inline with a study of the language used in satellite engineering in general, and that stemming from SSTL (Surrey Satellite Technology Limited) and Surrey Space Centre in particular. Both studies have an empirical basis. The observational study (mainly questionnaire-based) was designed to ask questions related to knowledge diffusion within the company during 2002-2005 period, as part of my doctoral research coverage. The questionnaire-based studies were not based on intuitions on how knowledge is managed, rather based on a set of empirical questions, partitioned under five sections namely:

1- Awareness and Commitment
2- External Environment
3- Information Technology
4- Knowledge Maintenance and Protection
5- Organizational Issues

We have investigated the diffusion of knowledge within SSTL, based on the practice within SSTL, as articulated through the questionnaire. There were two sets of questionnaire-based observations. The pilot study was conducted with managers and whereas the second run of the questionnaire was intranet-based, and more widespread. SSTL, is a small knowledge-based organization, for *minimalism*, a knowledge based organisation is one where knowledge is being the dependent input variable, as the need would exist for organisational resources to acquire such knowledge from physical entities (i.e. knowledge workers) and convert it as input for electronic storage medium (s), making it easier for retrieval and dissemination of information. Thus, knowledge (encompassing data and information) would be needed for creating and offering a product and service line mix, including that contained in individual employees and that in SSTL (as a collective entity, expertise accumulated over time). SSTL's principal assets are its engineers, its project managers and its researchers. Collectively, the engineers, managers and researchers are sometimes called *knowledge creating crew* (Nonaka *et al*, 1995). In a rapidly developing, high-technology field like satellite engineering, it is important to communicate, share and validate knowledge. We aim to describe in this paper our understanding of the nature of a specialist organization in a quantifiable manner, and the constructs of a knowledge management audit conducted through the observational study within a satellite manufacturing SME, based in the UK. We have examined how knowledge flows and is adapted between commercial and research types of corpora. One of the major results deduced from the observational study was that knowledge diffusion is paramount within the lifetime of an organization, and could be supported by information systems. Leading us to investigate on how knowledge diffusion takes place, in an empirical way. Our analysis shows that research papers (created within educational institution) and commercial documents (created within spin-offs of such higher education institution) can be distinguished rather on the basis of single word and compound terms. These two specialist lexis show the potential for identifying points of mutual interest in the diffusion of knowledge from the research institution to the commercialization process, thus to application(s) within a domain.

2 Method

Nonaka et al's (1995) knowledge conversion model is intuitive. It is based on long experience and judgement. Such model emphasizes the importance of practice, knowledge amongst knowledge workers. The case studies produced were between researchers, practitioners, and managers. There was transfer of knowledge from researchers to knowledge workers. Such has yielded a contingency table for the transfer of knowledge, so-called *knowledge conversion model* that generates four knowledge conversion modes. Such model is plausible but remains largely intuitive. Our interest is tacit to explicit knowledge conversion (externalization) and explicit to explicit knowledge conversion (combination). The reason we have studied an SME (Small to Medium Enterprise) because it would appear that knowledge would be shared because smaller groups would get together easily, i.e. no logistics involved. As well as it appears that in a SME knowledge bottlenecks which are characteristic of large organizations would not exist. Being in relation to the size of SMEs, managers are expected to interact with and understand needs and requirements of knowledge workers. Consider an organization like SSTL, Surrey Satellite Technology Limited, we focused on the interaction between knowledge engineers and knowledge practitioners, and were aiming to see how knowledge is shared. In order to investigate the gap in knowledge diffusion within SSTL, we did an observational study, and a study of language used in satellite engineering in general. Both studies have an empirical basis. A bimodal research method was followed within the specialist domain of satellite manufacturing applied within SMEs [Small to Medium Enterprise]. Inclusive of an Observational study: questionnaire and interview based and a Corpus-based study: analysis of text repositories. Thus, involving extraction and modelling of specialist terminology collated from: public domain publications (i.e. NASA, British Standards Institute – Terminology Specification, and BMP - Best Manufacturing Consortium database), and specialist domain publications (i.e. Surrey Space Centre and SSTL).

Figure 1 represents a relational view of the methodology, integrated within the possible set of agents for knowledge diffusion, being composed of a 2-tier process. Whereby knowledge is assumed to flow among or across from knowledge workers, to the organization, then to worldwide (horizontally), but the adaptation phase comes into place once knowledge is personalised and applied (vertically). However, such methodology was implemented in the specialist nature of the domain of investigation. Yin (1994) identified five components of research design that are important for case studies: the study's questions, its propositions, its unit (s) of analysis, the logic linking the data to the propositions, the criteria for interpreting the findings. The above components were integrated within the observational study, as guidelines to the formulation of the different stages involved within the conduct of this research, from the pilot run of the survey study, to the intranet-based survey and historical studies. In which the intranet-based survey seemed to

Knowledge Adaptation

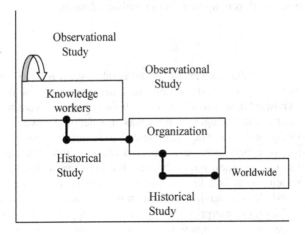

Knowledge Flow

Fig. 1 Knowledge diffusion through flow and adaptation processes

generate interest, impact and a set of internal actions. Supported as well by over 30% participation rate for the intranet-survey, and being composed mostly by middle to senior managers. Evidence of knowledge diffusion and support for it was manifested as outlined in the table below.

Table 1 shows how knowledge bottlenecks have been looked upon during the flow and adaptation of knowledge amid agents involved in its diffusion. Through the observational study and the historical study, behaviour between agents was modelled using the techniques prescribed below.

Table 1 Knowledge diffusion in the environment of a small organization

Agent A	Agent B	Artefact	Technique
Person	Person	Opinion, practice, know-how, organizational structures	Questionnaire-based study
Person	Organization	PhD Dissertation, Research Publications, technical reports	Text Analysis
Organization	Organization	Specialist documents (i.e. technical documents, technology-specific documents, missions documents)	Text Analysis
Organization	Worldwide	Specialist documents (i.e. technical documents, technology-specific documents, missions documents)	Text Analysis

The intranet-based survey study had 5 sections, stretching to cover possible areas of applications and implications for knowledge diffusion within an organization, namely - Knowledge Management Awareness and Commitment, External Environment, Information Technology, Knowledge Maintenance and Protection, and Organizational Issues. On the other hand, for our historical and special corpus, we followed where applicable and pertinent, Atkins et al's (1992:2), five principal stages for corpus building. Outlined in table 2:

Table 2 Stages for building a corpus (Atkins et al, 1992:2)

Stage	Description
Specification and design	Corpus type is identified taking into account sample size, language varieties and the time period to be sampled.
Computer Hardware and software	Hardware and software needs for the corpus project are estimated.
Data capture and mark-up	The data/texts are captured and transformed/transferred to electronic form, keyboarding, or audio transcription. The captured files are then marked-up with embedded codes containing text features.
Corpus processing	Includes basic tools, i.e. word frequency lists, concordance, and interactive standard query tools and tools for lemmatization, tagging, collocation etc.
Corpus growth and feedback	New materials may be added to the corpus or some of the old materials may be deleted according to feedback from previous analysis to reach a balanced and enhanced corpus.

Specification and design of a corpus and its processing are the most important steps in building the corpus and for any kind of subsequent study. Second and fourth stages are not so important due to the technological advances in computer hardware and software. The importance of the last stage depends on the nature of the study. Studying the state of the specialist terminology is considered important for the study of the language discourse. Corpus-based studies are empirical and depend on both quantitative and qualitative analytical techniques (Biber et al, 2002). Therefore to get results have an important effect, the corpus must be sampled and created carefully: "the decisions that are taken about what is to be in the corpus, and how the selection is to be organized, control almost everything that happens subsequently. The results are only as good as the corpus" (Sinclair, 1991:13).

3 Observational Study

The term knowledge management is used to articulate the concept that knowledge is an asset on a par with the tangible assets of any organisation - land, capital, plant and machinery. Management involves the management of assets; ergo knowledge should be managed from its inception through its nurturing to maturity to exploitation and to ultimate obsolescence. The term was also coined to indicate that knowledge within organisations is communicated not only through the typical organisational hierarchies but also through interaction between members of the organisations across the hierarchies and the different structures (divisions/departments and their functions, management style, communication culture, computer-mediated processes, practices and so forth) contained with an organization. The questionnaire portrays through its five sections, some of the concepts raised within the Knowledge Management field, outlined in Section 1. Two runs for the questionnaire-based study were conducted, a pilot study, and an intranet-based study. The majority of the respondents were knowledge practitioners (i.e.

team members). Over 80% from the intranet-based questionnaire were as such, like reported from the respective representative of the study onsite, head of Research and Development at SSTL. The key point was that the managers were more optimistic and confident about extent of knowledge sharing. Our analysis has been supported by the feedback received from one of the key managers cited previously. Our method is no more then holding a mirror to an organization and what is reflected is the management of knowledge within the organization when looked upon from the five different facets of the questionnaire sections (i.e. awareness and commitment, external environment, information technology, knowledge maintenance and protection, and organizational issues). The questionnaire study raises the need for a knowledge map through both the pilot and intranet-based observational studies, one that is specialist in nature. That can represent the domain language providing an environment for querying and validation for the knowledge worker, and thus containment of both elements of such knowledge (explicit and tacit). Allowing as well for the knowledge conversion modes (Nonaka, 1995) to take place, and hence knowledge to be created and utilized. This may act as basis for the research conducted on whether SMEs do create the dynamics of innovation, as such dynamics may need to encapsulate the sharing of the domain knowledge (touted and supported by knowledge workers), and thus embedded within the domain's *ontology* – referring to the explicit formal specifications of the terms in the domain and relations among them (Gruber, 1993). This part of the research (observational/introspective) has focused on the organizational structures (management hierarchies, attribution and validation of knowledge, and so forth) in place, enabling or facilitating the diffusion of knowledge. Our conclusions from this survey; based on the feedback and responses received, affirm that knowledge sharing is encouraged. As well as innovation being encouraged either through collective or individual effort(s), and facilitating knowledge sharing is possible through availability of knowledge maps and communication channels between multi disciplinary teams for specialist areas.

The above results, from either the pilot study or the intranet-based studies; have encouraged us to explore how a collection of specialist documents will facilitate knowledge diffusion and perhaps to construct knowledge map.

4 Text Analysis and Corpus-Based Studies

Text analysis should be taken to mean the analysis of text by algorithmic processing, and that may involve the computation of specialist lexis within a given or emerging domain. An algorithm may be defined as a step-by-step procedure capable of being run on a computer, hence rendered automatic or semiautomatic. Before it can actually be run, however, the algorithm must be coded in some computer language as part of a software program. The tool currently used for the purposes of this research is *System Quirk*, a computational linguistic software system providing a computer-mediated environment for text analysis created by the Artificial Intelligence Group, University of Surrey. The compound terms generated through System QUIRK/Ferret (Artificial Intelligence Group, University of Surrey), illustrate to a certain extent the composition and acceptance of frequent specialist words within a text repository. As well within the language of the domain and the domain knowledge; since latent clusters of concepts, may be represented by each of the compound terms (a hierarchy of concepts through morphological productivity of terms). Whereas each term's relevance to a collection of documents is erratic, it may be validated (*combined* and/or *externalized knowledge*) by the knowledge worker as it composes toward a given terminology. That signifies being of use to the individual knowledge worker or group of them. Thus, achieving acceptance based on consensus within an organization, and growing to be part of it and its external environment *(ontological spectrum)*. We have used these and other sets of compound frequent terms extracted based on a statistical criterion (in relation to the BNC); for comparative purposes, sometimes referred to as *"Lexical Signatures"* to index collections of text to be contained within a text repository. Frequent compound terms extracted from the collection of documents (Source: Surrey Space Centre corpus) listed below, illustrate the specialist nature of the organization and the domain of the knowledge within aspects of satellite technology encompassed by the research and possibly of the commercial activities of the organization. The compound terms selected below, from the corpus of SSTL and Surrey Space Centre is from a listing of over 50.000 compound terms within the corpus. Such contains all collated research publications of the organization (s) aforementioned.

Table 3 Ranking of select compound terms in SSTL/Surrey Space Centre corpora

Rank	Compound Term	Rank	Compound Term
1	low cost	32	doppler shift
4	propulsion system	33	swath width
6	remote sensing	33	satellite platform
7	surrey satellite technology ltd	33	narrow angle image
8	surrey space centre	34	sstl microsatellites
9	board computer	34	control system
10	low earth orbit	34	satellite missions
15	spectral bands	34	data products
17	disaster monitoring constellation	34	disaster monitoring
18	attitude determination	35	multiple satellites
19	earth observation	35	satellite engineering
21	launch site	35	radiation environment
22	remote regions	35	space science
23	ground station	35	board processing
24	launch vehicle	35	satellite design
28	board computers	35	mission lifetime
29	satellite communications	35	system design
30	synchronous orbit	36	satellite programme
30	satellite technology	36	synthetic aperture radar
30	band downlink	36	board data handling
31	solar panels	37	data storage
31	global coverage	38	space technology

The above table is illustrative of the morphological productivity (Bauere, 2001) of single word terms, like: cost, satellite, system, launch, sensing, et cetera. Whereas, their compound word formations may be representative of a morphological process based on which knowledge of the domain flows and adapts to the organizational setting in which it is created. Sometimes similar terms that were ranked differently have appeared within the collection of documents collated from the Swedish Space Corp satellite technology news corpus, as shown below in table 4. Being possibly illustrative of wider *ontological spectrum* (knowledge sharing), of the knowledge of the satellite technology domain and corresponding research and commercial activities. That could as well happen to be dependent on the source of authorship; thus biased. Implying in turn a wider *epistemological spectrum* (knowledge theory), as suggested by Nonaka et al (1995).

Table 4 Ranking of select compound terms in Swedish Space Corp corpora

Rank	Compound Term	Rank	Compound Term
2	launch vehicle	40	resolution images
3	geostationary orbit	40	proton launch
8	launch pad	41	reusable launch vehicle
10	geostationary transfer orbit	43	satellite manufacturing
15	shuttle mission	44	remote sensing
17	surrey satellite	46	satellite constellation
19	rocket boosters	48	meteorological satellite
20	shuttle missions	49	spy satellite(s)
22	satellite launch	50	satellite launched
24	remote sensing satellite	52	mobile satellite
27	manned spaceflight	53	manned spacecraft
29	satellite launcher	54	satellites launched
30	geostationary satellite launch vehicle	54	geosynchronous orbit
31	satellite launches	55	satellite payloads
32	launch initiative	58	remote sensing satellites
36	synchronous orbit	60	disaster monitoring constellation
38	launch vehicles	60	launch mission
38	remote manipulator	60	launched satellites

Examining compound terms within collections of PhD Theses, from Surrey Space Centre. These compound terms have appeared to present some dominant terms within, thus knowledge created and utilized. As shown in Table 5 below, select compound terms are listed, relating their frequency behaviour to the number of total compound terms generated from the corpus in percentage value (frequency / total number of compound terms found). The data presented in the table below (composing 20.96% of the total compound terms found) is presented as such to see lexical composition of such collection of documents – the extent to which each compound term contributes to the total number of compound term. Nonetheless, terms like: mobile satellite, satellite communication, satellite network, leo satellite, satellite constellation, remote sensing, and last but not least geostationary satellite orbit. Though all exist in satellite technology corpora, which were collated from sources prescribed previously. Some common ground is possibly available for such concepts to be shared across such specialist domain, and organizations within. This is assumed to facilitate the diffusion of knowledge within such domain (s). However, level of adaptation and further flow of the knowledge involved, is related to technological implications for the knowledge worker or organization.

Table 5 Compound terms within a listing of PhD theses titles

Compound Term	Relative frequency ratio	Compound Term	Relative frequency ratio
mobile satellite	2.94%	selective fading	0.37%
satellite communications	1.47%	satellite constellation	0.37%
processing satellites	1.47%	IP multicast	0.37%
satellite networks	1.10%	geomobile satellite	0.37%
IP telephony	0.74%	multicast strategies	0.37%
satellite constellations	0.74%	adaptive multiuser detection	0.37%
leo satellite	0.74%	thrust orbit	0.37%
mobile satellite communications	0.74%	geostationary satellite	0.37%
satellite multimedia	0.74%	noise amplifier	0.37%
geostationary satellite orbit	0.74%	satellite inertia matrix	0.37%
mobile communications	0.74%	orbit calibration	0.37%
gravity gradient	0.74%	satellite diversity	0.37%
orbit satellites	0.74%	sstl satellites	0.37%
remote sensing	0.74%	frequency bands	0.37%
novel orbit propagation algorithm	0.37%	ozone content	0.37%
satellite imaging	0.37%	satellite measurement	0.37%

The observational and historical studies carried out, have provided better understanding into the field of investigation. Such studies provided the basis and validation for inferences made. Based on Nonaka et al's (1995) terminology used within the *knowledge conversion model*, portraying creation of knowledge and corresponding conversion processes. It is believed that knowledge undergoes a combination and socialization conversion process (for knowledge flow) within an organization or across a (sub) domain (s), and undergoes an internalization and externalization conversion process (for knowledge adaptation) within an organization or across a (sub) domain (s).

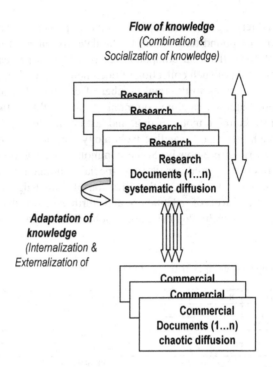

Fig. 2 Knowledge diffusion through the *knowledge conversion model (Nonaka et al, 1995)*

Figure above expands Nonaka et al's (1995) knowledge conversion model, to include consideration for how knowledge flows and is adapted within research and commercial documents. Within the case of an SME, Small to Medium Enterprise, such knowledge flow and adaptation through a *knowledge conversion model* may be a framework that could stimulate innovation through conversion of knowledge amongst the *Knowledge Creating Crew* (Nonaka et al, 1995) and stemming from an organization.

5 Conclusive Remarks

The case study is a method of learning about a complex instance through extensive description and contextual analysis. The product is an articulation of why the instance occurred as it did, and what may be important to explore in similar situations, in our case the specialist knowledge and its diffusion is the product. As the observational study laid the framework for the conduct of our research, it was focused on examining knowledge flow, and corresponding practices and information technology support in place. Results from the observational study have indicated that knowledge bottlenecks may exist, in particular were technological support could be needed. The transmutation of science into technology is a complex process when one sees unique ideas highlighting the past scientific landscape and beneficial technological artefacts in the present. The notion of satellite technology or space technology, with variable scope and scale, was an ostentatious idea that has led to a range of remote sensing and earth observation instruments for instance. The *unique idea* is a key reference point for forecasting how the idea will metamorphose into an artefact. Knowledge is communicated through so-called semiotic systems: written text, images, mathematical and chemical symbols, and so on. The knowledge of emergent domains is yet to standardize its symbol systems which simply add to the (creative) chaos inherent in such emergent systems. The analysis of change in written text, amongst the most changeable semiotic system at the lexical level at least, may reveal a consensus or dissension in the use of terms. Terms denote concepts and textually help us to understand how knowledge evolves in an emergent domain. The emergent domain of small satellite technology was studied as an exemplar. This is our attempt to establish a method, which covers a broad range of texts, research articles, commercially-driven documents and state-of-the-art papers representative of research and development conduced within an organization, to observe the emergence of a new domain.

We have by design focused on an innovative organization to establish our method which is driven by knowledge workers, document-based and guided by terminology utilized. The method will facilitate the construction of knowledge maps in an objective and systematic fashion. This method will help in establishing knowledge visualization studies in the realm of decision making focused on how research is exploited and how such a process can be facilitated, at lexical and knowledge worker levels. Whilst aiming to model sustainability of an organization through its continuous knowledge diffusion processes from persons composing such organization. It is an intuitive statement that research ideas and experimentation form the basis of new technologies, products, and practices. The research effort leads to the creation of new knowledge, and to the suspension of 'obsolete' knowledge, and this knowledge crosses over into technology. Perhaps a comparative analysis of the choice of terms (lexical signature) will indicate the extent of this cross-over. In this spirit of specialist knowledge still in the realms of research and not quite making it into the construction of artefacts and vice versa, we have compared the rank order of the most frequent words in the research corpus of SSTL/SSC papers with that of the Swedish Space Corp satellite technology news corpus, or between Surrey Space Centre PhD research theses and SSTL research publication, for instance.

Our analysis shows that research papers and commercial documents can be distinguished somewhat on the basis of single word and compound terms that were generated automatically. These two lexical signatures show the potential for identifying cross-over points in the diffusion of knowledge from the research arena to applications domain. The metamorphosis of science into technology is a complex process when one sees innovative ideas highlighting the past scientific landscape (i.e. in the form of PhD theses and state-of-the-art research papers) and beneficial technological artefacts in the present. The notion of satellite technology, with variable applications, was a unique idea that has led to a range of remote sensing devices for example. The innovative idea is a key reference point for forecasting how the idea will metamorphose into an artefact. Knowledge is communicated through so-called semiotic systems: written text, images, mathematical and chemical symbols, multimedia and so on.

The knowledge of emergent domains is yet to standardize their symbol system which simply adds to the (creative) chaos inherent in such emergent systems. The analysis of change in written text, amongst the most changeable semiotic system at the lexical level at least, may reveal a consensus or dissension in the use of terms. Terms denote concepts and textually help us to understand how knowledge diffuses in a domain. The specialist domain of satellite technology or space technology, specifically an organization in such a domain was studied as an exemplar. This is our attempt to establish a method, which covers a broad range of texts, PhD theses, journal articles, technical reports, and state-of-the-art review papers, to observe the emergence of a domain and hence specialist diffusion of knowledge.

References

Atkins, S., Clear, J., Ostler, N.: Corpus Design Criteria. Literary and Linguistic Computing 7(1), 1–16 (1992)

Bauere, L.: Morphological productivity. (Cambridge studies in linguistics; 95). Cambridge University Press, Cambridge (2001)

Biber, D., Conrad, S., Reppen, R.: Corpus Linguistics: investigating language struc-ture and use. Cambridge University Press, Cambridge (2002)

Burrows, J.F.: Computers and the Study of Literature, pp. 167–204. Blackwell, Oxford (1992)

Gruber, T.R.: A Translation Approach to Portable Ontology Specification. Knowledge Acquisition 5, 199–220 (1993)

Holmes, D.I.: The Evolution of Stylometry in Humanities Scholarship. Literary and Linguistic Computing 13, 111–117 (1998)

Liebowitz, J.: Building organisational intelligence: A knowledge management primer. CRC Press, Boca Raton (2000)

Nonaka, I., Takeuchi, K.: The Knowledge Creating Company: How Japanese Companies create the Dynamics of Innovation. Oxford University Press, Oxford (1995)

Sinclair, J.: Corpus, Concordance, Collocation. Oxford University Press, Oxford (1991)

Sperberg-McQueen, C.M.: Text in the Electronic Age: Textual Study and Text Encoding, with Examples from Medieval Texts. Literary and Linguistic Computing 6/1, 34–46 (1991)

Stemler, S.: An Overview of Content Analysis. Practical Assessment, Research & Evaluation. A Peer-Reviewed Elec-Tronic Journal 7(17) (2001),
http://ericae.net/pare/getvn.asp?v=7&n=17
(last accessed, 15/11/2002) ISSN 1531-7714

Surrey Space Centre (University of Surrey) – SSTL (Surrey Satellite Technology Lim-ited), George Edwards Library (University of Surrey), collection of text documents 1979-2002, http://www.surrey.ac.uk/Library,
http://www.ee.surrey.ac.uk/SSC/, http://www.sstl.co.uk

Swedish Space Corp, collection of text documents, historical corpus (1997-2002),
http://www.ssc.se

Yin, R.: Case study research: Design and methods, 2nd edn. Sage Publishing, Beverly Hills (1994)

Session E
Knowledge Transfer Partnership
Case Studies

Knowledge Exchange and Learning and Development in a Newly Formed SME: An Example from the Knowledge Transfer Partnership Scheme

Martin Wynn[1], Erin Lau[2], and Peter Maryszczak[2]

[1] Department of Computing, University of Gloucestershire, Cheltenham, GL50 2RH, UK
`mwynn@glos.ac.uk`
[2] Optimum Consultancy Ltd, 5 Court Mews, London Road, Cheltenham, GL52 6JQ, UK
`{erin.lau,peter.maryszczak}@optimum.uk.com`

Abstract. This paper focuses on the how the Knowledge Transfer Partnership (KTP) scheme has been used to introduce new information systems and advance learning and development at Optimum Consultancy Ltd, which was formed on 1st July 2008 via the amalgamation of Hama Ltd and J Orchard Consulting Limited. This new company now has 35 staff and turnover grew from £2.4m in 2008-9 to £3.1m in 2009-10. The knowledge base partner is the University of Gloucestershire, based in Cheltenham, UK. The KTP product is arguably the most used channel for effecting knowledge transfer between universities and local industries in the UK. The impact of the project is reviewed in terms of improved efficiencies, professional development, skills enhancement and organisational change. Learning and development were embodied in this major project to implement an integrated IT solution for the new company and rationalise and standardise the core business processes in the three offices situated at different locations in UK.

Keywords: Knowledge transfer, learning and development, IS strategy, systems integration, process change.

1 Introduction: The Key Business Requirements

1.1 Formulation of Project Objectives

The project was initially formulated in Spring 2008 when the Business Development manager from the University of Gloucestershire (UoG) met the Managing Director of Hama Ltd at a University event. The company's core business was project and cost management in the property, engineering and construction fields; and its customer base included major retailers (Harrods, Selfridges), rail operators

R.J. Howlett (Ed.): Innovation through Knowledge Transfer 2010, SIST 9, pp. 203–212.
springerlink.com   © Springer-Verlag Berlin Heidelberg 2011

(Docklands Light Railway, London Underground), Financial Sector (London Stock Exchange, Sumitomo Mitsui Bank Corporation), Business Relocations (Metronet, Tepnel) and Sustainable Development (Carbon Trust, Crown Estate). Hama Ltd was about to merge with one of their competitors, Orchard Consulting, and required learning and development (L&D) support in two main areas:

- The merging of their business systems and associated support processes. This was seen as absolutely critical to the success of the merger as neither company had reliable systems and without a rapid implementation of a new technical and informational infrastructure, the new business would not be capable of functioning as a unified entity.
- The establishment of common business procedures to underpin the expansion of the new company's market share. In particular, it was essential to unify and standardise business activities that impact on the merged customer base.

A number of options for supporting Hama's merger with Orchard Consulting were explored. It was decided to harness the government supported KTP scheme as the umbrella arrangement under which project based L & D training and embedding could be provided. The KTP scheme can be used for any project that provides bottom-line benefit to the company partner, but is often geared to projects that inject innovation and/or new technology into the operations and culture of the company (Wynn, 2009, Wynn et al, 2009). The UK government will fund over 50% of the employment, training and direct support costs of an experienced graduate - the 'Associate' – to lead a key change project, and also funds consultancy from the University for half a day a week, to bring transfer of knowledge from the University to the company.

A two year KTP project was deployed to review and establish the new business processes in Optimum (the new company) and then to evaluate, procure and implement new corporate information systems. At the same time, a shorter 40 week KTP was used within that 2 year period to address the specific problems of amalgamating three separate offices and standardising sales processes and support materials, including the website. Both projects were central to Optimum's new corporate strategy of growing market share in the project management services field through state-of-the-art information support services and slick efficient customer management. A strong theme running through both KTP projects was to maximise learning and development opportunities for Optimum staff, so that new systems and processes were fully embedded and could be exploited to maximum benefit for the company.

1.2 Ownership of the Business Objectives and Main Stakeholders

Managing the expectation of key stakeholders is critical to the success of any project or collaborative consultancy. The main stakeholders were the Project Board and project sponsor, Optimum company members, Project Team members, IT suppliers and Optimum clients. The Government KTP Advisor and the UoG were stakeholders via their roles on the Project Board and the Project team.

A power/interest grid can usefully be used to classify stakeholders and enable appropriate communication. Based on involvement and expectation, stakeholders are prioritised, focusing on their power and interest in the project outputs (Fig. 1). As the Project Board and team members were the most significant stakeholders, communication with them was done by weekly, monthly and quarterly meetings, reviewing progress against plan and checking main goals and associated plan milestones.

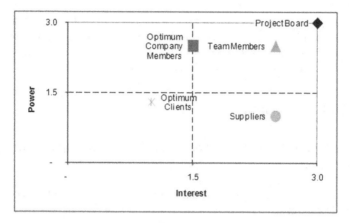

Fig. 1 Main stakeholders in KTP programme based on Power/Interest (based on scale of 0 = low; 3 = high)

1.3 Desired Outcomes and Link to Corporate Strategy

An internal review of systems and processes highlighted the fact that the company information systems were inadequate to support the company's current and future operations. Historically, the IT systems in Hama and Orchard Consulting were set up in an ad-hoc manner whenever a need arose. Separate software and hardware systems were purchased without detailed analysis of their impact on the overall IT and corporate strategies. The merger of the two companies in July 2008 posed greater challenges to combine and upgrade two different IT/IS architectures and in particular to align and standardise the business processes across three offices.

New integrated systems and standardised business processes would provide infrastructure support for steady growth and improved margins, without the stop-start addition of administrative overheads. In addition, as the new systems were implemented, there was an urgent requirement to implement a refreshed and refocused business development strategy that crystallised the different roles of the three Optimum offices (Cheltenham, London and Haywards Heath) and provided direction to the roll-out of the new Enterprise collaboration system. A major output was a new process and associated procedures for responding in a consistent and streamlined manner to customer enquiries across the organisation. This encompassed a review and evaluation of how Optimum's services and products could best be combined to meet varying customer needs and improve customer service. Overall, the desired impacts were:

- To reduce general administration time of fee earning project managers by at least 5% (i.e. 2 hours per week in searching for documents, collating reports from various spreadsheet and manual data re-keying)
- To improve the ratio of support staff to fee earners from 1:3 at the start of the project to 1:5 at project close
- To improve efficiency in reporting, forecasting, monitoring and controlling tools in all business activities and thereby achieve a growth in revenue from £2.4m in 2008/9 to £5m in 2012-13 (£3.1m turnover achieved in year ending June 2010), with a profit margin of circa 12%

2   Project Deliverables

2.1   Analysis of Problem Situation and Desired Impacts and Outcomes

The contracting stage took place in the summer of 2008. The Director of Hama Ltd (Peter Maryszczak) worked with staff from the UoG to design a project brief to satisfy L&D objectives that would gain financial support from central government under the auspices of the KTP scheme. The definition and scoping phase defined the purpose as well as the scope of the project. To do this, a multiple cause analysis (MCA) was adopted (Fig. 2). This process was subsequently repeated with the managing director of Orchard Consulting to develop the shorter 40 week project to address and support the change agenda in the selling practices and processes across the three offices of the new combined company.

The combined impact of the KTP initiatives was to provide Optimum staff with new skills and behaviours in three main interrelated areas:

- Establishment of an IS/IT function with the capabilities to support and develop a new systems platform to support sustained growth.
- New skill sets in the use of leading edge software as part of daily company operations
- Process understanding and ownership to facilitate on-going process improvement, particularly in the sales and operation areas.

In recent years, when new business was won, the lack of sound systems meant administrative and management staff was sucked in to support the delivery of key projects. This produced fluctuations in turnover, profit and staffing levels. New integrated systems and standardised selling processes would provide infrastructure support for steady growth and improved margins for the new combined company.

2.2 Financial Parameters, Timescales and Milestones

The finalised agreements for the two projects were embodied in contracts signed by Optimum and the UoG. For the major two year KTP project to implement a new systems platform, the budget for all UoG staff, travel and subsistence was set at £115K, with Optimum contributing 33% (£38K) and the Technology Strategy Board (a central government agency) providing 67% (£77K). In addition, Optimum agreed to invest a further £100K in hardware, software, staff time and training costs.

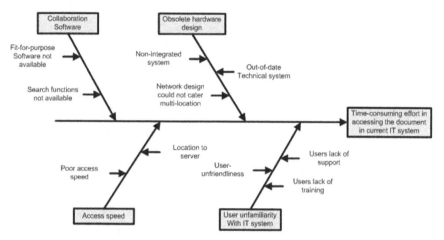

Fig. 2 Using Multiple Cause Analysis to help scoping and definition

For the shorter 40 week process alignment project, the staff consultancy budget from the UoG was set at £38K, with Optimum providing £15K and the Technology Strategy Board contributing £23K. Again, Optimum also committed to spend a further £8K on staff time and sales support materials. For the two projects combined, total investment over the two years will be £261K, with £100K being provided by the central government. The overarching two year project started in September 2008 and completed in September 2010, and the 40 week process alignment project started in October 2009 and completed in June 2010. Both projects have a series of milestones embodied in project plans. The key milestones – implementation of new systems, improved efficiencies and standardisation of customer centric processes have now been delivered.

2.3 Learning and Development Objectives

The design of the L&D initiatives was initially a three way process involving Optimum senior management, UoG staff and the government KTP adviser. The KTP funding and authorisation process requires a detailed project plan to be put together with clear benefits and deliverables. This was done for both the two year full KTP and the 40 week short KTP that ran in parallel. The generalised training goals concerned the improvements in capabilities noted in section 2.1 above, but these were weighted slightly differently according to job role. The three main training goals were:

- Learning to use and exploit the new collaboration software (Workspace)
- Learning to support the IS/IT function
- Learning to optimise business processes

Their relative importance for fee-earning and administrative/support staff is shown in Fig. 3.

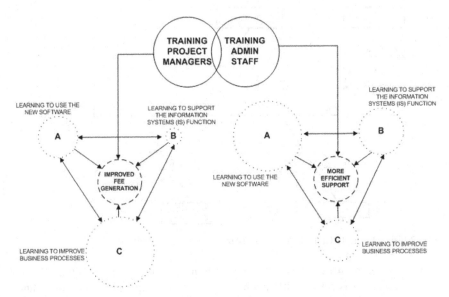

Fig. 3 Generalised training goals for Project Managers and support staff in the project

3   Project Benefits: Improved Efficiencies, Professional Development, Skills Embedding and Organisational Change

3.1   Delivery of L&D Initiatives

Within the three generalised training goals noted above (Fig. 3) were a range of L&D objectives concerning skills and competencies related to project management, systems selection and implementation, systems support, process change and organisational development. These included:

- The application of selected PRINCE2^{TM1} processes and components for managing the projects

---

[1] PRINCE2^{TM} is a Trade Mark of The Office of Government Commerce

- The use of PRISM Buy-Build methodology for selecting and implementing new software
- The application of process mapping tools and techniques for process reengineering
- The training of user champions who would become the standard bearers in the deployment of the new integrated Workspace systems and customer centric processes and procedures (Workspace was the new software acquired as part of the project)
- The application of ITIL Continual Service Improvement (CSI) concept for reviewing the project and sustaining improvement

To achieve these, a variety of training methods and learning formats was applied, following experience gained in a range of international training contexts (Wynn and d'Ayala, 1982, 1986). These were interspersed across the duration of the KTP projects and introduced by University staff as appropriate at different stages in the learning cycle. Methods used included:

- *formal training courses* to train users on the use of the new Workspace software were held on company premises and at the software house in Nottingham. Other courses to develop skills in PRINCE2 project management and service management principles were run at the University of Gloucestershire by qualified trainers.
- *structured workshops* to develop understanding of current systems issues, the PRISM Buy-Build methodology, process mapping analysis outcomes, and systems supplier selection options
- *brainstorming* to identify all possible requirements for new systems and risk identification and assessment
- *structured discussions and panel discussions* with the software suppliers, culminating a final panel discussion at the University's campus in London in April 2009
- *role plays, performance 'try-outs'* and *simulations* to allow systems users to try out new systems functions in a test environment
- *team tasks* were deployed on the three company sites to ascertain reporting requirements from the new systems and also to transfer project data from old systems into the new Workspace environment

3.2 Change Indicators

The impact of the overall project and L&D initiatives was assessed from a number of perspectives:

Professional development: Impact on the stakeholders (especially the company members) mainly lies in the changes in their normal work routine in IT systems. Ongoing training, coaching and support were needed to ensure that they got used to the new Workspace systems and processes as quickly as possible. Embedding the principles of PRINCE2 in the project management process also constituted an upgrade in professional practice in the systems area.

Organisational development: Certain company policies and procedures had to be modified, clarified or replaced to adequately support the new company. Business process flow charts were revisited to re-emphasise the role of the process owners, their responsibilities and new activities associated with each process. The IT/IS function was also significantly advanced, moving from the stage of end-user 'Contagion' to a stage of 'Control' and 'Integration', in Nolan's classic model of the evolution of the IT function (Nolan, 1979)

Resource commitment: To implement the recommended IT and process change solution, significant commitment and support from company staff was necessary. They needed to spend time on training & learning, and a range of tasks geared to the ushering in of a new way of working. Resources from the supplier were also needed for system support and configuration.

Skills Enhancement: The training of over 30 staff in the use of the new Workspace software through formal training sessions, supported by workshops and one-to-one coaching was a major embedding of new skills. This will be carried on via the selected four user champions. The embedding of new consistent sales processes to underpin the business development element of the system has also been a significant up-skilling in process understanding, ownership and management.

Change Management: Creating enthusiasm among staff for the new Workspace system was very important. A key to this is identifying and addressing common bottlenecks and causes of frustration and irritation in their job roles. End users will ultimately be responsible for making the system a continuing success. Thus, it is important to make them understand why the change is happening and what is needed to effect the change. For example, clear process flow diagrams, with defined roles and responsibilities, can improve communication and engender support for process change necessary to deliver project outcomes.

3.3 Impact on Optimum's Business Performance

For Optimum, the actual impacts of the project have been in improved efficiencies, time savings and the avoidance of additional administrative headcount. It was estimated that 5% of working time was being wasted due to inefficient IT systems and associated procedures. Removing this waste through improved efficiency of the IT systems portfolio has contributed up to £60K per year by reducing the administrative work of the fee-earning project managers, whilst avoidance of additional administrative headcount has provided an additional saving of £140K per year (4 extra administrative heads saved as the company has moved to a 1:5 admin-fee earning staff ratio). The Workspace software provides instant access to corporate information on overall company performance, forward workload and future prospects as well as full details for every job, including who is looking after it, the client, fee type, value, allocated costs and the margin that is being achieved has greatly improved the efficiency in all business activities. Moreover, the standardised enquiry response process is now being embedded through workshops and structured discussions with customer facing staff. This alone will drive an estimated additional increase in turnover of circa £200K per annum from 2010/11.

This will emanate from improved quality of customer response and the improved prospects of securing new business against competition. The standard components of customer response (project data sheets, accounting and insurance templates) are now in place across all three offices, allowing more time to focus on the customised elements of customer response (project specific approach an method). Different combinations of products and services are being fine-tuned for different market sectors and key customers. These benefits are now combining to enable the growth of the business and increased profit margins.

3.4 Stakeholder Impacts and Perceptions

The KTP programme focused on implementing an integrated approach to systems development and process change across the three offices of Optimum. The initial implementation addressed the key business information bottlenecks of document control, sales contacts and access to project information. The impact and perceptions of key areas of the Optimum business are summarised below:

People: This new collaboration software is able to integrate the management of time and resources and the recording of skills and training into the mainstream corporate database. Optimum staff benefit from having instant access to forward schedules and resource availability without a reliance on monthly paper/spreadsheet reports, which is very difficult and time-consuming to maintain.

Business Development: The new integrated system and sales procedures play a key role in business development, keeping track of enquiries, and underpinning sales and marketing campaigns. Any work done on prospective jobs is kept in the system. This allows the tracking and managing of the 'new work pipeline'. When Optimum wins a job, it gets migrated to a project record which holds all of that history. Before the new software and associated procedures were introduced, this data was held in spreadsheets in which a lot of information was being duplicated with no version control.

Finance: The existing accounting software (Sage) has now been linked with the new Workspace system. Senior management and team leaders now have instant access to information about overall company performance, forward workload and future prospects as well as full detail on every job (e.g. fee type, value, allocated costs and the margin that is being achieved). This leads to more accurate costing and invoicing and shortens the management reporting cycle. The link to the timesheet system ensures that costs are up to date and this eliminates what was a laborious process under the old way of working.

Operations: Management of projects has been made more efficient through the new Workspace system. Once a bid is won all the information is ready to be automatically transferred to the project record. This ensures continuity and reduces errors. The ability to find things more quickly will prove increasingly useful as project progress. Previously, field-based staff has been struggling to access the information they needed. The system will bring a complete picture of what the business is doing in one place and directors are better informed about projects. They can log into the system to access financial information on a project instantly.

4 Concluding Remarks

This case study has illustrated how the KTP scheme can be used to support an SME achieve key business objectives that revolve around a combination of multi-disciplinary learning and embedding of new skill sets and knowledge. Utterback (1994) has noted that 'a strong technological base is as critical to the prosperous survival of a firm as a good understanding of markets and a strong financial position', and knowledge transfer from Universities can play a crucial role in the technological advancement and organisational development of local industries. The UK Department of Trade and Industry (DTI) has specified a range of products for supporting local businesses (DTI, 2003), including the KTP scheme which provides direct support in excess of £25 million per annum for knowledge transfer projects in firms of all sizes, but particularly in SMEs of less than 250 staff (Wynn and Jones, 2006). The KTP scheme has a track record of benefits delivery to all parties, which suggests it is worth continued support in the current period of economic downturn and reduction in public expenditure budgets.

Acknowledgements. The Partnership received financial support from the Knowledge Transfer Partnership programme. KTP aims to help businesses to improve their competitiveness and productivity through the better use of knowledge, technology and skills that reside within the UK Knowledge Base. KTP is funded by the Technology Strategy Board along with the other government funding organisations.

References

1. Wynn, M.: Developing and implementing IS strategy in SMEs. Management Research News 32(1), 78–90 (2009)
2. Wynn, M., Turner, P., Abas, H., Shen, R.: Employing knowledge transfer to support IS implementation in SMEs. Industry and Higher Education 23(2), 111–125 (2009)
3. DTI, DTI Innovation Report, Competing in the global economy: the innovation challenge (December 2003)
4. Nolan, R.L.: Managing the crisis in data processing. Harvard Business Review (March-April 1979)
5. Utterback, J.M.: Mastering the Dynamics of Innovation (How companies can seize opportunities in the face of technological change). Harvard Business School Science and Innovation Policies, HMSO, Crown Copyright, 55 (1994)
6. Wynn, M., d'Ayala, P.G.: Human settlement management training: an approach to course design. Ekistics 49(292), 78–84 (1982)
7. Wynn, M., d'Ayala, P.G.: Handbook for the Design and Organisation of Courses. UNESCO Human Settlement Managers Training Programme and Man and the Biosphere (MAB) Programme, Revised Version, CIREA, Parma (1986)
8. Wynn, M., Jones, P.: Delivering Business Benefits from Knowledge Transfer Partnerships. International Journal of Entrepreneurship and Small Business 3(3/4), 310–320 (2006)

Providing e-Business Capability on a Legacy Systems Platform: A Case Study from the Knowledge Transfer Partnership Scheme

Rizwan Uppal[1], Martin Wynn[1], and Phillip Turner[2]

[1] Department of Computing, University of Gloucestershire, Cheltenham, GL50 2RH, UK
`{ruppal,mwynn}@glos.ac.uk`
[2] TPG DisableAids Ltd, Plough Lane, Hereford, HR4 OED, UK
`itmanager@tpg-disableaids.co.uk`

Abstract. This paper focuses on the how the Knowledge Transfer Partnership scheme has been used to develop and implement a technical strategy to support e-business trading by an SME dealing with the NHS and other public authorities. In this instance, the company (TPG DisableAids) decided against the introduction of new core systems but preferred instead to pursue a strategy of building e-business capabilities on legacy systems that were deficient both technologically and in terms of functional capacity. This resulted in a number of technical and business challenges that were addressed via the KTP project.

Keywords: Knowledge transfer, e-business strategy, legacy systems, systems integration, process change.

1 Introduction

1.1 Company Background

TPG DisableAids is a provider of equipment for the elderly and disabled and has grown steadily since 1984 to employ 47 staff today. The company assembles and distributes a wide range of products from primary manufacturers, such as Stannah who make a range of stair lift products. The company currently has an annual turnover of £4.3m (2009/10), with stair lift products generating about one-third of turnover but over 50% of profits.

TPG DisableAids' market can be divided into different segments (NHS, local authorities, district councils, residential & nursing homes, private individuals). Nationwide, this is a multi-billion pound market, which is growing as the age profile of the population increases. Competition comes from some of the national equipment dealers operating in the region (e.g. Stannah Lifts, who are also a supplier to TPG DisableAids) and one or two other smaller locally based companies

R.J. Howlett (Ed.): Innovation through Knowledge Transfer 2010, SIST 9, pp. 213–222.
springerlink.com   © Springer-Verlag Berlin Heidelberg 2011

with less than 5 staff each. The business opportunity is there to rapidly grow market share, particularly in the new market segments driven by public authority care management, insurance industry home equipment provision, and lifestyle products for the elderly. TPG DisableAids business plan is to double their turnover within 5 years to £8.5m in 2014/15 which is dependent on developing e-business capabilities in line with changes in NHS and public authority procurement practices. It is important that the company have the systems capability to respond to the equipment and service requirements of the NHS and related bodies at short notice as the elderly and disabled leave hospital and return to their homes. The NHS e-procurement initiatives require specific inter-organisational systems integration capabilities which the company has hitherto not had. This alignment is critical to the expansion plans of the company.

1.2   The Challenge and Opportunity of Legacy Systems

When packaged software first became widely available in the early 1990s, many companies moved quickly to procure and implement either a range of standalone packages (such as Sage, Manugistics, Peoplesoft) or integrated ERP suites (such as SAP, JD Edwards or Oracle). However, once the real costs and complexities of such projects became apparent, many companies began to look at alternatives that maximized the value of their investment in existing legacy systems. This trend was encouraged by the failure of some of the early ERP projects to deliver expected benefits. As Jeffrey and Morrison (2000) concluded, 'You don't have to go far to bump into lots of evidence that shows how ERP software has not delivered on the promises of vendors.' By the mid-1990s, the data warehouse was perceived by some as constituting an effective alternative strategy to wholesale replacement of old systems, by extracting data from new and legacy systems alike to provide timely aggregated management information, one of the main apparent benefits of new integrated ERP packages.

The data warehouse concept achieved considerable success and was onwards developed to utilize not only relational database technology, but also multi-dimensional spreadsheet type engines (the so called OLAP products – On-Line Analytical Processing) and subject specific mini warehouses often termed 'data marts'. By the late 1990s, however, the emergence of the web and the growth of the concept of information portal provided another possible option that could build on existing legacy systems and yet provide some of the benefits of across the board systems replacement (Wynn, 2000). This saw the emergence of the concept of middleware that could act as an information exchange between legacy systems and packaged software alike, and also provide a link through from in-house systems to the corporate portal or web-site.

1.3   The KTP Scheme

The UK KTP scheme attempts to harness the skills, knowledge and experience that exist in the higher education sector and apply them to projects in local industries.

As the Work Foundation recently noted 'universities are a valuable source of knowledge and innovation which can benefit.....existing businesses, whilst close linkages with businesses are also very valuable to universities' (Work Foundation, 2010). The KTP scheme has been in operation in one guise or another since the 1970s and the basics of the scheme are as follows:

- The university partners with a local company to deliver a project – typically of two years duration – of direct bottom-line benefit to the company
- The university and company design the project proposal which is then submitted the Technology Strategy Board – a UK Government organisation. If successful the UK Government provide circa 50% of direct costs for projects with small to medium sized enterprises.
- The cost subsidy applies to the salary of a full-time project manager or technical expert (recruited by the company and university together), a supervisor from the University working half a day a week on the project, and associated training, travel, equipment and support costs. The full-time project manager or technical expert (known as the KTP Associate) is recruited onto the University payroll, but works full time in the company.

KTP projects are multi-faceted and provide benefits for all parties. Knowledge transfer is at the heart of these schemes, with the Associate and university supervisor acting as conduits for a range of skills and knowledge that can be brought in to help the company move forward. The university benefits from involvement in real-world project delivery which often produces conference research papers and publications; and the Associate gets the opportunity to play a key role in a high profile company project, supported by a blue-chip training programme.

1.4   Project Objectives and Outputs

The project objectives were to underpin a transformation of the company from a traditional family business to a highly efficient e-business, operating electronically across its extended supply chain. Failure to enable electronic trading would cause significant damage to the company's ability to tender for upcoming supply contracts (and post sales services) and have a detrimental effect on efficiency. The KTP overarching objective was to optimise business processes and implement new cross-supply chain systems.

Outputs were targeted to include the following:

- Technology infrastructure upgrade to support cross supply chain information exchange.
- Top-level process maps for TPG DisableAids extended supply chain, identifying opportunities for process integration
- New information reporting capabilities providing improved communication and sharing of information in-house and with key clients and suppliers. This was seen as particularly significant in the tracking of large contracts covering several years transactions.

- New e-procurement/order capture capabilities to allow transaction processing with NHS and other key customers.
- New integrated systems incorporating bespoke elements for systems integration and web access.

2   Evolution of the KTP Project

2.1   Research Questions

From a research perspective, the key challenge was to determine if a technical strategy based on preserving old legacy systems and using a range of technologies could provide the e-business capabilities the company required to trade electronically with public authority client base. In essence, it was about finding an answer to what Laudon and Guercio Traver (2010) call the 'e-commerce site-building puzzle' (Fig. 1). This requires a systematic consideration of a number of key questions:

1. Could a data warehouse be constructed to extract, aggregate and summarise key performance data from the old Sybiz legacy financial systems?
2. Could an information portal be built that could sit 'on top' of the in-house legacy systems to allow electronic order capture and invoice posting?
3. Could bespoke 'middleware' be used to link these new technology elements together to function alongside the old legacy systems platform?
4. Could business processes be changed to support and exploit the business opportunities afforded by the technology innovation?

2.2   The Technology Challenge

From a technology perspective, the challenge was to understand the company's business processes and IT infrastructure, upgrade IT infrastructure and associated technical strategy to provide a solid platform (middleware) to build new capabilities to reduce cost and exploit new business opportunities. The key new strategic element was the need to develop an electronic trading capability through a web portal linked to the middleware infrastructure which fills a technology gap between the company's old legacy accounting system and new modern technologies possessed by key customers. Key project phases are shown in Fig.2.

To understand TPG DisableAids business needs, it was deemed essential to understand company business processes. This was not a simple task because of a com- plex and tight relationship between business processes and the company's bespoke accounting software package (Sybiz Vision). Often software packages are customised or adapted to fulfil the needs and requirements of an organisation, but here company growth has been a very gradual transition that has occurred around their accounting system. This made the task of studying and understanding business processes problematic. It was essential to ascertain possible capabilities and

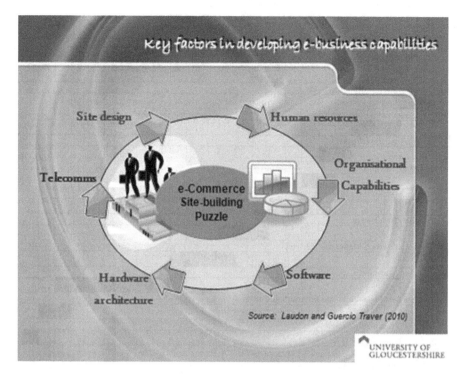

Fig. 1 The e-Commerce site-building puzzle
Source: Laudon and Guercio Traver (2010)

the current restrictions associated with adherence to the limited functionality of the existing accounting system.

2.3 IT Infrastructure Strategic Upgrade

There was a significant technology mismatch between the company's legacy accounting system and the new capabilities developed in modern technologies, so it was essential to upgrade the existing IT infrastructure to provide a solid foundation for those modern technologies. Information security was another concern for future development. Implementing information security controls on existing systems provided a secure environment for future development. A consistent IT strategy was also required to keep all technologies working and upgraded in line with business needs and requirements. TPG DisableAids decided to use open source/freeware support for in-house development which provides a secure, reliable and a flexible platform to develop in house systems capabilities. Implementing freeware technologies in a live business environment required significant research and knowledge transfer. MySQL Community Server and PostFix Email Server are two examples of secure and reliable open source technologies deployed in the project.

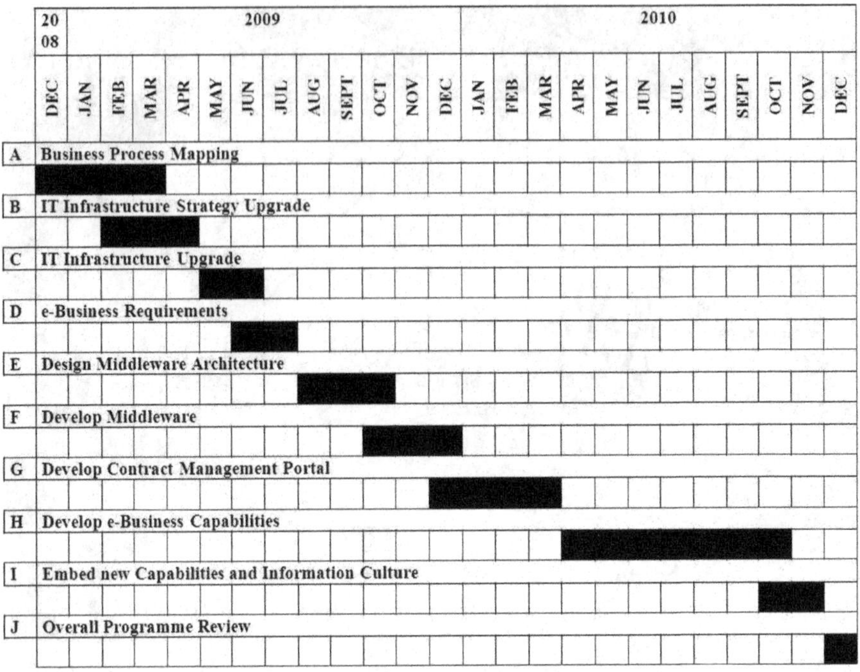

		2008	2009												2010										
		DEC	JAN	FEB	MAR	APR	MAY	JUN	JUL	AUG	SEPT	OCT	NOV	DEC	JAN	FEB	MAR	APR	MAY	JUN	JUL	AUG	SEPT	OCT	NOV	DEC
A	Business Process Mapping	■																								
B	IT Infrastructure Strategy Upgrade			■	■																					
C	IT Infrastructure Upgrade					■	■																			
D	e-Business Requirements							■																		
E	Design Middleware Architecture								■																	
F	Develop Middleware									■	■															
G	Develop Contract Management Portal											■	■													
H	Develop e-Business Capabilities															■	■	■	■							
I	Embed new Capabilities and Information Culture																						■	■		
J	Overall Programme Review																									■

Fig. 2 The project plan

2.4 Middleware Design and Development

This phase of the project had the most technological challenges. From design to development and implementation, at every stage, there were unexpected challenges due to the technical and functional shortcomings of the legacy accounting package and its limited integration capabilities.

There were two main phases of middleware development:

1. Development and synchronisation of a middleware database
2. Data synchronisation of the data mart containing contracts information

Converting the information from old file structures in the legacy accounting system to a modern RDBMS (Relational Database Management System) was the biggest challenge. Technology-wise there were only limited options available to convert the flat data files (DBF format) that existed in the legacy accounting system (Sybiz Vision). This challenge was accomplished by using further open source products to convert DBF files into modern SQL based information. Assuring quality and consistency of the data was another challenge. The overall performance of the above mentioned operations in terms of time was another challenge. Handling these challenges simultaneously amplified the overall difficulty of the task. Integration and well designed architecture were key to accomplishing this complex and critical phase of the project.

A further phase of middleware development centred on extracting data from the middleware database and transforming it into modern database objects which are the basic foundation for any modern software design pattern.

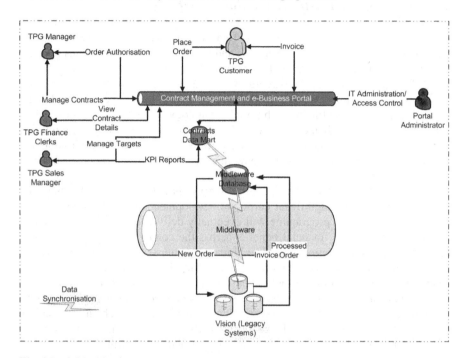

Fig. 3 TPG DisableAids systems architecture

2.5 Portal Architecture

This was the critical deliverable from the project from a business perspective and was dependent on the stability and reliability of the middleware synchronisation of the database, data mart and legacy systems (Fig.3). The portal architecture is based on the modern technology (MVC J2EE) design pattern whereas the design of the middleware database is necessarily based on the old legacy accounting systems file structures. To plug the gap between these two technologies, a modern data mart was created. A one-way synchronisation technique was used to extract data from the middleware database and populate the data mart on a regular basis. The main task of this synchronisation is to clean and transform primitive data from the middleware database into the technologically modern data mart objects.

Extracting data from the legacy accounting system to a modern MVC based portal was a collective effort utilising different open source products, freeware utilities and intelligent architecture design/development with tight integration. This allowed the production of capabilities and functionality not possible in the old legacy system.

3 Project Benefits

3.1 Delivery of e-Business Trading Capability

The company is now in a position to trade electronically with key customers including NHS Shared Business Services and local authority organisations responsible for the provision of disabled facilities grants and associated products and services. This may include trading via third party data transfer intermediaries or other similar agencies. Improved efficiencies can be seen throughout the order and sales processing procedures utilising web portal technology, whereby order information is accepted over the web and returned to the customer as an invoice, thus minimising the opportunity for human or machine error. The company's environmental impact has also been improved by removing the need to print paper documents and post to customers.

The introduction and transfer of explicit supply chain management thinking and knowledge has been a significant corporate development that will produce many ancillary benefits. The programme will eventually deliver technological, managerial, philosophical and financial benefits across the company's supply chain and business-to-business relationships. The blurring of inter-organisational boundaries through supply chain management techniques, controls and information flows will enable the companies in the supply chain to compete with the increasing number of large multi-nationals entering the consumer disabled-equipment sector. Demonstrating the practical benefits of electronically enabled supply chain automation technology and in-house knowledge facilitates the future evolution of organisational and business unit specialisation required for planned growth.

3.2 Research Findings and Change Indicators

The impact of the overall project can be assessed from a number of perspectives:

Technology application: The key research questions focussed on the possible application of middleware and data warehouse concepts to allow e-trading through a bespoke portal. The project manifestly demonstrated that this was achievable.

Organisational development: Certain company policies and procedures had to be modified, clarified or replaced to adequately support the new company. Business process flow charts were revisited to re-emphasise the role of the process owners, their responsibilities and new activities associated with each process.

Resource commitment: To implement the recommended solutions the company has utilised a number of obsolete servers, thereby extending the value of prior capital investment and reducing budgetary costs for implementation.

Skills Enhancement: The training of in-house workers to use the portal enhances very basic IT skills necessary for clerical work. For many involved, the use of the portal eases the use of ordering portals that are being developed by up-stream supply chain suppliers and manufacturers. Furthermore, the production of accurate financial figures allows staff to reason confidently and develop confidence in software systems. The legacy systems are widely acknowledged, within the company, to have poor data accuracy and misleading reports.

Process change: Creating enthusiasm among staff for new ways of working and the use of modern technology facilitates their expectations and adaptation to future wholesale replacement of legacy systems with modern software and best practice workflow systems. Of major benefit is the plugging of the gap historically surrounding the management of contracts. To date, managerial thinking and outlook and day-to-day management of contracts has been hit and miss, with, in particular, *a priori* adherence to Service Level Agreements with KPI reporting for the customer and in-house managers absent or hard to produce. The philosophical change from fire-fighting contract management to properly managed, KPI/dashboard based operations is the first fundamental shift in decision support and monitoring of operations within the company. Such changes are necessary to ease the cultural shift away from 'finger on the pulse' *ad hoc* decision making pertinent in the small company to the normal repertoire of techniques and tools of the medium sized SME. Finally, as a tool that is utilised by major stakeholders in each department, the portal and associated processes force into sharp focus the team effort viewpoint. In particular, effort in one department produces benefits for another, which forces employees and departmental managers to take a holistic view of effort and benefit.

3.3   Impact on TPG DisableAids' Business Performance

The actual impacts of the project have been in many areas. In addition to the impacts on company culture discussed above, the project has provided KPIs for contracts to date, reduction in errors in pricing for larger customers and errors during invoicing, and finally, and most importantly, in the day-to-day management of existing contracts and the creation of new contracts pricing structures.

Throughout the project, a number of additional changes have been necessary, which have resulted in (a) more reliable email systems, (b) more accurate and extended documentation and contact searching facilities, (c) increased information security, (d) development of middle-ware to allow future work on data-cleansing, (e) improvements in IT worker conditions, and (f) an increased awareness of the benefits of IT and information reporting. All these changes have directly or indirectly improved business performance across the whole company.

4   Concluding Remarks

The KTP scheme is central to UK government policy for re-invigorating and supporting British industry, not least in this period of economic downturn. It provides direct support of circa £25 million per annum for graduates – normally with several years' business experience - to undertake specific knowledge transfer projects in firms of all sizes, but particularly in SMEs of less than 250 staff (Wynn, 2009).

This KTP project brought TPG DisableAids significant technology based business benefit and competitive advantage. As Urwin (2000) has remarked, 'rapidity of response and ability to move quickly is an important advantage which small companies have over their bigger rivals, and the internet enables them to use it to the full'. In addition it has allowed the company to maximise the value of its investment in old legacy systems and to choose the appropriate time in its business cycle to replace them. This case study has illustrated how the scheme can be used to usher in new technologies in an evolutionary manner to support an SME achieve its key business objectives without significant cross-company upheaval.

Acknowledgements. The Partnership received financial support from the Knowledge Transfer Partnership (KTP) programme. KTP aims to help businesses to improve their competitiveness and productivity through the better use of knowledge, technology and skills that reside within the UK Knowledge Base. KTP is funded by the Technology Strategy Board along with the other government funding organisations. Grateful acknowledgement is also due to Alastair Gibbs, Managing Director and Mandy Harrold, Finance Director of TPG DisableAids for their on-going support and encouragement.

References

1. Jeffrey, W., Morrison, J.: ERP, one letter at a time. CIO Magazine (September 1, 2000), http://www.cio.com
2. Wynn, M.: From E-commerce to E-business at HP Bulmer – pioneering technologies in the drinks industry. Virtual Business 4(9) (September 2000)
3. Work Foundation, No city left behind? The geography of the recovery – and the implications for the coalition, Work Foundation (July 2010)
4. Laudon, K., Guercio Traver, C.: E-Commerce 2010, 6th edn. Pearson, London (2010)
5. Wynn, M.: Information systems strategy development and implementation in SMEs. Management Research News 32(1), 78–90 (2009)
6. Urwin, S.: The Internet as an information solution for the small and medium sized business. Business Information Review 17(3), 130–137 (2000)

Session F
Innovation and Enterprise

Firms' Nature and Characteristics and Their Attitude toward Publication: An Analysis of the Italian Biotech Sector

Rosamaria D'Amore and Roberto Iorio*

University of Salerno
(Department of Economics and Statistics)
riorio@unisa.it

Abstract. The aim of the paper is to analyse if the classification of the biotech firms according to the OECD criteria, together with another important firm characteristic, the size, are related to different behaviour of the firms toward the generation and circulation of knowledge; that is, if the specific activity of the firm, that implies its belonging to an OECD typology, has an effect on the number of publication done and on the number and quality of collaborations activated by that firm, even controlling for firm size. The empirical analysis shows that such relationship does exist.

1 Introduction

An increasing attention is being devoted by many scholars in the field of the economics of innovation to the phenomenon of the firms that publish papers in the "open" literature. Nevertheless, the relationship between the nature and characteristics of the firms and their different attitudes and behaviours toward publication and research collaboration has not been probably enough investigated. This paper tries to increase the knowledge in this direction, treating this issue with reference to the Italian biotech sector.

The biotech sector is characterized by an high level of knowledge intensity and a pervasive nature of innovation. It has therefore an high production of new knowledge, that is often disclosed trough the scientific publication or by the introduction of innovations. Besides, new knowledge creation is rarely a solitary activity: knowledge creation, sharing and innovative activities take place within networks and come out from collaborations.

On the other side, this sector is characterized by an high level of complexity because of its multidisciplinarity and pervasive nature of innovation; boundaries of industries that can be included under the umbrella of "biotechnology" are blurring. This causes a great heterogeneity inside the sector, up to the point that it becomes

---

* Corresponding author.

R.J. Howlett (Ed.): Innovation through Knowledge Transfer 2010, SIST 9, pp. 225–233.
springerlink.com   © Springer-Verlag Berlin Heidelberg 2011

difficult to give a definition and to precisely identify it. We considered the Italian situation: indeed, the Italian Statistical Office (ISTAT) does not identify a biotechnology sector, therefore it does not give a definition of it. An attempt to manage such complexity has been done by the OECD, which on the contrary provides a definition of the whole sector and, on the basis of the kind of activity mainly conducted, tries to identify the different typologies of the firms inside the sector.

We therefore face up a sector characterized by high heterogeneity and high propensity to publish. Basing on the work done by d'Amore and Vittoria (2006, 2008, 2009), who identified the existing Italian biotech firms and classified them according to the OECD criteria, and crossing their data with data on publication and firms characteristics, we try to investigate the relationship between these two aspects, that is to verify if the different characteristics of the firms (different activities that generates a different OECD categorization, together with another relevant aspects like size) are relevant in explaining different attitudes to make publications in the "open" literature and to make publication in collaboration.

The paper is so articulated: the next section contains a literature review about the reasons why firms make research in collaboration and why they publish papers in scientific journals. The following section briefly describes the typologies of biotech firms identified by the OECD. A section with a description of data follows, than a section with the results of the empirical analysis. Some final considerations conclude the paper.

2 Why Do Firms make Research in Collaboration and Why They Publish the Results?

The biotech sector is characterized by an high degree of collaboration between innovative agents (firms, universities, research centres, hospitals) and an high propensity of firms to involve themselves in the practice of open knowledge: many firms publish scientific papers in the reviews read by the scientific community. The two phenomena go often together, in the sense that the papers are frequently the results of collaborative research.

Indeed, the phenomenon of publications by firms is not obvious, because there is a clear incentive for firms to keep the results of their research secret or to patent them, in order to transform them in commercially useful results.

Nelson (1990) argued that firms have many good reasons to publish (selected) results of their research endeavors of low competitive value: to maximize visibility and link up to the scientific community, but also to establish intellectual claims and legal rights. Hicks (1995) points out that the corporate research papers in the open literature may also signal R&D capabilities to (potential) partners and suppliers. Cockburn and Henderson (1998) claims that firms publish in order to increase their absorptive capacity. According to Stern (1999) firms might also publish with universities to reduce labour costs: collaboration through joint publication allows the firm to use the human capital of the university scientists without having to pay him/her a high wage, as their reward may be the publication itself.

This "open science" mechanism produces a pool of knowledge that can be used freely by the international scientific community from which corporate researchers draw very heavily (Jaffe, 1989). Zucker et al. (1998) report that, especially in periods when there is a shift in technological paradigm to one closely linked to science, publications by the leading firms are crucial for mobilizing relevant in-house research and external research to make a successful transition.

If firms do decide to publish, many of these papers are likely to be co-authored with researchers working in other institutions Particularly in the biotechnology sector, characterized by an high level of innovation, one way to produce innovation is the collaboration in publication between firms and research institutions (universities, research centres, hospitals). Knowledge transfer becomes a crucial point in the sector. The inter-institutional co-authorship of research articles is a fundamental form of knowledge transfer and creation. Inter-institutional co-authorship, regardless of the type of the organizations involved, occurs when at least two different co-authors of a scientific paper have different affiliations. This type of interaction entails the tacit transfer of information and knowledge as a result of personal contacts between the authors, even where the process is scantly formalized: tacit knowledge is more easily exchanged via direct collaboration (Rosenberg, 1990). Collaborations with universities are particularly frequent. The perspective of the university as a key contributor to wealth generation and economic development has increased in recent decades (Mansfield and Lee, 1996; Etzkowitz and Leydesdorff, 2000).

3 The OECD Classification of the Biotech Sector

As we said in the previous sections, one of the main characteristics of the biotech sector is its multidisciplinarity. Its definition comprise a broad range of knowledge fields; in fact there are many different definitions existing in the literature and are given by internationally influential bodies. The most frequently used definition is given by OECD: "Biotechnology consists in the use of scientific and engineering principles (based on microbiology, genetic, biochemistry, chemical and biochemical engineering) to transform materials using biological agents (such as micro organism, enzyme, animal or vegetable cells) with the purpose to obtain goods and services" (OECD, 1989).

The OECD Statistical Framework for technology also defined biotech activities, identifying six classes. The main distinction is between production and service activities. Among production activities, it distinguishes between active, innovative and dedicated biotech firms, in order to identify activities more or less focused on biotech. In particular, a biotechnologically active firm (BAF) is defined as a firm engaged in key biotechnology activities, like the application of at least one biotech technique to produce goods or services and/or the performance of biotechnology R&D; a dedicated biotech firm (DBF) is a BAF whose *predominant* activity involves the application of biotech techniques to produce good or services and/or the performance of biotech R&D; an innovative biotech firm (IBF) is defined as a BAF that applies biotech techniques for the purpose of implementing new products or processes.

Among service activities, it distinguishes R&D, market and other service oriented firms. In particular a biotechnology R&D firm with no product sales is classified by the Italian national statistical offices into the R&D service industry category; targeted firms include firms classified as wholesalers, for instance local operations of large foreign pharmaceutical firms, whose local affiliate performs biotechnology research, but acts mainly as a wholesale distributor; other types of services firms are included if they are using biotech techniques for the purpose of providing a services (for example waste management and environmental remediation firms).

4 Description of Data

In order to build a database of scientific publications in the biotech sector it has been done an intersection of three databases: *i*) RP Biotech data base; *ii*) ISI Web of Science; *iii*) *Analisi Informatizzate delle Aziende* (AIDA).

RP Biotech data base. It is a collection , created by D'Amore and Vittoria (2006, 2008, 2009), of (potentially all) the Italian firms belonging to the biotech sector, in activity at the end of 2005, according to the OECD definition and classified according the OECD typologies described before. This database collects 865 firms; 501 of them are for profit firms, 364 are no profit firms. We focus our attention on the life-science for-profit firms, whose total number is 371.

ISI databases, especially the Science Citation Index®, and the web-based version Web of Science® (WoS), is a detailed bibliometric database of journal articles and citations of worldwide research literature that contains 14 000 international peer-reviewed scientific and technical journals. Each journal is attributed to one or more WoS-defined Journal Categories. The bibliographic record of each publication contains information on the authors' addresses and their institutional affiliations. We assumed that a publication refers to a firm when, among the addresses of the article, there was the address of that firm; the number of institutions whose the authors belong to is assumed to be given by the number of addresses.

AIDA (from *Bureau van Dijk*). It contains balance sheets information of all firms operating in Italy. We used data on the annual number of employees.

Data from ISI and AIDA are collected from 2001 to 2005.

Data in AIDA on the annual number of employees were found for 212 of the 366 life-science for-profit firms; this is the sample we considered for the following analysis. In this sample the firms are so distributed among the different OECD categories: among the "production" firms, there are 18 active firms, 27 dedicated firms and 62 innovative firms; among the "services" firms: 19 R&D firms, 37 targeted firms, 10 other services firms. 39 firms are not easily included in any category (the so-called "out" firms).

5 Results of the Empirical Analysis

Firms belonging to different OECD typologies have different characteristics and goals. We can infer a different behaviour towards publications of scientific papers. In fact there are significant differences in the propensity to publish across the typology of firms. The values let to think to a division of the firms, according to the

OECD criteria, in two groups: the first one, composed by innovative (on average 2.41 publications a year), R&D firms (2.26 publications), dedicated (1.64 publications), that is the group that publish more frequently; the other one, composed by firms belonging the other categories – active (1.03 publications), other services (0.28 publications), targeted (0.27 publications), and "out" (0.18 publications)- for whom publication is, on average, a quite rare event. This distinction is almost co-incident with the distinction between production and services firms.

Nevertheless, many other characteristics of the firms, besides their main activity that determines the OECD typology, may influence their publication attitude. We focus our attention on firm size. There is a positive and significant correlation between the number of employees and the number of publications (significant overall and between correlation, not significant within). This result may be explained through a "direct resource effect" (larger firms have greater internal research resources and this increases the quantity and quality of research) and an "indirect resource effect" (larger firms activate larger networks and this generates more and better research) (Iorio, Labory and Paci, 2007).

It is possible to analyse more precisely, in a multivariate context, the effect of OECD characteristics and dimensions of the firms on the number of publication through a regression analysis.

We use a panel database whose dimensions are firms (individuals) and years (time); the dependent variable is the number of publications made by each firm each year; the independent variables are the number of employees (as a proxy of firm size) and the OECD categories (included as dummy variables) the firms belong to. Because the dependent variable, the number of publications, is a count data, the more suitable technique is the negative binomial regression. We run a panel regression with random effects. The results are shown in Table 2.

Table 2 Determinants of the number of publications. Results of the negative binomial regression on panel data; years 2001-2005

COVARIATES	Coefficients
Employees	0.00041***
OECD category (benchmark: Innovative)	
Targeted	-1.35404***
R&D	0.92437*
Active	-0.61875
Dedicated	0.02831
Other services	-1.07445
Out	-1.81635***
_constant	0.992535***

*Significant at 90 % level; **Significant at 95 % level; ***Significant at 99 % level;
Notes: Dependent variable: number of papers published by each firm each year
Number of observations: 797 (212 groups); WaldChi2(7)hood: 43.22 (prob>chi2: 0.0000);
Log Likeihood: 778.37268.

These results show that firms belonging to targeted and "out" publish less then the innovative firms (these result are significant at 99% level of significance), while R&D firms publish more (at 90% level of significance). The employees variable is positive and significant.

Therefore we can confirm what we saw in the previous statistic analysis, but in this case we can conclude that it's true also *ceteris paribus* (equal employees): the belonging of a firm to an OECD category rather then the other one is a determinant of the propensity to publish and that also the firm size has a positive effect on publications. It has to be underlined that the ranking of typologies of firms is not the same in the bivariate and multivariate analysis: the innovative firms on average publish more, but R&D firms show a stronger propensity to publish if we "control" for the firm dimension: this is consistent with the nature of the R&D firms, particularly interested in basic research; innovative firms are, on average, the largest firms of our sample: controlling for firm size, they loose their primacy in publication.

Now we turn to the analysis of the determinants of the number of collaborations.

The data we collected on firm publications confirm what we sustained in Section 2: collaborations are frequent in the biotech sector. In fact about 83% of the publications done by the Italian biotech firms are made in collaboration with other partners.

We considered the collaborations on an institutional point of view, that is taking into account the institutions of affiliation of the authors: for our purpose there is a collaboration in publication (a co-autorship) if a publication is done by one or more authors belonging to a biotech firm and one or more authors belonging to one or more other institutions. We observed that the co-autorships happen more frequently with universities, than with hospitals and research centres, while collaborations in publications among firms are quite rare.

Our aim is to explore in which way the belonging to different classes of OECD influences the propensity to collaborate in publications. As for the number of publications, we want to make this analysis in a multivariate context, controlling for firm size, assuming that this one has an impact on the propensity to collaborate. In other words, trough a regression analysis on the cross section data, we explore if the belonging to a one OECD class rather then to another is relevant for the propensity to collaborate, even "controlling" for the dimension of the firms, measured like in the previous analysis with the number of employees.

We use a cross-section database, whose individuals are the single publications; the dependent variable is the number of institutions to whom the authors of the publication i belong; the independent variables are the characteristics (like in previous analysis: employees and OECD category) in year t (the year of the publication) of the firms that published the publication i.

Also in this case the dependent variable (the number of collaborations for each publication) is a count variable, then the negative binomial regression is the best technique to adopt.

The results of the estimation are shown in Table 3.

Table 3 Determinants of the number of collaborations. Results of the negative binomial regression on cross-sectional data; years 2001-2005

COVARIATES	Coefficients
Employees	0.00010***
OECD category (benchmark: Targeted)	
Innovative	-0.28112**
R&D	-0.41678***
Active	-0.01208
Dedicated	-0.25596*
Other services	-0.54484*
Out	-0.25319
_constant	1.10171***

Significant at 90 % level; **Significant at 95 % level; ***Significant at 99 % level;
Notes: Dependent variable: number of collaborating institutions for each publication
Number of observations: 1150; Pseudo R^2 (McFadden): 0.0059; LRChi2(7): 27.05
(prob>chi2: 0.0003); Log Likeihood: -2263.1215.

We can say that the targeted firms (the benchmark category) have more collaborations respect the other OECD classes also *ceteris paribus*, considering the dimensional variable. In particular, this is significant (at at least 90% level of significance) for 4 OECD classes: R&D, innovative, dedicated, other services; in the other cases the values are not significant.

These results are consistent with the goals of each typologies of firms. In fact, if we look to the targeted firms, as said in the section 3, their main goal is to find firms engaged in key biotechnology activities, wherever they are currently classified. So, it is obvious that this is the typology of firm with the highest number of collaborations. Also the kind of collaboration should be observed: targeted firms have frequent collaborations with hospitals; this fact can be justified by the core business of this typology of firm, that is the sale of the biotech products and, considering that the main costumers are the hospitals, it sounds as obvious that they have many collaborations with them. On the other side, R&D firms, more devoted to the basic research, collaborate very frequently with universities and research centres.

The multivariate analysis also shows that the dimensional variable is significant, as the number of collaborations grows with the dimension of the firms, but the difference in the number of collaboration among the different OECD classes may not be exclusively attributed to the different average dimensions of the different typologies firms, as this difference holds even controlling for firm size.

6 Conclusions

This work is focused on the analysis of a sector that is unanimously considered as leading in the contemporary knowledge driven economy. More specifically, this paper aims to explore the complexity of the biotech sector, characterized by an high level of knowledge intensity, an high degree of heterogeneity and an high level of dynamism. We tried to manage this complexity trough the analysis of the propensity to disclose knowledge by the different typologies of firms belonging to this sector.

We based on a previous work (d'Amore and Vittoria, 2006, 2008, 2009) that, moving from the problems of definition and classification of the Italian biotech sector, ends up with the creation of an original database, including all the Italian biotech firms, classified according to some typologies, defined by the OECD, that should underline different characteristics of the biotech firms, mainly their fundamental activity. We wanted to know if and how these characteristics influence the behaviour of the biotech firms. To give an answer to this question we analyzed a relevant theme in a knowledge intensive and science based sector, that is the propensity of firms to publish a scientific article and to collaborate with research institutions or with other firms to make such publications; the considered period is 2001-2005.

Our first analysis on the propensity to disclose the knowledge, based on the number of publications made by each firm, showed that the propensity to publish is different according to the different OECD typologies. In particular, we may identify two groups: the first one is composed by biotech firms that are interested to the basic research (innovative, dedicated and R&D) and that show an high propensity to publish; the second one is composed by firms that are more far from basic research and so have a low propensity to publish. Thanks to the econometric analysis, we can observe this behaviour also *ceteris paribus*, considering the dimensional variables, that is also very important in determining the propensity to publish.

We also analyzed the propensity to collaborate, considering the number of collaborations in publications. Also this analysis shows that the firms have a different behaviour according to their typology; indeed this analysis shows a different behaviour of the firms respect to the previous analysis. In particular, we note that firms, like the targeted ones, have a low propensity to publish but have the highest propensity to collaborate. We also observed that different kind of firms, with their different goals, develop different kind of research, therefore have different partners.

These results have some consequences also in terms of policy. A knowledge based economy and particularly a knowledge based sector, like biotechnology, requires fine tuned policies to implement innovative capacity. A key topic of a modern innovation policy is surely the increase of the incentives to collaborate in research and to diffuse the knowledge achievement. Our analysis about the different approach inside the biotech sector to these issues induced us to think that the policies to adopt in relation to the collaboration in research and the dissemination of its results should be different in relation to the different typologies of firms. A complex and differentiated sector requires differently modulated policies.

References

D'Amore, R., Vittoria, P.: Le Biotecnologie in Italia. Ricerca per la costruzione di un Data Base generico per le analisi di settore e di un Repertorio per le policy. Quad. Ric. DISES (Univ. Salerno) 23 (2006)

D'Amore, R., Vittoria, P.: Il nuovo settore delle imprese biotecnologiche. Fonti informative, indicatori statistici e ambiti di policy. L'ind 2, 329–346 (2008)

D'Amore, R., Vittoria, P.: Assessing statistical standards for emerging industries. Applying OECD statistical codes to Italian biotech population lists. World Rev. Sci. Technol. Sust. Dev. 6, 233–243 (2009)

Cockburn, I., Henderson, R.M.: Absorptive capacity, co-authoring behaviour and the organisation of research in drug discovery. J. Ind. Econ. 46(2), 157–182 (1998)

Etzkowitz, H., Leydesdorff, L.: The dynamics of innovation: from National Systems and "Mode 2" to a Triple Helix of university–industry–government relations. Res. Policy 29, 109–123 (2000)

Hicks, D.: Published Papers, Tacit Competencies and Corporate Management of the Public/Private Character of knowledge. Ind. Corp. Chang. 4, 401–404 (1995)

Iorio, R., Labory, S., Paci, D.: The determinants of research quality in Italy: empirical evidence using bibliometric data in the biotech sector. Work Pap. DISES (Univ. Salerno) 3/190 (2007)

Jaffe, A.B.: Real effects of academic research. Am. Econ. Rev. 79, 957–970 (1989)

Mansfield, E., Lee, J.Y.: Intellectual property protection and U.S. foreign direct investment. Rev. Econ. Stat. 78, 181–186 (1996)

Nelson, R.R.: Capitalism as an engine of progress. Res. Policy 19, 193–214 (1990)

OECD, OECD, Biotechnology: economic and wider impact, Paris (1989)

Rosenberg, N.: Why do firms do basic research (with their own money)? Res. Policy 19, 165–174 (1990)

Stern, S.: Do scientists pay to be scientists? NBER Work Pap. 7410 (1999)

Zucker, L., Darby, M., Brewer, M.: Intellectual human capital and the birth of US biotechnology enterprises. Am. Econ. Rev. 88, 290–306 (1998)

Networks of Co-autorship in the Publications of the Italian Biotech Firms: The Role of Different Institutions

Rosamaria D'Amore[1], Roberto Iorio[1,*], and Agnieszka Stawinoga[2]

[1] University of Salerno
  (Department of Economics and Statistics)
  riorio@unisa.it
[2] University of Naples-Federico II

Abstract. In this paper we analyse, through the instrument of the social network analysis, the network of co-authorships in the publications of the firms belonging to the Italian life-science biotech sector. We identify the kind of institutions the authors of the publications are affiliated to (we divide the institutions in universities, research centres, hospitals and firms), then we observe the role of the different institutions inside such network. The analysis shows the central role of the universities but also the importance of hospitals, frequent partners in publications, and of the research centres for their "bridging" role between different institutions.

1 Introduction

Biotech is a strongly science-based sector, where the production of new knowledge and new products is absolutely usual. As nowadays it commonly happens, such production of new knowledge happens as the result of collaborations, among firms or between firms and the institutions devoted to the "production" of science (universities, research centres, etc.). Another characteristic that is nowadays common to many science-based sectors is that at least part of the new knowledge produced by the firm is frequently disclosed through the instrument of the "open science" (publications on scientific journals, conferences, etc.). As a joint result of this two points, firms often publish co-authored papers.

In this paper we analyse the network of co-authorships in the publications of the firms belonging to the Italian life-science biotech sector. We take into consideration the institutions the authors of the papers belong to, classifying them into four categories: universities, research centres, hospitals and firms. The aim of the paper is to analyse the role of the different institutions inside the network of publications, through the instrument of the Social Network Analysis (SNA).

---

* Corresponding author.

R.J. Howlett (Ed.): Innovation through Knowledge Transfer 2010, SIST 9, pp. 235–243.
springerlink.com   © Springer-Verlag Berlin Heidelberg 2011

In this way we try to look inside the structure of collaborations and knowledge exchange of the biotech sector, whose relevance is undoubted, because of its high level of research and innovativeness .

The paper is so articulated: the following section introduces the theme of generation and exchange of knowledge in the biotech sector that often take the form of co-authored papers; in the third section we focus on the use of the SNA to study the phenomenon of the co-authorship; the fourth section describes the sources of the data and shows some descriptive statistics about publications and publishing institutions; the fifth paragraph introduces some methodological principles of the SNA; the sixth paragraph illustrates the results of the SNA; some final considerations conclude the paper.

2 Research Collaborations and Firm Publications in the Biotech Sector

The biotech sector is characterized by a complex knowledge base, where the sources of expertise are widely dispersed. Network relations are frequently used to access this knowledge. As Powell at al. (1996) argue, the locus of innovation will be found in networks rather than in individual firms. Biotech rely mostly on inter-organizational collaborations. There are many organizations where it is possible to find the knowledge, the expertise useful for the firm: it is possible to find it in the universities, in the research centres, in the hospitals. According to the triple helix vision, there are intensive scientific collaborations between universities, industrial organization and government agencies (Etzkowitz and Leydesdorff, 2000; Etzkowitz et al. 2000) and particularly universities may increasingly function as a locus of national knowledge intensive network.

Biotech sector is not only multi-disciplinary, but it is multi–institutional as well. In fact, in addition to research universities, both start-up and established firms, government agencies, non profit research institutes and leading hospitals play a key role in conducting and funding research (Powell et al., 1996). Notwithstanding this articulated institutional framework, universities keep a key role: a large fraction of biotechnology firms originated from universities or at least depend on linkages with universities for their competitive success (Audretsch and Stephan, 1996; Powell et al., 1996; Zucker et al., 1998; Stuart and Sorenson, 2003; Owen-Smith and Powell, 2004; Stuart et al., 2007).

The new knowledge generated by these collaborations not only takes the form of industrial innovations, but it is often disclosed trough the scientific publications: research collaborations often generate co-authored publications. Over two-thirds of even formal alliance partners in this field also appear as partners in scientific publications (Gittelman, 2006) and there is a close link between successful patents and scientific publications in this field (Gittelman and Kogut, 2003; Murray and Stern, 2007).

Given the importance and the frequency of publications done by firms, if the aim is to study the dynamics of the knowledge exchanges and of the innovative networks inside a technological field, considering that data on publications are usually of high quality and easy to access, it is possible to study the publications of the firms.

3 The Use of the Social Network Analysis to Study the Co-author Relationship

The SNA is a tool useful to analyse, in many situations, how individuals or organizations are related. It is a multidisciplinary methodology, developed mainly by sociologists and researchers in social psychology in the 1960s and 1970s. The SNA is based on the assumption of the importance of relationships among interacting units or nodes. Trough the shape of a SNA we can determine a network's usefulness to its individuals, understand the linkages among social entities and the implications of these linkages.

One way of studying such networks in academic research communities is to conduct co-citation analysis, where the links are established through the way authors refer to one others' research and publications (Horn et al., 2004; Lin, 1995). Another good way to study similar networks was observed by Newman (Newman, 2001a, b) who studied co-authorship networks and research collaborations within academic research communities to understand collaboration network patterns and characteristics.

In this paper we adopted the co-authorship analysis rather than co-citation analysis, because the co-authorship more directly reflects the nature and structure of formal relationships among members of a research community (Newman, 2004). The study of scientific collaboration helps to establish groups and work networks that can be analyzed and evaluated through bibliometric techniques and represented in what some authors call co-authorship networks or bibliometric maps. These analyses, applied to the study of co-authorship and collaborative relationships between institutions for scientific publications, allow the existing relations between the social agents responsible for the publications to be identified and represented graphically, setting out the number of members in the network, the intensity of the relationships existing between them and who the most relevant members are with respect to a wide range of measures or indicators.

The peculiarity of our study is that it is conducted at an institutional level: an example of an empirical study, trough the SNA, of co-autorships networks at an institutional level may be found in Chinchilla-Rodriguez et al. (2008)

4 The Data: Sources and Some Statistics

In order to build a database of scientific publications in the biotech sector we made an intersection of two databases: *i*) RP Biotech data base; *ii*) ISI Web of Science.

RP Biotech data base. It is a collection of Italian firms belonging to the biotech sector according to the OECD definition, created by D'Amore and Vittoria (2006, 2008, 2009). This database collects, at the end of 2005, 865 firms. 501 of them are for profit firms, 364 are no profit firms. For this analysis we considered only the 306 life-science for- profit firms[1].

ISI databases, especially the Science Citation Index®, and the web-based version Web of Science® (in the following pages WoS) provide the best source of information to identify the basic research activity across all countries and fields of science. It is a detailed bibliometric database of journal articles and citations of worldwide research literature, that contains 14000 international peer-reviewed scientific and technical journals.

We obtained information about publications of the selected firms, across the period 2003-2005. The record of each publication in ISI-Web of knowledge reports, among other kinds of information, the name of the authors and the name of the institutions the authors belong to. We extracted all the publications where the name of at least one biotech firm appeared among the institutions of affiliation. Then, in order to develop our analysis at the institutional level, we divided the institution in five categories (universities, research centres, hospitals, Italian biotech firms, other firms) and established what category each institution belonged to.

115 of the considered firms made at least one publication during the period 2003-2005. The total number of publications is 1053.

The total number of the affiliation institutions of the authors is 900; besides the 115 Italian biotech firms, we identified 218 universities, 289 hospitals, 134 researcher centres and 114 other firms

The institutional co-operation in publication is very frequent: in 918 on the total number of 1053 publication (87.2%) the authors belong to more than one institution; in the others 135 publications the only institution of affiliation is one of the Italian biotech firms. The average number of institutions per paper is 3.43. There are only two firms which did not write any paper in collaboration.

5 Methodology for the Social Network Analysis

The primary step to represent our data as a network was creating an affiliation network, where the set of actors were composed by the set of publications done by authors affiliated to the Italian biotech firms and the institutions (firms, universities) all the authors of the publications are affiliated to.

We used a categorical variable "Type of institution" to classify nodes in the five institutional categories described in the previous section. The aim of the analysis is to describe the scientific collaboration network among Italian biotech firms at an institutional level. The research design is quite simple: we calculate

---

[1] In the following, for the sake of brevity, we refer to them as Italian biotech firms.

some metrics to quantify the centrality and/or connectivity of each node (institution), then we calculate the average value for each of the five kind of institutions.

The most commonly used metrics in the analysis of centrality are: the degree centrality, the closeness centrality and the betweenness centrality (Freeman 1979).

The degree centrality considers nodes with the higher number of adjacent edges (higher degree). In a collaboration network, if we consider a binary matrix, degree is equal to the number of collaborators an author has. In the case of a valued matrix the degree is equal to the number of collaborations. The number of collaborations is greater than the number of collaborators, as it is possible to collaborate many time with the same collaborator. As the valued degree measures the number of interactions, it seems to give a more interesting information than the "non-valued" degree, and this is the reason why we give more attention to the valued degree in our analysis; anyway to compare values of centralization indices they must be calculated on binary matrix. It must be underlined that, being our analysis at an institutional level, our nodes are not the authors but the institutions the authors belong to and the collaborations happen among institutions.

Closeness centrality is a global metric based on the average length of the paths linking a node to others and reveals the capacity of a node to be reached. Since our whole network is disconnected, we could not obtain the value for this index for whole graph, so we decided to obtain it for the main component of the network.

Betweenness centrality is a metric based on geodesic distances counts; it represents the nodes ability to influence or control communication in the network. The betweenness centrality focuses on the capacity of a node to be an intermediary between any two other nodes.

To understand if the different groups of institutions collaborate within and between each other we examined the homophily of ties in the network. We used the E-I index, which is based on comparing the numbers of ties within groups and between groups. Values of this index can range from -1, when all ties are within members of the group, to 1 when all ties are external to the group. The E-I index can be applied at three levels: the whole network, each group, and each node

To study and represent networks we used following software for network analysis: Ucinet 6.221 (Borgatti, Everett and Freeman, 2002) and NetDraw 2.089 (Borgatti, 2002).

6 The Results of the Social Network Analysis

In this section we present some results of our analysis.

Figure 1 shows the undirected network of co-authorships (900 nodes and 4729 ties) with evidence of different types of institutions and links among. There are five different colours and figures for different types of institutions: Red Circle-Italian biotech company, Blue Square-University, Purple Up Triangle-Hospital, Yellow Box-other company, Green Down Triangle-Research Centre.

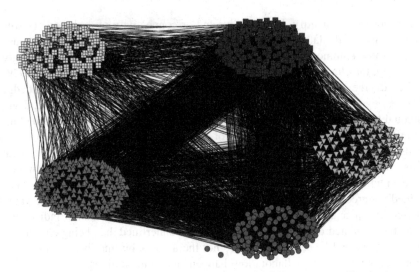

Fig. 1 Undirected network of co-autorships. Evidence of different types of institutions and links among.

For the whole network, the mean degree centrality overall is 10.509 (each subject co-authored papers on average with 10.509 subjects), with a maximum of 132, belonging to the Public University of Milan (Università Statale di Milano). The mean valued degree centrality overall is 15.127 (each subject has done on average 15.127 co-authorships), with a maximum value of 313, always belongs to the Public University of Milan. Considering the institutional categories, the highest value of mean degree belongs to the universities (23.16), even higher than the average value for the Italian biotech firms (22.41), that was expected to be rather high, as in every publication there is at least one of the Italian biotech firms, because our database is based on their publications. Therefore we can say that the universities cover a central, key-role in networks of publications. At a certain distance we find the hospitals (12.61), followed by research centres (10.93) and other firms (6.11). It is interesting to observe that the total number of the hospitals (289) is higher than the total number of universities (218), while the order in terms of mean degree is the opposite: a lot of hospitals collaborate with firms, but the collaboration with every single hospital appears as more occasional than the collaboration with every single university. In fact three universities have the highest values of degree in binary network (the public University of Milan is followed by the University of Turin and University of Rome "La Sapienza"); the first hospital (Hospital San Raffaele, Milan) comes at the fourth place. In the valued network the hospital San Raffaele takes the second place after University of Milan and it is followed by University of Rome "La Sapienza" and University of Turin.

These values and considerations underline, therefore, the prominence of the universities in the biotech system of innovation, respect to other locus where research is done (research centres and hospitals), while collaboration between firms (especially if they belong to different sectors) is quite unusual, at least in publications.

We also calculated, for the whole network, the values regarding betweenness. Considering the average values for each kind of institution, we observe the highest value for the Italian biotech firms (normalized value: 0.566), followed by universities (0.422), research centres (0.077), hospitals (0.074) and other firms (0.018). We can therefore observe two inversions (universities/biotech firms; hospitals/research centres) respect to the order for the mean degree: while co-authorships with research centres are less frequent than with hospitals, research centres are more "able" than hospitals to "bridge" different partners.

With regard to closeness centrality, University of Milan is the closest node to other institutions; among the biotech firms Bracco Imaging is the closest. Considering the average values for each kind of institution, we observe that the mean values of closeness centrality for the five different kind institutions are very similar.

The property of the research centres to bridge different institutions emerges from another kind of statistics, the E-I index: the average value for research centres is 0.711, for hospitals is 0.087: we can interpret such result in the sense that, if a paper is co-authored by a biotech firm with a research centre, there are frequently one or other kind of authors too (low level of homophily); the opposite happens if there is a co-authorship with an hospital (the presence of partners of another institutional type is much less frequent: high level of homophily). The value for this index is next to the maximum (0.995) for the Italian biotech firms: this derives from the very low degree of inter-firm collaborations; this can also explain the high value of this index for other firms (0.634); universities have, on average, a value of 0.271.

7 Conclusions

In recent years Italy has known a process of rationalization, or even drastic cuts, of the public expenditure in the fields of education and research. There is therefore the need to understand in depth the points of strength and weakness of the research and innovation system, in order to operate reasonable choices. In this paper we tried to analyse an important aspect of the Italian system of research and innovation: through the instrument of the SNA we investigated the networks of co-authorships in the publications of the firms belonging to a highly innovative sector, the life-science biotech. We analysed the role and importance of the different kinds of institutions that constitute the Italian system of innovation (Universities, research centres, hospitals and firms). This analysis reveals unambiguously the central role covered by the universities, particularly by the great universities in the Northern Italy (Milan, Turin) and in Rome. There is likely a relationship between this observation and the high number of biotech firm situated in the region of Milan (Lumbardy), Turin (Piedmont) and Rome (Lazio).

The other institutions have an important role too: a lot of hospitals make research in collaboration with firms, publishing the results; the research centres often participate in large and heterogeneous networks, having the role to bridge different institutions; a point of weakness seems to be the infrequent collaborations among firms. On a policy point of view, we may conclude that each kind of institution has its peculiar and fundamental role in the system of innovation, therefore it seems that should be avoided to valorise some institutions penalising others. This kind of study, focused on a country, like Italy, that shares with many advanced countries the condition of a high level of technology, but not at a leadership level, may be for many countries an useful example of such attempts to understand in depth important parts of the innovation system.

References

Audretsch, D.B., Stephan, P.E.: Company-Scientist locational links: the case of biotechnology. Am. Econ. Rev. 86, 641–652 (1996)

Borgatti, S.P.: Netdraw Network Visualization. Analytic Technologies, Harvard (2002)

Borgatti, S.P., Everett, M.G., Freeman, L.C.: Ucinet for Windows: Software for Social Network Analysis. Analytic Technologies, Harvard (2002)

Chinchilla-Rodríguez, Z., Moya-Anegón, F., Vargas-Quesada, B., Corera-Álvarez, E., Hassan-Montero, Y.: Inter-institutional scientific collaboration: an approach from social network analysis. In: PRIME International Conference (2008)

D'Amore, R., Vittoria, P.: Le Biotecnologie in Italia. Ricerca per la costruzione di un Data Base generico per le analisi di settore e di un Repertorio per le policy. Quad. Ric. DISES (Univ. Salerno) 23 (2006)

D'Amore, R., Vittoria, P.: Il nuovo settore delle imprese biotecnologiche. Fonti informative, indicatori statistici e ambiti di policy. L'ind 2, 329–346 (2008)

D'Amore, R., Vittoria, P.: Assessing statistical standards for emerging industries. Applying OECD statistical codes to Italian biotech population lists. World Rev. Sci. Technol. Sust. Dev. 6, 233–243 (2009)

Etzkowitz, H.A., Leydesdorff, L.: The dynamics of innovation: from National Systems and "Mode 2" to a Triple Helix of university–industry–government relations. Res. Policy 29, 109–123 (2000)

Etzkowitz, H.A., Webster, C., Gebhardt, B., Terra, R.C.: The future of the university and the university of the future: evolution of ivory tower to entrepreneurial paradigm. Res. Policy 29, 109–123 (2000)

Freeman, L.C.: Centrality in Social Networks: Conceptual Clarification. Soc. Netw. 1, 215–239 (1979)

Gittelman, M.: National institutions, public-private knowledge flows, and innovation performance: A comparative study of the biotechnology industry in the United States and France. Res. Policy 35, 1052–1068 (2006)

Gittelman, M., Kogut, B.: Does good science lead to valuable knowledge? Biotechnology firms and the evolutionary logic of citation patterns. Manag. Sci. 49, 366–382 (2003)

Horn, D.B., Finholt, T.A., Birnholtz, J.P., Motwani, D., Jayaraman, S.: Six degrees of Jonathan Grudin: a social network analysis of the evolution and impact of CSCW research. Paper presented at the Computer Supported Cooperative Work (2004)

Lin, C.H.: The cross-citation analysis of selected marketing journals. J. Manag. 12, 465–489 (1995)

Murray, F., Stern, S.: Do formal intellectual property rights hinder the free flow of scientific knowledge: an empirical test of the anti-commons hypothesis. J. Econ. Behav. and Organ. 63, 648–687 (2007)

Newman, M.E.J.: Scientific collaboration networks. I. Network construction and undamental results. Phys. Rev. E 64, 0161311–0161318 (2001a)

Newman, M.E.J.: The structure of scientific collaboration networks. Proc. Natl. Acad. Sci. 98, 404–409 (2001b)

Newman, M.E.J.: Coauthorship networks and patterns of scientific collaboration. Proc. Natl. Acad. Sci. 101(S1), 5200–5205 (2004)

Owen-Smith, J., Powell, W.: Knowledge networks as channels and conduits: the effects of spillovers in the Boston biotechnology community. Organ. Sci. 51, 5–21 (2004)

Powell, W., Koput, K., Smith-Doerr, L.: Interorganizational Collaboration and the locus sof innovation: networks of learning in biotechnology. Adm. Sci. Q 41, 116–145 (1996)

Stuart, T.E., Sorenson, O.: The geography of opportunity: Spatial heterogeneity in founding rates and the performance of biotechnology firms. Res. Policy 32, 229–253 (2003)

Stuart, T.E., Ozdemir, S.Z., Ding, W.W.: Vertical alliance networks: the case of university-biotechnology pharmaceutical alliance chains. Res. Policy 36, 477–491 (2007)

Zucker, L., Darby, M., Brewer, M.: Intellectual human capital and the birth of US biotechnology enterprises. Am. Econ. Rev. 88, 290–306 (1998)

Stitching an Organisation's Knowledge Together–Communities of Practice as Facilitator for Innovations Inside an Affiliated Group

M.A. Weissenberger[1] and Dominik Ebert[2]

[1] University of Kassel, Chair for Innovation and Technology Management, Germany
[2] Robert Bosch GmbH, Stuttgart, Germany

Abstract. A major challenge for innovation inside an affiliated group, such as the Robert Bosch GmbH (Bosch), is to enable an effective knowledge exchange between its subsidiaries. One of the driving forces in Bosch's internal knowledge transfer is communities of practice (CoP). During the last fifteen years the company has established these experience exchange groups around its most important technical topic areas. By conducting a qualitative study we wanted to find out how these groups influence the handling of knowledge. In addition to the expected results on the identification and the transfer of knowledge, we also discovered that communities of practice advance innovation in an indirect way. Bosch files 15 patents per working day. To establish an evidence, whether communities of pratice contribute to this knowledge creation, we conducted a second study. Within the study, we analysed the relationship between inventions and communities of practice quantitatively and found a significant correlation.

1 Chances and Challenges in an Affiliated Group

The challenge today's companies face is to ensure the exchange of knowledge between their employees to fulfill the need to innovate. Especially in affiliated groups with a diversified product portfolio and widely autonomous associate companies, proper functioning of knowledge management is an essential success factor. Bosch is one such highly diversified multinational company. Besides its established core business as the world's largest automotive supplier the company has captured leading market positions for power tools, renewable energy and

R.J. Howlett (Ed.): Innovation through Knowledge Transfer 2010, SIST 9, pp. 245–252.
springerlink.com

security systems, to name but a few The company has subsidiaries in more than 50 countries and approximately 260 locations worldwide. It had a turnover of about 38 billon euro in the year 2009.

The diversity of the Bosch group is both an opportunity and a challenge. Overcoming the difficulties of cross-divisional knowledge transfer can lead to highly innovative products. With more than 3,800 new patent applications in 2009 the company is Germany's largest patent applicant and one of the most innovative enterprises in the world. However, as an innovative company it further needs tools that stitch together the organization's knowledge in an adequate way.

It is a known fact that the knowledge developed for a specific context can not be passed directly to another context. This is called the 'transfer problem' (Weissenberger-Eibl 2005). Following are some of the potential barriers leading to the transfer problem:

- *Lack of transparency:* In affiliated groups it is often difficult to get an overview about who is working on which topics. This may lead to the duplication of work in different divisions.
- *Cognitive barriers:* In some cases employees prefer to develop a new solution instead of taking an existing one ('not invented here syndrome'). In other cases local optimization may reduce the global result ('it's not my job phenomena'). Sometimes individuals or departments even refuse a knowledge transfer to protect their status as experts ('knowledge is power attitude').
- *Missing abilities:* The ability of employees to recognize the value of information available in one business unit, to assimilate it, and to apply it to their business need. This is called the absorptive capacity (Cohen and Levinthal 1990).

Therefore effective knowledge management tools have to be used to reduce these barriers, thereby benefitting from the diversity of a company. A community of practice (CoP) is one such knowledge-transfer tool that helps to reduce the transfer problem.

2 Communities of Practice as Knowledge Bridges

In its original conception, communities of practice are groups where people can learn in a social context. The idea behind the concept is that the transfer of implicit knowledge can succeed through personal interaction. Such groups have been existing in many cultures for a long time (Lave and Wenger 1991). Brown and Duguid (1991) took this idea and adopted it to business context. They regard CoP as an instrument which amalgamates working, learning and innovation. In companies these groups connect employees from different business divisions and therefore enable cross-divisional exchange of knowledge (see Figure 1).

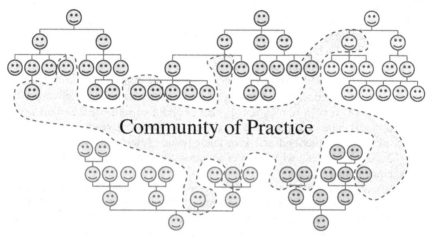

Fig. 1 Overcoming organizational boundaries through CoP

In the following years, communities of practice attract more and more attention in the upcoming field of knowledge management. Therefore, there are an increasing number of publications giving different views. However, sometimes there are contradicting recommendations for the implementation of CoP in companies. While the first extensive conception about communities of practice describes them as completely informal groups (Wenger 1998), a later work suggests to 'cultivate' them in an organisation by establishing adequate surrounding conditions (Wenger et al. 2002). Today, there are CoPs with clear objectives and targets which they are expected to fulfil. This is termed as a best practice in a business context (McDermott and Archibald 2010).

By now, CoP is a popular knowledge management instrument not only in single companies, but also in affiliated groups. An expert survey shows that CoP is an often applied and an appropriate method to make different forms of knowledge in company networks accessible (Weissenberger-Eibl 2006). As literature already gives different recommendations for organising the expert groups, the diversity of communities of practice in companies is even bigger.

Bosch belongs to the group of pioneers where first set of communities of practice were established in business context in the mid nineties. While some groups were found bottom-up, others have been installed by management. The groups have different degrees of formalization, different targets and different ways to organise the experience exchange. Today there are more than 150 active CoP which are built on different roots, pursue different goals and use different organisational designs. Therefore, they are of interest for the study to find out relationships between the knowledge transfer through CoP and innovation.

3 Empirical Research on Communities of Practice

We have conducted an empirical study, to explore the functioning of successful communities of practice and their effects on intra-organisational knowledge-based collaboration. The current state of research does not offer more than a couple of ideas suggesting possible changes in an organisation's knowledge culture. However, they do not explain wherefrom the changes come (Oliver and Kandadi 2006, Sollberger 2008, Liebowitz 2008). Therefore, we decided to start with a qualitative survey to investigate the effectiveness of CoPs.

We have conducted 22 semi-structured interviews with experts from Bosch's communities of practice. The respondents include predominantly spokespersons and members of some communities of practice and also some of their colleagues, supervisors and experts that accompany the groups for many years. The interviews were recorded, transliterated and kept anonymous. Afterwards we have categorised the statements topic wise when conducting a combined qualitative and quantitative content analysis.

In line with our expectations we found a strong relationship between communities of practice and knowledge transfer. From the beginning it is believed that CoP enables the informal exchange of knowledge in a social context (Lave and Wenger 1991). The main reason that triggers this exchange is the personal relationships among the CoP members which are build during CoP meetings. This helps to reduce cognitive barriers since the persons seeking information learn the importance of others' work. In general, groups that spent more time on socialising events were more successful in exchanging knowledge beyond the meetings. In some cases, the community members develop a strong sense of belongingness, which was the base for a broad cross-divisional knowledge exchange.

Besides the direct know-how transfer, the prime benefit of having a community of practice is to know whom to contact for which problem. It is also found out that the outperforming communities of practice spend noticeable amount of time for discussions during the meeting. Instead of cramming the agenda with too many agenda points, they normally focused on one topic that was of interest.

Interestingly we also found a couple of long term changes in the field of knowledge management which helped better usage of knowledge across the organisation. For example, the collaboration of community members from different business units helped to standardise technical terms that had been used differently across the organisation. Therefore, it is now easier to understand the views and findings of experts across business units through relevant documents such as research and development reports. Moreover, the CoP members improved their ability to gauge the relevance of sources and to find the required knowledge within the company. Overall, communities of practice seem to enlarge the absorptive capacity of their members. However, it is difficult to assess the relation between communities of practice vis-à-vis innovations.

We did a second study to test whether there is a relationship between knowledge transfer through CoP and its members' ability to innovate in their ordinary business environment. This time, we analysed the invention disclosures of 405 engineering departments as well as their participation in CoP. For each department we calculated two metrics, the number of inventions per capita and the amount of memberships in CoP per capita. As we did not know whether the assumed correlation is linear or not, we decided to calculate an ordinal scale correlation. The data did not meet the demand of equidistant spaces between the measuring points which is a requirement for the calculation and meaningful interpretation of Spearman's rho (Bortz et al. 2008). Therefore, we decided to use the more conservative rank correlation coefficient Kendall's tau (Kendall 1938).

As a result of the study we calculated a tau value of 0.121 on a level of significance of 0.01. This means that there is a small correlation between the participation in CoP and innovation. With a likelihood of 99%, we can be sure that the measured correlation is not random. The effects of communities of practice on a department's capacity to innovate may be very different depending on the way the experience exchange groups are organised. Hence, we will give some recommendations about how to use them as catalysers for new ideas.

4 Practical Implications

The empirical research shows that CoP is an instrument that can trigger innovation inside affiliated groups. However, there is big variance between the practices of each CoP leading to large innovations in one department and less innovations in another department. An organisation has several options to set the course for innovation-creating communities of practice.

First of all, the community members have to define the targets they want to achieve. This might be in terms of experience exchange, resolving practical problems or a cross-divisional activity such as internal process standardisation. In the latter cases it may be useful to officially assign target responsibility to the community of practice (McDermott and Archibald 2010). Depending on the groups task there are various ways to steer the group.

A key player of every community of practice is its spokesperson. He is critical for the group's success as he can govern the community of practice into different directions (Weissenberger-Eibl and Ebert 2010). One of his instruments is planning of the agenda. We have identified about 15 different possible program items for the community meetings and their impact on the handling of knowledge. The creation of new ideas gets fuelled by agenda items that make running activities and new ideas transparent. Possibilities to do so are regular reports of conferences where community members participated, a status report about running activities in the business units or an experience exchange about new interesting scientific papers. One of the CoP groups that we studied even had a special agenda item at the end of their meeting where new internal and external inventions were discussed.

Sometimes it may be interesting to invite external guests such as university professors or suppliers. In that case, an organisation can support its communities of practice by creating and communicating adequate surrounding conditions. As most CoP members will come with very limited knowledge about data security issues, they need clear guidelines for the external knowledge exchange for example when dealing with employees of research centres. Moreover, in affiliated groups there may be uncertainty about the knowledge transfer between the mother company and its joint ventures. Communicating the official company rules will make it significantly easier for the CoP members to use external knowledge sources.

A further aspect that our research revealed is the critical role of the communities of practices' members' supervisors. They influence the priority for participation in the experience exchange groups. Therefore, an official commitment of the company's top management is necessary which directs the cross-divisional knowledge exchange. At Bosch the engineering executive management approved a document that allows the associates to spend part of their working time for the cross-divisional knowledge transfer.

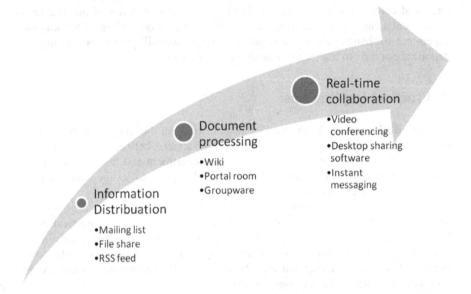

Real-time collaboration
•Video conferencing
•Desktop sharing software
•Instant messaging

Document processing
•Wiki
•Portal room
•Groupware

Information Distribuation
•Mailing list
•File share
•RSS feed

Fig. 2 Supporting information and communication technology

Another possibility for an organisation to enable CoP is to provide an innovation-supporting information and communication technology infrastructure. From 'media richness theory' we know that the communication tools that a group needs, depends on the complexity of its tasks (Daft et al. 1987). Some communities of practice might need frequent face-to-face meetings encouraged by complex collaboration tools, such as portal rooms, wikis or groupware applications. Other groups that focus more on information distribution might need a possibility to start in-house RSS feeds or simply a well-structured file sharing system. International working communities of practice may be supported best with easy to handle video conferencing equipment and desktop sharing software. Figure 2 shows some possible tools for communities of practice depending on the groups' tasks.

5 Conclusion

We have investigated the functioning and the effects of CoP inside an affiliated group. Our research shows that communities of practice are not only a tool that helps to transfer implicit knowledge but also to trigger innovation. The creation of new ideas is mainly based on a better transparency of the existing knowledge inside the company and a couple of long-term changes in the use of knowledge in the day-to-day work of the CoP members. These long-term effects include the development of a common language for technical terms inside the company network, a better understanding of the relevance of sources and improved abilities to find relevant knowledge inside the affiliated group. For the first time in community of practice research, we proved quantitatively a relationship between the participation in those groups and the number of a department's inventions. Therefore, we consider communities of practice as an important tool for knowledge creation.

Referring to the empirical research result at Bosch we can give a couple of recommendations for the successful organisation of communities of practice in an organisation. First of all, the stakeholders of the CoP should have clarity on the objectives and expectations of the CoP. In some cases it may be necessary to assign explicit objectives to the group. Afterward, the spokesperson of the CoP can orient the group towards its aims for example by the composition of the meeting agenda. The task of the organisation's knowledge management is to ensure adequate surrounding conditions. Important aspects of these surrounding conditions are a clear communication of the rules for external knowledge exchange, the commitment of the top management for intra-organisational knowledge exchange and a mixture of information and communication tools.

References

Bortz, J., Lienert, G.A., Boehnke, K.: Verteilungsfreie Methoden in der Biostatistik. Springer, Berlin (2008)

Brown, J.S., Duguid, P.: Organizational Learning and Communities-of-Practice: Towards a Unified View of Working, Learning and Innovation. Organ. Sci. 2(1), 40–57 (1991)

Cohen, W.M., Levinthal, D.A.: Absorptive Capacity: A New Perspective on Learning and Innovation. Adm. Sci. Q 35(1), 128–152 (1990)

Daft, R.L., Lengel, R.H., Trevino, L.K.: Message equivocality, media selection, and manager performance: Implications for information systems. MIS. Q 11(3), 354–366 (1987)

Kendall, M.G.: A new measure of rank correlation. Biom. 30(1-2), 81–93 (1938)

Lave, J., Wenger, E.: Situated learning. Legitimate peripheral participation. Cambridge University Press, Cambridge (1991)

Liebowitz, J.: Think of others in knowledge management: making culture work for you. Knowl. Manag. Res. Pract. 6(1), 47–51 (2008)

McDermott, R., Archibald, D.: Harnessing Your Staff's Informal Networks. Harv. Bus. Rev. 88(3), 82–89 (2010)

Oliver, S., Kandadi, K.R.: How to develop knowledge culture in organizations? A multiple case study of large distributed organizations. J. Knowl. Manag. 10(4), 6–24 (2006)

Sollberger, B.A.: Wissenskultur. Erfolgsfaktor für ein ganzheitliches Wissensmanagement. Haupt Verlag, Bern (2006)

Weissenberger-Eibl, M.A.: Wissensmanagement in Netzwerken für Klein- und Mittelbetriebe. In: Ciesinger, K.-G., Holwaldt, J., Klatt, R., Kopp, R. (eds.) Modernes Wissensmanagement in Netzwerken, Perspektiven, Trends und Szenarien, DUV, Wiesbaden (2005)

Weissenberger-Eibl, M.A.: Wissensmanagement in Unternehmensnetzwerken. Konzepte, Instrumente, Erfolgsmuster. Cactus Group Verlag, Kassel (2006)

Weissenberger-Eibl, M.A., Ebert, D.: Die kritische Rolle des Community-of-Practice-Sprechers. Wissensmanagement 12(5), 24–26 (2010)

Wenger, E.: Communities of practice. Learning, meaning, and identity. Cambridge University Press, Cambridge (1998)

Wenger, E., McDermott, R., Snyder, W.: Cultivating communities of practice. A guide to managing knowledge. Harvard Business School Press, Boston (2002)

Living Labs Are Innovation Catalysts

Maurice D. Mulvenna[1, *], Birgitta Bergvall-Kåreborn[2], Brendan Galbraith[1], Jonathan Wallace[1], and Suzanne Martin[1]

[1] TRAIL Living Lab, University of Ulster, Newtownabbey, UK
md.mulvenna@ulster.ac.uk
[2] Luleå University of Technology, 971 87 Luleå, Sweden

Abstract. Living labs are increasingly facilitating new ways to stimulate innovation. They offer the possibility to catalyse how innovation can be carried out, focusing on user communities supported by information technology. However, living labs are poorly understood by the business community, in particular by small to medium companies who arguably have the potential to benefit most from accessing the services provided by living labs. This position paper sets out the context for the rising popularity of living labs, explaining how public-private-academic partnerships offer new ways or carrying out innovation activities that are increasingly user-orientated. The paper also discusses the issues and opportunities arising from this new approach.

1 Introduction

In the economy, new products and services are created and existing ones are changed to meet the needs of the marketplace. Knowledge and technology transfer activities involving universities support and add value. While this innovation 'eco-system' has been established for some time and at first glance seems to be well understood, there are major forces at work that are changing the ecosystem beyond recognition. These forces and the impact of them on knowledge transfer and innovation service provision are discussed in this paper. In section 2, the models and processes of support for companies are outlined and how this support has evolved in the post-industrial information society is described. In section 3, how innovation has evolved towards more network friendly cyclical models of innovation is described and user-driven innovation is explained. In section 4, living labs are described, their history, methods, processes, services, policy background and philosophy is outlined. Section 5 discusses the issues and opportunities arising from living labs before the paper offers conclusions in section 6.

---

* Corresponding author.

R.J. Howlett (Ed.): Innovation through Knowledge Transfer 2010, SIST 9, pp. 253–264.
springerlink.com   © Springer-Verlag Berlin Heidelberg 2011

2   The Evolution of Support for Companies

The mode of provision of support to companies who are trying to increase their capability and capacity for innovation in product and/or service development is typically facilitated through a regional or national economic development agency. The support is normally available though thematic programmes aimed at increasing research and development activity in companies, ranging in size from large multi-nationals to Small-to-Medium Enterprises (SMEs) (EU, 2003).

Support available may include reduced costs for office space, and more sophisticated provision including units in science parks and business incubators, where additional supporting services are available. These can include infrastructure support, for example, access to low-cost, super-fast broadband, as well as access to mentoring for business development functions such as, for example, marketing development or accessing venture funds. Normally new companies, often high-tech based, who show potential for high-growth are the target.

At both national and regional levels, economic development agencies have used the kind of support outlined above to stimulate the development of an economy based upon the creation of wealth around using and manipulating information or knowledge, variously called the 'information society' (Machlup, 1962) or the 'knowledge economy' (Drucker, 1969).

These models of support rely on using physical locations with accompanying services, where the governance is usually provided through an economic development agency or its subsidiary. Supporting services are bought in and provided to the on-site client companies. This does help companies establish their 'bricks-and-mortar' presence, and provides a safe harbour for new companies as they seek to develop and then prove their business model and products or services.

However, the information society in which these companies seek to thrive has very different characteristics to those that existed a relatively short time ago, for which the science parks and incubators were created. The biggest difference is the scale of opportunity arising from globalisation of markets and economies. Another characteristic that has changed is the use of technology. Information and Communications Technology (ICT) has revolutionised our world as technological progress in computers, networks and new media computing impact on society and business.

In the information society, electronic information, products and services can be developed collaboratively using the Internet. The resulting digital information, products and services can be bought and sold and electronically delivered to the customer.

Where a company is located may not be so important, and may even be a burden for some businesses (Rifkin, 2000). Even concepts such as geo-located clusters of unusual competitive success recognise the value of partnerships (Porter, 2007). What is important is that the company has access to the human capital (Becker, 1994) that it requires in order to deliver value to the customer at a profit to the business. A second item of prime importance is that the company is responsive to the market, and to the needs of its customers.

In effect the company becomes a kind of innovation engine that facilitates collaboration of human capital into the production of digital information, goods and services that meets the needs of the other set of people with whom it need to collaborate, its customers.

This concept is not new, and draws upon the work on innovation adoption and diffusion and innovation life cycles (Rogers, 2003). The 'crossing the chasm' concept (Moore, 1998) further developed the innovation diffusion idea by developing ideas on how to cross the chasm between early adopters and mainstream market users for high-tech products.

This extension of the innovation diffusion paradigm predates the ubiquity of the Internet and it therefore does not take into account the ease with which companies can market electronically to customers, customers can talk to companies about their products and services, and most importantly, customers can talk to customers about company's offerings using the Internet as an open discussion forum.

This social connectedness arising from the evolution of our information society is called the 'network economy' (Kelly, 1998) and highlights the networked interconnectedness that ICT affords our new knowledge economy, especially using social media. In the networked economy, the network is the channel for many business functions for conventional products and services and potentially all business functions for information-based electronic products and services.

The value in the network economy is inherent in the network, not within individual companies, and it is the network economy and the underlying ICT that facilitates new economies of scale that foster new business models and processes that are often employed to get an offering to market rapidly, and to grow sales or equivalent measures of success exponentially; example, "Over 1,000,000 calls placed from Gmail in just 24 hours!" (Google, 2010).

In the network economy, companies converse with their customers, suppliers and all the other network stakeholders in a more normalised space where all voices are more or less equal. In this space, a single customer voice can grow into a significant groundswell of complaint that can cause significant problems for a company, while a senior company officer may have to earn respect from a company employees or customers.

The key impact of the network economy in our information society age, however, is the realisation that network connectivity affords significant business advantage. The more inter-connected your business is with your customers, suppliers and staff, then the more value your business can derive from the network.

Companies therefore need to be open to the networks in which they operate, at many levels. They are moving away from proprietary systems, processes and software that act as barriers to open communication, towards new concepts of engagement with their customers and other stakeholders.

This evolution in the provision of the type of support for companies is just one of a number of areas of changes wrought in the business environment. Another area of change is around the concept of innovation and how that concept is utilised, and this is discussed in the next section.

3 Innovation

Innovation is a "change in the thought process for doing something, or the useful application of new inventions or discoveries" (McKeown, 2008). Historically, innovation has been characterised as a linear process, driven and controlled by the industrial developers of products for the marketplace. In the information society, it is increasingly seen as a catalyst for growth and competitiveness and has been enthusiastically promoted at regional, national and international level and included in new policy formulation. However, it has evolved from a linear process more towards a network model involving partners supporting innovation, often focused on cycles of innovation activity. These partnerships of interaction can take many forms but one model that is increasingly being used is a triple-helix model of engagement (Etzkowitz, 2003), where the three types of stakeholders are industry, government and academia, often also called academic-public-private partnerships. This model and its variants works well within the concept of network economy, facilitating *ad hoc* or permanent partnerships as required, focused on problem solving and commercial exploitation of intellectual property and know-how arising from the partnerships. The most interesting facet of these kinds of models for engagement is the active participation of academia, cementing a role for entrepreneurial universities in innovation activities that are becoming increasingly influenced by network economy concepts.

However, arguably the greatest change in how we should consider innovation in the context of open innovation, where it is claimed that innovation can thrive when a company utilises a network of partnerships beyond its traditional internal resources (Chesbrough, 2003). The partnerships can facilitate technology development, licensing of existing intellectual property, access to external capital as well as sales and marketing partnerships. A typology for open innovation is emerging, encompassing different strategies; for example, innovation seeker, innovation provider, intermediary and open innovator (Gianiodis et al. 2010). However, while there is a significant volume of academic publishing activity that embraces open innovation as a new paradigm to help describe innovation in our networked knowledge economies, there are also those that assert that open innovation is 'old wine in new bottles' (Trott and Hartmann, 2009). In their paper, they argue that while closed innovation principles are indeed limited, companies today no longer adhere to these closed innovation principles but rather have long ago changed their mindsets to think beyond their company's borders. These 'closed' principles are:

1. "The smart people in our field work for us.
2. To profit from Research and Development (R&D), we must discover, develop, produce and ship it ourselves.
3. If we discover it ourselves, we will get it to market first.
4. If we are the first to commercialize an innovation, we will win.
5. If we create the most and best ideas in the industry, we will win.
6. We should control our Intellectual Property (IP) so that our competitors do not profit from our ideas."

Contrast these closed principles with the equivalent open innovation principles, which are:

1. "Not all of the smart people work for us so we must find and tap into the knowledge and expertise of bright individuals outside our company.
2. External R&D can create significant value; internal R&D is needed to claim some portion of that value.
3. We don't have to originate the research in order to profit from it.
4. Building a better business model is better than getting to market first.
5. If we make the best use of internal and external ideas, we will win.
6. We should profit from others' use of our IP, and we should buy others' IP whenever it advances our own business model."

Trott and Hartmann argue that Chesbrough has created a "false dichotomy by arguing that open innovation is the only alternative to a closed innovation model. We systematically examine the six principles of the open innovation concept and show how the Open Innovation paradigm has created a partial perception by describing something which is undoubtedly true in itself (the limitations of closed innovation principles), but false in conveying the wrong impression that firms today follow these principles."

More useful is their observation that open innovation is still inherently a linear concept, although technology and ideas can 'move' in and out at all stages. They argue further: "modern innovation models should once and for all get rid of the notion of linearity in the innovation process" and adopt cyclical concepts of models such as Cyclic Innovation Model (Berkhout et al. 2007). In this model, explicit feedback paths are added as well as feed forward options. By harnessing these paths in a cyclical architecture, a dynamic system is created to model an organisation or network and its innovation activities.

This section on innovation has described how innovation has evolved and described how more network friendly cyclical models of innovation offer promise. The use of models such as triple-helix explicitly recognises the value of partnerships and the different stakeholders and their roles in facilitating and supporting innovation. However, there is one other stakeholder who has occasionally been fully involved in innovation processes around product and service creation and development, but is only now becoming recognised as perhaps the ultimate stakeholder in these processes. That stakeholder is the user, and the following subsection describes user-driven innovation.

3.1 User-Driven Innovation

The importance of users in the design process for product and service innovation has long been recognized. It is natural to involve users, and indeed the resulting quality and appropriateness of a product or service suffers in some way if users are not involved in some way in the processes that together make up the design stages.

User Centred Design (UCD) is an approach that puts the customer or user at the centre of the design process (Rubin, 1994). UCD has been successfully used in many product designs and is supported by standards (ISO-13407, 1999). The key aim in UCD is to learn what product or service is best suited to meet the needs of the user, and the intended benefit arising from the application of the approach is better usability in the resulting designed product or service. There is a long tradition of user-orientated, experience-based approaches developed to realise these aims and benefits, including user experience (Norman et al. 1995), contextual design (Beyer and Holtzblatt, 1998), action research (Lewin, 1946), and cooperative (participatory) design (Bødker et al. 1993).

There are also fresh approaches emerging such as crowdsourcing (Howe, 2006) where design challenges can be opened out to a broad population of people or the wisdom of crowds (Surowiecki, 2004) where it is posited that groups of free thinking people are likely to make certain types of decisions better than an individual. But arguably the most interesting is the lead user concept (von Hippel, 1986). This concept stems from research finding that it is often the user who can realise a commercially successful product or service, rather than the producers (von Hippel, 1988), and that a particular type of user, the lead user, may be responsible for the majority of the innovative thinking (Urban and von Hippel, 1988).

Many of these new approaches in user-centred innovation are facilitated by ICT, and can thrive in a network economy society. The developers of products and services now have extremely powerful, useful and potentially profitable techniques and approaches that are centred on ICT-supported innovation processes that embrace the customer, citizen or user. However, while there are models for engagement in innovation partnerships, such as triple-helix, until recently the support has been focused on science parks, business incubators and other activities more related to supporting fledgling new companies than partnerships that support research and development and innovation activities around new ideas tested with users. A new paradigm of support has emerged that extends the triple-helix model to involve users, and indeed its name reflects its philosophy to create a research laboratory wherever the users are testing products and services; in effect, a living lab.

4 Living Labs

The architect and academic, William J. Mitchell, created the concept of living labs. Mitchell, based at MIT, was interested in how city dwellers could be involved more actively in urban planning and city design (Mitchell, 2003). The ideas of citizen involvement in the design process was subsequently taken up and developed further in Europe by various research communities. A small number of living labs, created across Europe in 2005, primarily from the Computer Supported Cooperative Working (CSCW) research community, formed the European Network of Living Labs (ENOLL) in 2006. Successive waves of new living labs have since been created and, in 2010, there are, for example, 15 living labs in the UK and over 250 living labs across Europe and beyond.

The ENOLL living labs recognise, as did Mitchell, that technology, in particular ICT plays a powerful catalytic role in user engagement and most of them are focused on using technology to support user engagement, research novel ways of engaging with users, and communicate findings rapidly and accurately using low-cost, mass-adopted tools such as social networks.

Living labs are "collaborations of public-private-civic partnerships in which stakeholders co-create new products, services, businesses and technologies in real life environments and virtual networks in multi-contextual spheres" (Feuerstein et al. 2008). A simpler definition is "a collection of people, equipment, services and technology to provide a test platform for research and experiments" (FarNorth, 2010). Some position living labs as a kind of technological test-bed (Ballon et al. 2005) while others classify them as "innovation methodologies" (Kallai and Bilic-ki, 2008).

It is apparent from an examination of the living labs that many have a particular niche in which they operate. Some labs are region-based, others focus on a particular product family for example, automotive design, while others seek to address particular societal needs in, for example, healthcare. However, the use of technology to engage and support users as early as possible in product and service development is the common denominator for all of them.

Many living labs, and indeed other research organisations, are experimenting with variations of innovation techniques. In Arizona State University, the InnovationSpace is a research and development lab that "seeks to commercialize product design concepts that are progressive, possible and profitable" (Rothstein and Wolf, 2005). The lab also aims to institutionalise trans-disciplinary collaboration and focus on knowledge transfer activities with private partners to commercialise design concepts. The innovation model used in InnovationSpace is integrated innovation, and a key stage in this model is satisfying consumer demand. Another model of innovation is networked innovation, which "involves combining ICT with explicit collaboration in consortiums through an innovation broker" and "ICT driven innovation that involves connecting organizations, knowledge and resources in collaborative structures and consortiums with the specific aim to deliver individual and collective value." (Van Buuren et al. 2009).

How living labs actually work centres on methods, processes and services utilised to translate the philosophy into engagement. The methods encompass approaches, tools and techniques that often make use of advanced and innovative application of ICT to create and sustain dialogues with users, for example analysis of system logs or automatically collected behavioural data, ethnographic research, questionnaires, focus groups, and observation (Følstad, 2008). The processes are varied but can be described along a development spectrum from the creation of ideas, engagement with user communities and other stakeholders, collection of data using a variety of methods usually facilitated by ICT, and the evaluation of results as well as the methods employed. These can be summarised as co-creation, exploration, experimentation and evaluation (Pallot, 2009).

Another useful perspective on innovation process is the innovation value chain (Hansen and Birkinshaw, 2007). The innovation value chain is viewed as an end-to-end process encompassing three main stages: idea generation, conversion and diffusion, with conversion including both selection of ideas and the subsequent development of them. While the innovation value chain described is generally seen as being controlled by a commercial organisation, the concept can arguably be said to stronger if the value chain is comprised of a variety of triple-helix stakeholders, each bringing their organisation's strengths to the process. In addition to these phase processes, other aspects that make living labs different from traditional research and development innovation labs have been described. These include openness, influence, realism, value and sustainability (Bergvall-Kåreborn et al. 2009). The concept of openness is valuable in living labs as it promotes open communication within and without the stakeholder groups in all development phases. However, there are problems in promoting openness while retaining, for example, IP rights. The principle of realism is a critically important one as it relates to the promotion of concepts where consensus is reached between stakeholders.

If living labs are to be understood by the broad community that comprises all the varied stakeholders and users (and to be successful, they must be clearly understood), then what they do has to be presented to that community in a transparent way. This is particularly true as each living lab has its own particular niche and the message of engagement may be drowned out by the use of language that can be new and potentially confusing to some of the stakeholders (co-creation, networked open innovation, harmonisation cubes, participatory design, etc.).

Services are a useful way of presenting the stakeholder with a set of 'competencies' with which the living lab is familiar. A living lab can be seen as a "service providing organization in the topic of R&D and innovation" with a set of resources including: areas of competency, local partners and stakeholders, ICT infrastructure, operational methodology and administrative resources (Molinari, 2008). Services in living labs have been listed as co-creation, integration and data preparation (Feuerstein et al. 2008).

Living labs operate within a policy framework in Europe. This framework has evolved, supported by ENOLL and the European Commission (European Commission, 2010), who see a role for living labs, particularly as components in the Competitiveness and Innovation Programme (CIP, 2010). There is a broader policy framework in which living labs could be situated, termed Territorial Cohesion (European Commission, 2008). This is defined as "a situation whereby policies to reduce disparities, enhance competitiveness and promote sustainability acquire added value by forming coherent packages, taking account of where they take effect, the specific opportunities and constraints there, now and in the future" (Faludi, 2009). There is an argument for giving living labs a role in a transversal policy where they facilitate a user-centred, open research capability in any engagement or initiative, rather than the current role where living labs are funded much like science parks or incubators using a sectoral policy philosophy (Marsh, 2008). This "territorial innovation" would be a move away from building specialised research centres towards integrating research with local and regional development stakeholders and municipalities, involving citizens from all areas of life to address problems affecting the territory.

5 Discussion

Living labs offers unparalleled opportunity to drive user-orientated innovation in partnerships between users, research organisations and universities, the public sector and private companies and organisations. A formal governance model exists

(ENOLL) to support the living labs in existence and the European Commission has explicitly supported the concept of living labs in its CIP funding regime. There are a wide variety of methods that can be brought to bear to solve problems in partnership.

So, why are living labs not more successful? Why are we not witnessing an explosive growth in the number of living labs being set up to solve a range of problems affecting society? If living labs have the support of policy makers such as the European Commission as well as grass-roots citizens and users, why do we not hear about living labs each and every day?

Perhaps one day we will see a living lab approach to community problem solving as a ubiquitous process, with methods and services clearly understood by all stakeholders. On the other hand, living labs may fade away as new paradigms of social engagement supported by ICT become commonplace.

The challenges and issues that are faced by living labs are seen as related to potential problems in each of the areas of infrastructure, methods, tools and policy (Feuerstein et al. 2008). Another issue relates to how the research organisations or universities can develop agendas for inter-disciplinary research (Mulvenna et al. 2009), which is beneficial to living labs but not to universities that seek to specialise in particular areas of expertise.

Other challenges have been identified as relating to collaboration, standardization and efficiency (Molinari, 2008). Each living lab has to develop its competencies in user-centred methods and engage with the stakeholders. In terms of standardization, living labs often carry out very similar practices of engagement but because many living labs have developed from different areas of science, there is no common and agreed vernacular. Efficiency may be an issue in living labs but the use of ICT to aid communication between stakeholders and users must help to reduce engagement budgets. In general, each of these issues is an issue of living lab standardisation and the presence of a governance organisation means that the living labs should have access to harmonisation and standardization roadmaps.

Living labs do not have strong, explicit and clear business models and determining usable models that facilitate profitable partnerships for all the stakeholders may prove difficult to resolve; for example, who owns the intellectual property of the product or service? On the other hand, they can fire the imaginations of the stakeholders and create a strong *esprit de corps* between the stakeholders, transcending traditional barriers between different groups.

Living labs may create many new opportunities for engagement between users as citizens, patients, and service or product users and all types of private and public stakeholders ranging in size from SMEs to multi-nationals and from local councils to the European Commission.

6 Conclusions

While living labs offer much promise in engaging with users to create new products and services, they are not widely understood outside some of the academic departments in which the concepts developed. In an attempt to address this issue, the backdrop against which living labs have developed has been described in this paper.

Our economies have moved from sectoral policies of making capital available for the creation of, for example, science parks and incubator units to innovation and investment philosophies that originate from concepts such as the information society and the knowledge economy and promote, for example, networking partnerships and access to innovation services. This change has been facilitated by the explosive growth and uptake of the Internet by people as well as businesses. The perception is of a business as a kind of innovation engine, consuming capital and producing output for global markets, often available electronically.

How companies innovate is also the subject of radical change, as they move to build *ad hoc* partnerships, value chains and networks to exploit their intellectual property and know how maximally and globally. Models such as the triple-helix model of academic-public-private partnership offer access to resources and R&D capacity that cannot be accessed internally by a company. This availability of open innovation partnerships accelerates the ability of a company to act globally, and the active involvement of users offers the potential for improved product or service design.

Living labs offer a collaborative partnership framework in which user-centred, innovation activities can take place. They offer methods to garner data and evidence on design, processes that develop ideas, oversee engagement with users and how data is evaluated, and services that package all the constituent components that make up a living lab into coherent offering that can be understood by the core stakeholder groups comprising users, businesses, civic partners and research organisations such as universities.

There are significant issues with living labs that primarily relate to standardisation of their offerings, which should be addressed through the existing governance organisation, ENOLL. However, they offer an exciting means to engage with users while aiming for commercial results and perhaps they offer a significant partnership resource for SMEs in Europe to retain their competitiveness.

References

Ballon, P., Pierson, J., Delaere, S.: Open Innovation Platforms For Broadband Services: Benchmarking European Practices. In: Proceedings of the 16th European Regional Conference by the International Telecommunications Society (ITS), Porto, Portugal, September 4-6 (2005)

Berkhout, A.J., van der Duin, P., Hartmann, D., Ortt, R.: The Cyclic Nature of Innovation: Connecting Hard Sciences with Soft Values. In: Advances in the Study of Entrepreneurship, Innovation and Economic Growth, vol. 17, Elsevier, Amsterdam (2007)

Becker, G.S.: Human Capital: A Theoretical and Empirical Analysis, with Special Reference to Education. University of Chicago Press, Chicago (1964)

Bergvall-Kåreborn, B., Ihlström Eriksson, C., Ståhlbröst, A., Svensson, J.: A Milieu for Innovation-Defining Living Lab. In: 2nd ISPIM Innovation Symposium, New York (2009)

Beyer, H., Holtzblatt, K.: Contextual Design: Defining Customer-Centered Systems. Morgan Kaufmann, San Francisco (1998)

Van Buuren, R., Haaker, T., Janssen, W.: Networked Innovation. Novay, The Netherlands (2009)

Bødker, S., Christiansen, E., Ehn, P., Markussen, R., Mogensen, P., Trigg, R.: The AT Project: Practical research in cooperative design, DAIMI No. PB-454, Department of Computer Science, Aarhus University (1993)

Chesbrough, H.: Open Innovation: The New Imperative for Creating and Profiting from Technology. Harvard Business School Press, Boston (2003)

Collier, D.A., Meyer, S.M.: A service positioning matrix. International Journal of Operations & Production Management 18(12), 1223–1244 (1998)

Drucker, P.: The Age of Discontinuity: Guidelines to Our Changing Society. Harper and Row, New York (1969)

Etzkowitz, H.: Innovation in Innovation: The Triple Helix of University-Industry-Government Relations. Social Science Information 42(3), 293–337 (2003)

EU Recommendation 2003/361/EC (2003),
http://ec.europa.eu/enterprise/policies/sme/facts-figures-analysis/sme-definition (last accessed September 11, 2010)

European Commission, Green Paper on Territorial Cohesion, Turning territorial diversity into strength, SEC, 2550 (2008)

European Commission, Living labs for User-Driven Open Innovation (2010),
http://ec.europa.eu/information_society/activities/livinglabs/index_en.htm (Last accessed September 11, 2010)

European Commission, Competitiveness and Innovation Programme (2010),
http://ec.europa.eu/cip/ (Last accessed September 11, 2010),

FarNorth, FarNorth Living Lab (2010), http://farnorthlivinglab.no/ (last accessed September 11, 2010)

Faludi, A.: Territorial Cohesion under the Looking Glass (2009),
http://ec.europa.eu/regional_policy/consultation/terco/index_en.htm (last accessed September 11, 2010)

Feuerstein, K., Hesmer, A., Hribernik, K.A., Thoben, K.-D., Schumacher, J.: Living Labs: A New Development Opportunity. In: Schumacher, J., Niitamo, V.-P. (eds.) European Living Labs-A New Approach for Human Centric Regional Innovation (2008)

Følstad, A.: Living Labs for Innovation and Development of Information and Communication Technology: A Literature Review. Electronic Journal for Virtual Organizations and Networks 10 "Special Issue on Living Labs" (August 2008)

Gianiodis, P.T., Ellis, S.C., Secchi, E.: Advancing a Typology of Open Innovation. International Journal of Innovation Management 14(4), 531–572 (2010)

Google tweet (August 26, 2010),
https://twitter.com/google/status/22199802288
(last accessed September 8, 2010)

Hansen, M., Birkinshaw, J.: The Innovation Value Chain. Harvard Business Review 85(6), 121–130 (2007)

von Hippel, E.: Lead user: a source of novel product concepts. Management Science 32(7), 791–805 (1986)

von Hippel, E.: The Sources of Innovation. Oxford University Press, New York (1988)

Howe, J.: The Rise of Crowdsourcing, Wired (2006),
http://www.wired.com/wired/archive/14.06/crowds.html
(last accessed September 11, 2010)

ISO-13407, Human-centred design processes for interactive systems (1999)

Kallai, T., Bilicki, V.: Innovation Commercialization in a Rural Region: A Case Study. In: Schumacher, J., Niitamo, V.-P. (eds.) European Living Labs: A New Approach for Human Centric Regional Innovation (2008)

Kelly, K., New Rules for the Wired Economy, Forth Estate (1998)

Lewin, K.: Action research and minority problems. J. Soc. 2(4), 34–46 (1946)

Machlup, F.: The Production and Distribution of Knowledge in the United States. Princeton University Press, Princeton (1962)

Marsh, J.: Living Labs and Territorial Innovation. In: Cunningham, P., Cunningham, M. (eds.) Expanding the Knowledge Economy: Issues, Applications, Case Studies. IOS Press, Amsterdam (2008) ISBN: 978-1-58603-924-0

McKeown, M.: The Truth About Innovation. Prentice Hall, London (2008)

Mitchell, W.J.: Me++: the cyborg self and the networked city. MIT Press, Cambridge (2003)

Molinari, F.: Services for Living Labs. In: Schumacher, J., Niitamo, V.-P. (eds.) European Living Labs: A New Approach for Human Centric Regional Innovation, pp. 978–973 (2008) ISBN 978-3-86573-343-6

Moore, G.A.: Crossing the Chasm: Marketing and Selling Technology Products to Mainstream Customers, Capstone (1998)

Mulvenna, M.D., Galbraith, B., Martin, S.: Enriching the Research & Development Process Using Living Lab Methods: The TRAIL Experience. In: Cunningham, P., Cunningham, M. (eds.) eChallenges 2009 Conference Proceedings. IIMC International Information Management Corporation (2009)

Norman, D., Miller, J., Henderson, A.: What You See, Some of What's in the Future, And How We Go About Doing It: HI at Apple Computer. In: Proceedings of CHI 1995, Denver, Colorado, USA (1995)

Pallot, M.: Engaging Users into Research and Innovation: The Living Lab Approach as a User Centred Open Innovation Ecosystem. Webergence Blog (2009), http://www.cwe-projects.eu/pub/bscw.cgi/715404?id=715404_1760838 (last accessed September 11, 2010)

Porter, M.E.: Clusters and the New Economics of Competition. Harvard Business Review (July 7, 2007)

Rifkin, J.: The Age of Access: The New Culture of Hypercapitalism, Where All of Life is a Paid-for Experience. Penguin, Putnam (2000)

Rogers, E.M.: Diffusion of Innovations, Simon & Schuster International, 5th edn (2003) (revised)

Rothstein, R., Wolf, P.: Re-Energizing Product Development: Innovation Space at Arizona State University. Design Management Institute Review 16(2) (Spring, 2005)

Rubin, J.: Handbook of Usability, How to plan, design and conduct effective tests. John Wiley &Sons, New York (1994)

Ståhlbröst, A., Bergvall-Kåreborn, B.: FormII- an Approach to User Involvement. In: Schumacher, J., Niitamo, V.-P. (eds.) European Living Labs-A New Approach for Human Centric Regional Innovation (2008)

Surowiecki, J.: The Wisdom of Crowds: Why the Many Are Smarter Than the Few and How Collective Wisdom Shapes Business, Economies, Societies and Nations Little, Brown (2004)

Trott, P., Hartmann, D.: Why 'Open Innovation' Is Old Wine In New Bottles. International Journal of Innovation Management 13(4), 715–736 (2009)

Urban, G., von Hippel, E.: Lead User Analyses for the Development of New Industrial Products. Management Science 34(5), 569–582 (1988)

How to Transfer the Innovation Knowledge from Craft Art into Product Design

A Case Study of Character Toys

Yang-Cheng Lin[1,*] and Chun-Chun Wei[2]

[1] Department of Arts and Design, National Dong Hwa University, Hualien, 970, Taiwan
[2] Department of Industrial Design, National Cheng Kung University, Tainai, 701, Taiwan

Abstract. How to design highly-reputable and hot-selling products is an essential issue in product design. Product designers design a product by considering physical elements or characteristics of the product, while craft artists create their works relying largely on their own particular expertise or experience. In order to clarify the innovation/creation process between craft art and product design, we conduct an experimental study on character toys using the Kansei Engineering approach and the Quantification Theory Type I analysis. The result of the experimental analysis shows that the innovation knowledge models built in this study can help product designers understand consumers' emotional feelings to transfer the innovation knowledge from craft design into product design. This approach provides an effective mechanism for facilitating the new product design process.

1 Introduction

Whether consumers choose a product depends largely on their emotional feelings of the product image [5]. Product designers need to comprehend the consumers' feelings or needs in order to design successful products in an intensely competitive market [7]. However, the way that consumers look at product image is usually different from the way that product designers look at product elements or characteristics [4]. On the other hand, product designers design a product by considering physical elements or characteristics of the product, while craft artists create their works (or crafts) relying largely on their own particular expertise or experience, which is regarded as something of a black box [1]. It is quite difficult to describe clearly by quantitative models or mathematical formulas, because it is concerned with the creation and innovation of human nature [6]. There is a gap between product designers and craft artists, due to the difference of the creation purpose and target. Consequently, it is a real challenge to transfer the innovation knowledge from craft art into product design.

R.J. Howlett (Ed.): Innovation through Knowledge Transfer 2010, SIST 9, pp. 265–274.
springerlink.com   © Springer-Verlag Berlin Heidelberg 2011

In order to clarify the innovation/creation process between craft art and product design, we conduct an experimental study on character toys (dolls, mascots, or called "公仔" in Mandarin), due to the great popularity in eastern Asia (particularly in Taiwan, Japan, and Hong Kong). In addition, the character toy is not only a craft work but a commercial product. The character toy is well suitable to be an object to illustrate how the innovation knowledge can be used. Furthermore, in order to explore the relationship between consumers' emotional feelings and product form elements, Kansei Engineering is adopted in this study to design highly-reputable and hot-selling products [4]. Kansei Engineering is as an ergonomic consumer-oriented methodology and design strategies for affective design to satisfy consumers' psychological requirements [8].

2 Quantification Theory Type I

The QTTI can be regarded as a method of qualitative and categorical multiple regression analysis method [3], which allows inclusion of independent variables that are categorical and qualitative in nature, such as product form elements and quantitative criterion variables within Kansei Engineering. The QTTI consists of the followings six steps [10]:

Step 1: Define the Kansei relational model associated with the Kansei measurement scores of experimental samples with respect to an image word pair.
Step 2: Calculate the standardized regression coefficients and the standardized constant in the model.
Step 3: Determine the matrix CCR of correlation coefficient of all variables.
Step 4: Calculate the multiple correlation coefficient R that is regarded as the relational degree of external criterion variable and explanatory variables.
Step 5: Calculate the partial correlation coefficients (PCC) of design elements to clarify the relationships between product form elements and a product image.
Step 6: Determine the statistical range of a categorical variable (product form element) by the difference between the maximum value and minimum value of the category score. The range of the categorical variable indicates its contribution degree to the prediction model with respect to a given product image.

3 Experimental Procedures of a Case Study

We conduct an experimental study using the concept of Kansei Engineering in order to collect numerical data about the character toys.

3.1 Experimental Samples of Character Toys

In the experimental study, we investigate and categorize various character toys with local and aboriginal cultures in Taiwan. We first collect 179 character toys and then classify them based on their similarity degree by a focus group that is formed by six subjects with at least two years' experience of craft and product design. The focus group eliminates some highly similar samples through discussions. Then the hierarchy cluster analysis is used to extract representative samples of character toys. Fig. 1 shows the 35 representative character toy samples.

3.2 Morphological Analysis of Character Toys

The product form is defined as the collection of design features that the consumers will appreciate. The morphological analysis [11], concerning the arrangement of objects and how they conform to create a whole of Gestalt, is used to explore all possible solutions in a complex problem regarding a product form.

The morphological analysis is used to extract the product form elements of the 35 representative character toy samples. The five subjects of the focus group are asked to decompose the representative samples into several dominant form elements and form types according to their knowledge and experience. Table 1 shows the result of the morphological analysis, with seven product design elements and 24 associated product form types being identified. The form type indicates the relationship between the outline elements. For example, the "width ratio of head and body (X_2)" form element has three form types, including "head $>$ body", "head=body", and "head $<$ body". A number of design alternatives can be generated by various combinations of morphological elements [2].

3.3 Emotional Feelings of Character Toys

In Kansei Engineering, emotion assessment experiments are usually performed to elicit the consumers' psychological feelings or perceptions about a product using the semantic differential method [9]. Image words are often used to describe the consumers' feelings of the product in terms of ergonomic and psychological estimation. With the identification of the form elements of the product, the relationship between the image words and the form elements can be established. The procedure of extracting image words includes the followings four steps [8]:

Step 1: Collect a large set of image words from magazines, product catalogs, designer, artists, and toy collectors. In this study, we collect 110 image words which are described the character toys, e.g. vivid, attractive, traditional, etc.
Step 2: Evaluate collected image words using the semantic differential method.
Step 3: Apply factor analysis and cluster analysis according to the result of semantic differential obtained at Step 2.
Step 4: Determine three representative image words, including "cute (CU)", "artistic (AR)", and "attractive (AT)", based on the analyses performed at Step 3.

Fig. 1 The 35 representative character toy samples

Table 1 The morphological analysis of character toys

	Type 1	Type 2	Type 3	Type 4	Type 5
Length ratio of head and body (X_1)	$\geqq$ 1:1	1:1~1:2	<1:2		
Width ratio of head and body (X_2)	head▯ body	head=body	head▯ body		
Costume style (X_3)	one-piece	two-pieces	robe		
Costume pattern (X_4)	simple	striped	geometric	mixed	
Headdress (X_5)	tribal	ordinary	flowered	feathered	arc-shaped
Appearance of facial features (X_6)	eyes only	partial features	entire features		
Overall appearance (X_7)	cute style	semi-personified style	personified style		

Table 2 The morphological analysis of character toys

No.	X_1	X_2	X_3	X_4	X_5	X_6	X_7	CU	AR	AT
1	3	2	1	1	4	3	3	73	61	64
2	1	1	1	1	1	2	1	72	45	43
3	2	2	1	3	3	1	1	70	64	71
4	2	3	2	4	2	2	2	63	52	54
5	2	2	1	1	4	2	1	68	59	55
6	2	2	2	4	3	2	2	65	66	69
7	2	2	2	4	5	2	2	52	66	61
8	2	3	2	4	4	2	2	53	61	60
9	2	2	3	2	2	2	2	63	59	59
10	2	2	1	3	2	2	2	55	63	65
11	1	1	2	3	4	2	1	70	69	67
12	1	1	3	2	2	2	1	57	54	61
13	3	3	2	4	4	3	3	48	69	76
14	3	3	1	4	4	3	3	62	68	78
15	3	3	2	2	2	3	3	54	63	68
16	3	3	1	2	3	3	3	62	74	72
17	3	3	2	4	2	3	3	55	68	66
18	2	3	3	2	2	2	2	71	65	61
19	2	2	1	1	2	3	3	41	52	75
20	2	2	2	1	1	3	3	39	53	63
21	2	2	2	2	3	3	3	41	50	58
22	2	2	2	3	2	3	2	44	74	62
23	2	2	2	1	2	3	3	43	59	74
24	2	2	1	3	2	3	1	54	60	62
25	2	2	2	2	2	1	1	63	52	62
26	1	2	2	2	4	3	2	58	71	68
27	1	2	1	2	4	3	2	57	61	66
28	1	1	2	2	1	1	1	62	56	73
29	1	1	1	3	5	3	2	76	67	74
30	1	1	1	3	3	3	2	68	59	65
31	1	1	3	2	2	3	2	71	60	70
32	1	1	1	4	4	1	1	61	49	51
33	1	1	1	4	5	1	1	72	59	57
34	2	3	2	4	2	2	2	38	48	49
35	1	1	1	3	5	2	1	78	59	79

To obtain the assessed values for the emotional feelings of 35 representative character toy samples, a 100-point scale (0-100) of the semantic differential method is used. 150 subjects (70 males and 80 females with ages ranging from 15 to 50) are asked to assess the form (look) of character toy samples on a image word scale of 0 to 100, for example, where 100 is most attractive on the AT scale.

The last three columns of Table 2 show the three assessed image values of the 35 samples. For each selected character toy in Table 2, the first column shows the character toy number and Columns 2-8 show the corresponding type number for each of its seven product form elements, as given in Table 1. Table 2 provides a numerical data source for building an innovation knowledge model, which can be used to develop a design support system for the new product design and development of character toys.

4   Innovation Knowledge Models for New Product Design

In this section, we present the result of applying the QTTI analysis in order to build an innovation knowledge model for transferring consumers' emotional feelings into product form design.

4.1   The QTTI Analysis and Results

We use the QTTI analysis to examine the relationship between the seven product form elements and three product images. In this paper, seven independent variables (i.e. the seven product form elements) and three dependent variables (i.e. the cute, artistic, and attractive product images) are used. The result of QTTI analysis is given in Table 3. In Table 3, the partial correlation coefficients indicate the relationship between the seven product form elements and each product image.

The highest variable of the partial correlation coefficient in the CU image is the "overall appearance" form element ($X_7 = 0.73$), meaning that "overall appearance" primarily affects the CU image of the product, followed by the "costume pattern" form element ($X_4 = 0.71$) and the "length ratio of head and body" form element ($X_1 = 0.70$). This implies that the product designers should focus their attention more on these most influential elements, when the objective of designing a new character toy is to achieve a desirable CU image.

In the last second row of Table 3, R means the correlation between the observed and predicted values of the dependent variable, and R^2 is the square of this correlation. R^2 ranges from 0 to 1. The category grade (form type grade) shown in Table 3 indicates the preference degree of the consumers' emotional feelings on the each category of independent variables. If the grade is positive, the consumers' emotional feeling leans towards the "specific" image strongly (e.g. cute, artistic, or attractive). On the contrary, the negative grade indicates that the consumers' emotional feeling favors the "counter-specific" image (e.g. un-cute, in-artistic, or un- attractive). For example, the category grades of 3 selected values of "costume style (X_3)" in the AR image are -4.40, 1.87, and 9.04 respectively. The result shows that the consumers' emotional feeling prefers the AR image if the "costume style (X_3)" is "robe" or "two-pieces", and feels the "inartistic" while "costume style (X_3)" is "one-piece".

Table 3 The result of QTTI analysis

Form element	Form type	CU image Form type grade (Category grade)	CU image Form element grade (Partial correlation coefficient)	AR image Form type grade	AR image Form element grade	AT image Form type grade	AT image Form element grade
X_1 Length ratio of head and body	$\geqq$ 1:1	-1.57		1.96		5.76	
	1:1~1:2	-5.26	0.70	-5.02	0.73	-3.74	0.34
	< 1:2	18.05		10.29		-0.91	
X_2 Width ratio of head and body	head > body	1.71		-8.63		-6.15	
	head = body	-2.14	0.29	3.30	0.76	2.43	0.29
	head < body	1.91		3.73		2.51	
X_3 Costume style	one-piece	-0.47		-4.40		-2.29	
	two-pieces	-1.55	0.46	1.87	0.64	0.42	0.36
	robe	7.97		9.04		6.89	
X_4 Costume pattern	simple	12.19		0.72		-2.88	
	striped	-1.35	0.71	-3.56	0.69	-1.84	0.59
	geometric	1.63		8.04		9.38	
	mixed	-7.14		-2.95		-3.75	
X_5 Headdress	tribal	-6.00		-2.39		-2.01	
	ordinary	-3.76		-3.36		-1.87	
	flowered	6.59	0.65	2.55	0.59	-0.05	0.34
	feathered	0.49		1.64		0.96	
	arc-shaped	8.30		6.66		5.95	
X_6 Appearance of facial features	eyes only	7.33		1.26		5.47	
	partial features	0.76	0.49	-1.76	0.28	-0.56	0.30
	entire features	-2.74		0.97		-1.18	
X_7 Overall appearance	cute	4.59		0.86		-5.45	
	semi-personified	6.42	0.73	2.53	0.48	-2.71	0.63
	personified	-14.04		-4.49		9.80	
	Constant	59.40		60.43		64.51	
	R	0.88		0.85		0.71	
	R^2	0.78		0.72		0.51	

4.2 Innovation Knowledge Models and Product Design Support System

As the result of the QTTI analysis, Models (4.1), (4.2), and (4.3) indicate the relationship between product form elements and the given product images. We can use these three models to input the values of seven product form variables, and then output the prediction values of theses three product images. These three models can help the product designers understand consumers' emotional feelings to transfer the innovation knowledge from craft design into product design.

CU: $\hat{y} = 59.4 - 1.57X_{11} - 5.26X_{12} + 18.05X_{13} + 1.71X_{21} - 2.41X_{22} + 1.91X_{23} - 0.47X_{31}$
$- 1.55X_{32} + 7.97X_{33} + 12.19X_{41} - 1.35X_{42} + 1.63X_{43} - 7.14X_{44} - 6X_{51} - 3.76X_{52} + 6.59X_{53}$
$+ 0.49X_{54} + 8.3X_{55} + 7.33X_{61} + 0.76X_{62} - 2.74X_{63} + 4.59X_{71} + 6.42X_{72} - 14.04X_{73}$ (4.1)

AR: $\hat{y} = 60.43 + 1.96X_{11} - 5.02X_{12} + 10.29X_{13} - 8.63X_{21} + 3.30X_{22} + 3.73X_{23} - 4.4X_{31}$
$+ 1.87X_{32} + 9.04X_{33} + 0.72X_{41} - 3.56X_{42} + 8.04X_{43} - 2.95X_{44} - 2.39X_{51} - 3.36X_{52} + 2.55X_{53}$
$+ 1.64X_{54} + 6.66X_{55} + 1.26X_{61} - 1.76X_{62} + 0.97X_{63} + 0.86X_{71} + 2.53X_{72} - 4.49X_{73}$ (4.2)

AT: $\hat{y} = 64.51 + 5.76X_{11} - 3.74X_{12} - 0.91X_{13} - 6.15X_{21} + 2.43X_{22} + 2.51X_{23} - 2.29X_{31}$
$+ 0.42X_{32} + 6.89X_{33} - 2.88X_{41} - 1.84X_{42} + 9.38X_{43} - 3.75X_{44} - 2.01X_{51} - 1.87X_{52} - 0.05X_{53}$
$+ 0.96X_{54} + 5.95X_{55} + 5.47X_{61} - 0.56X_{62} - 1.18X_{63} - 5.45X_{71} - 2.71X_{72} + 9.80X_{73}$ (4.3)

The innovation knowledge models enable us to build a character toy design support database that can be generated by inputting each of all possible combinations (4860, 3×3×3×4×5×3×3) of product form elements to the innovation knowledge models individually for generating the associated image values. Product designers can specify a set of desirable image values for a new character toy form design, and the database can then work out the optimal combination of form elements. The design support information helps product designers to find out the optimal combination of product form elements in terms of a given sets of product images. In addition, the design support database can be incorporated into a computer-aided design system to facilitate the product form in the new character toy development process.

5 Conclusion

In this paper, we have conducted an experimental study on character toys to demonstrate how a consumer-oriented design approach can be used to transfer the innovation knowledge from craft art to product design. The consumer-oriented design based on the process of Kansei Engineering has used the QTTI technique to build three innovation knowledge models (i.e. the CU, AR, and AT models). The result has shown that the innovation knowledge models can help product designers determine the optimal form combination of product design for a particular design concept of product image. Furthermore, the consumer-oriented design approach has been built a character toy design support database, in conjunction with the computer-aided design system, to help product designers facilitate the product form in the new product development process.

Acknowledgments. This research is supported in part by the National Science Council of Taiwan, ROC under Grant No. NSC 98-2410-H-259-064.

References

[1] Brown, S.L., Eisenhardt, K.M.: Product development: past research, present findings, and future directions. Academy of Management Review 20, 343–378 (1995)

[2] Cross, N.: Engineering Design Methods: Strategies for Product Design. John Wiley and Sons, Chichester (2000)

[3] Komazawa, T., Hayashi, C.: In: de Dombal, F.T., Gremy, F. (eds.) A Statistical Method for Quantification of Categorical Data and its Applications to Medical Science, North-Holland Publishing Company, Amsterdam (1976)

[4] Lai, H.-H., Lin, Y.-C., Yeh, C.-H., Wei, C.-H.: User Oriented Design for the Optimal Combination on Product Design. International Journal of Production Economics 100, 253–267 (2006)

[5] Lai, H.-H., Lin, Y.-C., Yeh, C.-H.: Form Design of Product Image Using Grey Relational Analysis and Neural Network Models. Computers & Operations Research 32, 2689–2711 (2005)

[6] Lin, Y.-C., Yeh, C.-H., Hung, C.-H.: A Neural Network Approach to the Optimal Combination of Product Color Design. In: NCM 2008, pp. 53–57. IEEE Computer Society, Los Alamitos (2008)

[7] Lin, Y.-C., Lai, H.-H., Yeh, C.-H.: Consumer-oriented product form design based on fuzzy logic: A case study of mobile phones. International Journal of Industrial Ergonomics 37, 531–543 (2007)

[8] Nagamachi, M.: Kansei engineering: A new ergonomics consumer-oriented techology for product development. International Journal of Industrial Ergonomics, 15, 3–10 (1995)

[9] Osgood, C.E., Suci, C.J.: The Measurement of Meaning. University of Illinois Press, Urbana (1957)

[10] Wang, C.-C.: Development of an Integrated Strategy for Customer Requirement Oriented Product Design. Ph.D. Dissertation, Department of Industrial Design, National Cheng Kung University, Tainan, Taiwan (2008)

[11] Zwicky, F.: The Morphological Approach to Discovery, Invention, Research and Construction, New Method of Though and Procedure: Symposium on Methodologies, Pasadena, pp. 316–317 (1967)

'Center for Global' or 'Local for Global'? R&D Centers of ICT Multinationals in India*

P. Vigneswara Ilavarasan

Indian Institute of Technology Delhi, India
vignesh@hss.iitd.ac.in

Abstract. Existing work on the information and communication technology (ICT) industry in India provides inadequate information on the R&D centers of multinationals. The present paper attempts to fill this gap through a content analysis of secondary data on R&D centers of ICT multinationals and examines their impact on the local science and technology systems. The analysis finds that the centers are almost equally divided between those who execute the designs of the headquarters (center for global) and those who collaborate in design making and execution along with their headquarters (local for global). It also deduces that the nature of the linkages between the centers and the local universities, public research laboratories and local firms are inadequate to effect knowledge transfer to India. The paper suggests a few policy recommendations.

1 Introduction

The present paper aims to achieve two objectives: to understand the nature of research and development (R&D) centers of information and communications technologies (ICTs) of multinational enterprises (MNEs) in India, and to examine the impact of these centers on the Indian science and technology systems. Despite the impressive growth of the Indian ICT industry in software services exports, it is criticized for the lack of innovation indicated by visible global ICT products or patents filed (D'Costa and Sridharan, 2003). The existing literature on the role played by the ICT MNEs in India offers contradictory views – MNEs have contributed to the technological upgrading of domestic firms (Patibandla and Petersen, 2002) and there are no technological spillovers (Athreye, 2002; D'Costa, 2003). Extant scarce studies on MNEs in India appear to be weak with very small sample sizes or no mention of sample sizes (for instance, Patibandla and Petersen, 2002). Also, the existing literature on the internationalization of R&D predominantly focuses on the perspective of the home country (De Meyer and Mizushima,

---

* The paper is based on a research study funded by the Technology Information and Assessment Council, New Delhi. Thanks to Dr. Jyoti for the research assistance.

R.J. Howlett (Ed.): Innovation through Knowledge Transfer 2010, SIST 9, pp. 275–282.
springerlink.com © Springer-Verlag Berlin Heidelberg 2011

1989), and views from the host countries, mostly developing, are inadequately studied.

Systematic data on the Indian ICT industry is almost absent and statistical frameworks of the government are being updated to address this problem (Chandrasekhar and Ghosh, 2008; Parthasarathi and Joseph, 2002). Hence we do not know either the number of R&D centers or the number of people working and the amount of investment. Given the lack of official data on the number of R&D centers, trade press reports provide varying numbers in the range of 77 to 230 (Ilavarasan and Malik, 2010). As the government is continuing its policy initiatives to attract investment by multinationals, an understanding of R&D centers and their impact on the local system will benefit policy makers.

In order to understand the nature of R&D centers, we used the conceptual framework given by Archibugi and Pietrobelli (2003) which broadly classifies R&D centers of MNEs in host countries into three categories. To quote (p. 878):

"Center-for-global
... a single 'brain' located within the company headquarters concentrates the strategic resources: top management, planning, and the technological expertise. The 'brain' distributes impulses to the 'tentacles' (that is, the subsidiaries) scattered across host countries.

Local-for-local
Each subsidiary develops its own technological know-how to serve local needs. The interactions among subsidiaries are, at least from the viewpoint of developing technological innovations, rather weak.

Local-for-global
... rather than concentrating their technological activities in the home country, [multinationals] distribute R&D and expertise in a variety of host locations. This allows the company to develop each part of the innovative process in the most suitable environment: semiconductors in Silicon Valley, automobile components in Turin, software in India. ..."

This framework is comprehensive and incorporates the ideas expressed in previous studies (for instance, Kuemmerle, 1997).

2 Methodology

We generated an exhaustive list of R&D centers of ICT MNEs by sieving information available in the trade press, newspapers, periodicals and government databases. We have included only those firms that have clearly mentioned that the firm or center is pursuing R&D-related activities. In order to understand the nature of work undertaken in these centers, we used online search engines like Google, site.securities.com, and Ibef.com. Further, we searched the website of individual companies. After deleting three firms for which no information was available either on the address or employees, a total of 160 centers was finalized. These 160 firms formed the population of the study.

We collected information on the following variables for the firms: name of the company, location in India, home country, number of professionals/ employees, amount of investment in India, technical domain, and nature of activities performed.

3 Nature of R&D Centers

Using the collated information on the firms, we attempted to understand the nature of R&D centers on the following variables: home country, technological domain, geographical location, number of people employed, finance and type of center.

We identified the home country of the MNEs by analyzing newspaper articles and other secondary material. In terms of population, the US is the leading country with the highest presence at 72 percent, followed by France, Germany, South Korea and the UK at 4 percent each. Japan constitutes 3 percent. One each from Austria, Brazil, Canada, China, Finland, Israel, Singapore, Taiwan, and Turkey together constituted 9 percent of the population.

Historically, in the ICT domain India is closer to US markets. The US is the leading export destination for the Indian IT industry, at 63 percent in 2007-2008 (NASSCOM, 2009). Firms that first explored the Indian market for offshoring were from the US. Also, during the 1960s-1990s, high-skilled Indians migrated to the US for higher studies and returned to start firms in India (Saxenian, 2006; Sharma, 2009).

In the population, more than half of the firms perform R&D-related activities in the domain of software services (63 percent), followed by telecommunications (5 percent). Fewer firms operate in other technical domains of the IT sector. The Indian ICT sector is predominantly a software services industry with USD 52 billion compared to USD 12 billion in hardware in 2008-2009 (NASSCOM, 2009). An analysis of the telecom equipment industry by Mani (2005) offers some insights into how the local manufacturing industry declined due to lack of market integration policy initiatives despite the technological capabilities of government research labs.

The Indian ICT industry is located in six major clusters anchored in the following cities: Bangalore, Hyderabad, Chennai, National Capital Region (NCR, composed of New Delhi, Gurgaon, and Noida), Pune and Mumbai. Our analysis showed that Bangalore tops the list of clusters with 56 percent, followed by Pune (15 percent) and NCR (13 percent) in hosting R&D centers.

Earlier research (Parthasarathy and Aoyama, 2006) showed that the Bangalore cluster presents the most mature ecosystem for the IT industry among the six clusters. Bangalore leads due to positive agglomeration effects (Markusen, 1996) created by the presence of a mix of government research labs, universities, a healthy mix of large and small firms, the abundance of high-skilled labor, the availability of complementary support systems like suppliers, and adequate infrastructure.

Centers might be located in Hyderabad, Pune, Chennai and Delhi due to negative agglomeration effects in Bangalore. Bangalore's physical infrastructure is unable to meet the demand, resulting in higher costs and pollution levels apart from rising labor costs.

We attempted to collect information on the investment and people employed by the ICT MNEs while constructing the database. Despite our best efforts, for 32 percent of the firms investment data was not available. Also, data for all the firms in the same year was unavailable. For some firms, data was available only for the years 2002 or 2003. Trade press releases mention the firms' future investment amount rather than the actual investment; we were unable to find out whether or not proposed investments were actually executed. Similar problem was faced in enumerating the number of employees. Hence, the paper does not present data on these variables.

To classify the centers on the basis of Archibugi and Pietrobelli (2003), we analyzed the information collated on each of the firms in the database. The content analysis of the news articles was done manually. Based on our understanding through the literature, we looked for keywords or phrases to understand the nature of work performed and the orientation of the centers. We were able to deduce whether the R&D centers were established to adapt their new product/technology to local needs or to develop new products/technology for global markets.

Firms that stated that the quality of manpower available in India would be used to support headquarters are classified as center-for-global. Activities performed in the center typically cater to the global market.

The second set of centers are established to tap into the huge local market. They adapt their product to Indian needs like developing software in Indian languages or providing services to telecom players. These centers use local talent to come up with products for the local market. These centers are noted as local-for-locals.

The third set of centers are established as part of a strategic decision and complement the innovation initiatives of the MNEs. These centers work on 'mainstream research' of the parent firms. Some companies even say that their India R&D center now contributes to the development of all major products in their areas of business. Such R&D centers are classified as local-for global.

In our population, the R&D centers of ICT MNEs are almost equally divided, with 50 percent serving as center-for-globals and 46 percent as local-for-globals. A small number of centers (4 percent) are established as local-for-locals.

The type of R&D center indicates the level of importance attached to India in the global strategy of MNEs. Center-for-globals operate on the labor cost arbitrage model among high-skilled labor. Control of the project activities still lie with the parent firms. whereas in the case of local-for-globals, the Indian center becomes an equal partner which is likely to share responsibility for product development activities. The transition from center-for-globals to local-for-globals indicates the growing importance of India as an important location of technological activity. National technological capabilities are enhanced multifold when the value of activities performed in the R&D centers increases. In the long run, MNEs expertise is disseminated locally. As the number of local-for-global centers increases, opportunities to serve the global market in the product segment opens up for the Indian workforce. The third set of centers, local-for-locals, uses the local workforce to develop products for the local market which is yet to be exploited.

4 Impact on the Local Science and Technology System

4.1 University-Industry Linkages

Linkages between local universities and industries are crucial to foster innovation in any industrial sector. Advancements in the research laboratories form the base for growth in applied technology areas of industry. The growth of Silicon Valley is repeatedly associated with its strong university-industry linkages (Saxenian, 2006).

We looked for the following possible relationships: (1) Funding for research projects / collaborative research – R&D centers will provide research grants to universities to undertake mutually beneficial research projects; R&D centers will also have their scientists working in the academic laboratories. (2) Joint teaching – scientists from the R&D centers will teach in the universities; (3) Specialized program sponsorship – The university in collaboration with the R&D center will introduce a certificate or diploma program for which industry will provide adequate support for laboratories, manpower and other inputs; (4) Student internships – Final-year students in undergraduate or master's programs from the university will spend one semester or summer in the R&D centers for which they are paid; (5) Faculty fellowships – Faculty members from the university will spend a certain amount of time in the R&D centers; (6) Campus placement – R&D centers do not have any relationship apart from visiting the campus to hire eligible students; and (7) Guest lecturers – People from R&D centers visit the university to deliver guest lectures.

These activities were grouped into two broad categories: research-related and training-related. Research-related associations between university and industry are long term and have dense interactions with subsequent intense knowledge transfer. In training-related activities, the relationship between industry and the university is relatively short term and beneficial to industry. Here industry prepares the workforce required for the industry. Although industry-supported programs do not bind the trainees or guarantee employment, the size of the labor pool increases for industry.

We found that 24 percent of the population is involved in research-related activities compared to 16 percent in training-related activities. More than half of the centers (60 percent) do not report any relationship with universities. In terms of importance, training-related activities are equal to research-related ones, as industry is dependent on the labor pool. However, research-related activities are important if the country wants to be a technology initiator rather than a follower in the long term.

4.2 Linkages with Government Research Laboratories

Public research laboratories play a significant role in any national innovation system. For instance, in Carnie Mellon Survey on industrial R&D (Cohen, Nelson &

Walsh, 2002), semi-conductor firms reported that public research is a source of new project ideas.

We tried to understand the relationship between R&D centers of IT MNEs and government research labs under three headings: (1) collaborative research projects – Centers will have joint collaborative research projects with government labs to develop a product; (2) sub-contracting – Centers will outsource part of the work to government labs; and (3) others – There is a possibility that scientists from government labs are working in R&D centers on a fellowship.

The content analysis indicates that only 3 percent of the centers are engaged in some form of collaboration with government labs. We did not find any related information for the rest of the firms. Earlier research on Indian bio-technology (Sardana and Krishna, 2006) indicated that public-funded research work is oriented towards publishing rather than commercialization.

4.3 Inter-firm Linkages

Host countries tend to benefit in multiple ways from MNEs. Apart from adding to employment and investment, MNEs also diffuse advanced technologies, both directly and indirectly (Meyer, 2004). In the Chinese and Irish ICT industries, MNEs played a significant role in diffusing global knowledge (Arora, Gambardella, & Torrisi, 2006).

We examined the relationship with local firms in three areas: (1) joint technical collaborations (JTC) – R&D centers will enter into a collaboration with the local firm; (2) joint sales collaborations – Local firms work as re-sellers to the products of IT MNEs, but are associated with the R&D centers; and (3) others – this includes R&D centers collaborating with other MNEs in India.

Our population of R&D centers of IT MNEs shows that one-quarter of the centers have joint technical collaborations (19 percent), followed by 11 percent sales-based collaborations and 6 percent of other kinds of relationships. More than half of the centers (64 percent) do not report any kind of collaboration with local firms. This indicates poor inter-firm linkages in industrial R&D in the IT sector in India.

5 Policy Recommendations

The lack of sub-sector level and R&D data on the ICT industry is a major handicap for researchers. The existing statistical framework followed by the Government of India is not aligned with international standards. *Frascati Manual* can be adopted to collect data on R&D expenditure, investment and personnel. Internationally comparable data will help researchers in comparative studies to learn from the experiences of developed countries.

Interaction between the R&D centers and universities is limited to training, and research collaboration with public laboratories is well below the desired level. Policy initiatives should encourage MNEs to collaborate with universities and public laboratories in research.

There are no linkages between IT MNEs and local firms, resulting in poor direct technology or knowledge transfer. The policy framework should provide an incentive structure for MNEs to partner with domestic firms in R&D.

Despite being known for its software industry, India does not have a world-class research center for software research. Such a center should be established, with researchers from academia and industry working together, either full time or part time. This center would serve as an important node in linking universities and industry.

References

Archibugi, D., Pietrobelli, C.: The globalization of technology and its implica-tions for developing countries: Windows of opportunity or further burden? Technological Forecasting & Social Change 70, 861–883 (2003)

Arora, A., Gambardella, A., Torrisi, S.: In : the footsteps of the Silicon Valley? Indian and Irish software in the international division of labour. SIEPR Discussion Paper No. 00-41. Stanford Institute for Economic Policy Research (2001)

Athreye, S.S.: Multinational firms and the evolution of the Indian software industry. Economics Study Area Working Papers 51, East-West Center, Washington (2003)

Chandrasekhar, C.P., Ghosh, J.: How big is IT. Business Line, March 11(2008),
http://www.thehindubusinessline.com/2008/03/11/stories/
2008031150170900.htm (accessed on December 10, 2008)

Cohen, M., Nelson, R.R., Walsh, J.P.: The influence of public research on in-dustrial R&D. Management Science 48(1), 1–23 (2002)

De Meyer, A., Mizushima, A.: Global R&D Management. R&D Management 19(2), 135–146 (1989)

D'Costa, A.P.: Uneven and combined development: Understanding India's software exports. World Development 31(1), 211–226 (2003)

D'Costa, A.P., Sridharan, E. (eds.): India in the Global Software Industry Inno-vation, Firm Strategies and Development. Palgrave Macmillan, London (2003)

Ilavarasan, P.V., Malik, P.: Study on the trends in public and private investments in ICT R&D in China, India and Taiwan. Draft report submitted to The Institute for Prospective Technological Studies, Brussels (2010)

Mani, S.: The dragon vs. the elephant: comparative analysis of innovation capability in the telecom industry of China and India. Economic and Political Weekly 40(39), 4271–4283 (2005)

Markusen, A.: Sticky places in slippery space: a typology of industrial districts. Economic Geography 72(3), 293–313 (1996)

Meyer, K.E.: Perspectives on multinational enterprises in emerging economies. Journal of International Business Studies 35(4), 259–276 (2004)

NASSCOM, The IT-BPO Sector in India: Strategic Review, New Delhi (2009)

Sardana, D., Krishna, V.V.: Government, university and industry relations: The case of bio-technology in the Delhi region. Science, Technology & Society 11(2), 351–378 (2006)

Saxenian, A.: The New Argonauts: Regional Advantage in a Global Economy. Harvard University Press, Cambridge (2006)

Sharma, D.S.: The Long Revolution: The Birth and Growth of India's IT Industry. Harper Collins, Noida (2009)

Parthasarathi, A., Joseph, K.J.: Limits to innovation set by strong export orien-tation: The experience of India's information and communication technology sector. Science, Technology and Society 7(10), 33–49 (2002)

Parthasarathy, B., Aoyama, Y.: From software services to R&D services: Local entrepreneurship in the software industry in Bangalore, India. Environment and Planning, 38(7), 1269–1285 (2006)

Patibandla, M., Petersen, B.: Role of transnational corporations in the evolution of a high-tech industry: The case of India's software industry. World Development, 32(3), 561–566 (2002)

Session G
Knowledge Transfer Case Studies

An Examination of an Innovation Intermediary Organisation's Methodology Using Case Studies

Ben Tura and Caroline Bishop

InnovationXchange, Birmingham Research Park, Vincent Drive, Birmingham B15 2SQ
Caroline.Bishop@ixc-uk.com

The intermediary organisation the InnovationXchange (IXC) has utilised a methodology to assist innovation through knowledge transfer for a range of organisations. A brief review of technology intermediary organisations introduces the emerging field of open innovation intermediaries before looking in more detail at the methodology of the InnovationXchange. The service users in two successful commercial transactions were interviewed to examine the methodology used to facilitate the transfer of knowledge and technology from UK academic research into global organisations. This paper will elucidate the mechanisms used to create value for services users based on a typology of innovation intermediaries.

1 Introduction

Open innovation encapsulates a process based on knowledge transfer both within and beyond the boundaries of an organisation to develop new products, processes and markets. The phrase 'open innovation' was coined and popularised by Henry Chesbrough (2003) to contrast with the closed model of innovation. The theory is discussed in detail elsewhere (Chesbrough *et al.* 2006).

The development of the Open Innovation paradigm has spawned the emergence of a number of intermediary organisations or innovation intermediaries. They act on behalf of organisations to facilitate movement of knowledge and technology across the organisational boundary. The innovation intermediary can take many guises and operate under a number of business models.

The appearance of intermediaries stretches back to the 16th century with middlemen reporting on innovations in the textile industry (Smith 2002). Such a 'traditional' approach is still found in consultants with deep knowledge of a particular industry sector acting as intermediaries (Fincham *et al.* 2008). A contemporary high profile business model is the internet-based intellectual property market place. Examples include NineSigma with solution providers in

R.J. Howlett (Ed.): Innovation through Knowledge Transfer 2010, SIST 9, pp. 285–295.

135 countries[1] and a reported 1 600 projects completed[2]; yet2.com, who report in excess of 120 000 users in 2010[3] and Innocentive who report 200 000 solvers and a total of 685 successes[4].

The UK's Knowledge Transfer Networks (KTNs) also perform a number of intermediary type functions. Funded by the Technology Strategy Board, KTNs were originally established as Faraday Partnerships in 1998 and exist with the remit to assist knowledge exchange and promote innovation in specific technology communities. At the time of writing 16 KTNs exist with memberships ranging from 300 to nearly 4000 individuals[5]. The Technology Strategy Board itself can be seen as an intermediary organisation, with a mission to 'Connect and Catalyse' (Technology Strategy Board 2008) it has both set an agenda and facilitated knowledge transfer through its competitions for collaborative funding.

NESTA (National Endowment for Science Technology and the Arts) has been endowed with £325 million National Lottery funding and the remit to improve innovation in the UK. Along with a number of other projects NESTA established a programme, Corporate Connect, to assist collaboration between small to medium enterprises and larger organisations (NESTA 2010). A spin out, 100%Open, has been launched based on the successes of Corporate Connect.

The scope of intermediary organisations covers large area cross-sector projects such as the European Enterprise Network through to local technology parks and organisations focussed on a specific technology area, Electronics Yorkshire is one such example.

Such a richness and diversity of intermediary organisations has proved difficult to classify and manage. However academic research has looked to codify intermediary approaches. Howells (2006) combined secondary research with an investigation of 22 technology intermediary organisations to formalise the functions of intermediaries. Ten functions were identified and detailed. Recently the functions were expanded (Lopez et al. 2010) and classified into three groups; Connection, Collaboration and Support and Technological services as shown in Table 1 Further research is ongoing into the selection and management of open innovation intermediaries (Mortara et al. 2010).

---

[1] NineSigma website: Our Network
   http://www.ninesigma.com/OurNetwork/OurNetwork.aspx Accessed 6 September 2010.
[2] NineSigma website: An interview with Andy
   Zyngahttp://www.ninesigma.com/News/AnInterviewWithAndyZynga.aspx
   Accessed 6 September 2010.
[3] Yet2.com website: Press Release 19 March 2010
   http://www.yet2.com/app/about/about/press?page=press82 Accessed 6 September 2010.
[4] Innocentive website: Facts and Stats
   http://www2.innocentive.com/about-innocentive/facts-stats Accessed 6 September 2010.
[5] Technology Strategy Board website
   https://ktn.innovateuk.org/web/guest/networks Accessed 6 September 2010.

Table 1 The functions of open innovation intermediaries as identified from the literature and field research (Howells 2006; Lopez *et al.* 2010).

Group	Function
Connection	Gatekeeping and brokering
	Middlemen between science policy and industry
	Demand articulation
Collaboration and support	Foresight and diagnostics
	Scanning and information processing
	Knowledge processing, generation and combination
	Commercialisation
Technological services	Testing, validation and training
	Accreditation and standards
	Regulation and arbitration
	Intellectual Property
	Assessment and evaluation

Lopez *et al.* (2010) further identify four types of innovation intermediaries; consultants, traders, incubators and mediators. By examining business models, strategies, sources of ideas and paths taken, the researchers drew up a typology to define the intermediaries (Table 2). Consultants therefore utilise their internal knowledge to provide a service, traders (e.g. NineSigma) leverage external intellectual property (IP), whilst incubators such as science parks provide physical spaces for interaction between start ups. Innovation mediators are exemplified by corporate open innovation projects such as those run by Lego, Nokia and Siemens who provide facilities and environments to draw in ideas and knowledge from users and external organisations.

Table 2 Intermediaries can be classified by the mechanism by which they create value for clients and the sources of ideas (Lopez *et al.* 2010).

Innovation Type	Intermediary Source of ideas / paths taken	Value creation
Consultants	Internal	Services
Traders	External	Services
Incubators	Internal	Infrastructure
Mediators	External	Infrastructure

According to this typology InnovationXchange would function as an innovation trader, however the company's business model differs substantially from traders in that InnovationXchange staff are co-located with clients and instead of generating revenue from posting challenges and solving problems a flat fee structure is used.

2 InnovationXchange Methodology

The InnovationXchange is a technology intermediary organisation that embeds multi-disciplinary scientists into client organisations. The ethos and business model of the InnovationXchange was set out by John Wolpert of IBM's Extreme Blue (Wolpert 2002). An intermediary was envisaged who would be able to facilitate knowledge transfer without revealing sensitive information on motives, applications or identity. Rather than focus on one industry the organisation takes clients with technology needs and capability across a range of sectors. The InnovationXchange approach was first trialled and tested in Australia in 2004 under the auspices of the Australian Industry Group and spun out into a not-for-profit company limited by guarantee in 2006. A hub of the InnovationXchange was established in the UK in the same year with a third hub based in Malaysia launched in 2009.

Three phases exist in the InnovationXchange process: intention, opportunity and connection. The intention phase has been analysed in detail by Wolpert (2006). Briefly an intention arises within an organisation 'to overcome a problem or capture a perceived opportunity'. This intention can then be assessed and managed by InnovationXchange Intermediaries.

The opportunity phase is based on the assessment of the intention and identification of gaps. These gaps can be thought of in terms of holes in the structure of the network (Burt 2001) that the InnovationXchange Intermediary can seek to resolve thereby brokering a flow of information between two or more parties. Theoretically the Intermediary can leverage InnovationXchange's internal network which has complete closure (Burt 2001). In a network with complete closure all nodes are connected to all other nodes, thus all clients´ intentions are be connected to all other clients' intentions in the network through the Intermediaries. In practice though, in this emerging field, connections are made from both within the InnovationXchange client base and from the wider connections and knowledge the InnovationXchange cultivates and maintains beyond its immediate client portfolio.

In the connection phase the intermediary acts as the *tertius iungens* – the third who joins (Obstfeld 2005). *Tertius iungens* behaviour describes a third party who introduces disconnected individuals or facilitates new co-ordination between connected individuals. Intermediaries may act in a classic "non-partisan" role (Simmel 1950) particularly in the first instance where the parties may not see the mutual benefit of a connection and the adversarial tension that characterises the non-partisan role is evident. InnovationXchange methodology departs from social capital theory in that the InnovationXchange acts as a *tertius* with the interests of the client at heart rather than seeking reward for themselves.

3 Research Methodology

Two case studies were chosen from the InnovationXchange UK hub for investigation. Recent projects that are currently being developed into products / processes were chosen to assist recall and to allow focus on value generation. A semi structured interview was conducted by the researcher, with a manager or academic involved in the project. The discussion was focused on identification of the role of InnovationXchange in one particular project rather than analysing the broader remit. Howells' functions (2006) with additions from Lopez *et al.* (2010) were used to structure discussion around the role of InnovationXchange (see Appendix), the ability of the service user to carry out intermediary functions, the value of the functions to the service user and further factors in the success of the project were also explored.

4 The Case Studies

Due to commercial concerns the identities of the services users have been kept anonymous. All service users were clients of InnovationXchange at the time the connections were made. In both case studies technology was sought by organisations with a global presence and provided by a UK university. The products being developed as a result of the connections are a novel safety product and a process for identifying novel compounds. The combined value of the products to the companies is estimated to be in the order of several million pounds.

5 Results

From semi structured interviews with the managers and academics involved in the case studies the functions of InnovationXchange Intermediaries in two successful cases were identified. The majority of activities fall in the connection and the collaboration and support groups with fewer activities in the technological services group (Fig. 1). Looking in closer detail the activities performed are in gatekeeping / brokering and scanning / information processing (Fig. 2). Activities may have more than one part and so can be scored multiple times for each interviewee, as an example the gatekeeping / brokering activity is further divided into two parts; matchmaking / brokering and contractual advice (see Appendix).

The activities perceived varied depending on whether the service user was a provider or seeker of technology. Technology seekers identified five functions being delivered (Fig. 3) while technology providers saw three (Fig. 4).

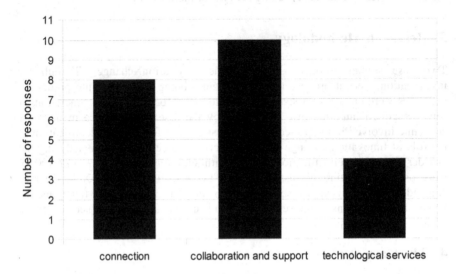

Fig. 1 InnovationXchange Intermediary activities fall into all three groups

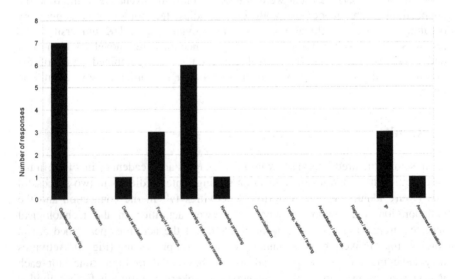

Fig. 2 InnovationXchange delivers most of its services in gate keeping / brokering and scanning / information processing

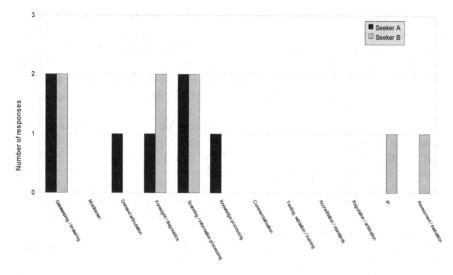

Fig. 3 Seekers of technology report five functions performed in a connection

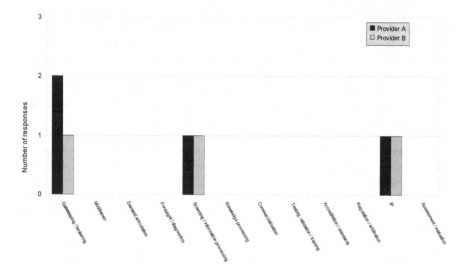

Fig. 4 Providers of technology report three activities performed in the connection

The most valued activities, as identified by the service users were scanning / information processing (three users) and gatekeeping / brokering (one user). Key benefits emphasised were; understanding the technological benefits rather than the commercial benefits and translating that to the scientists in question. A seeker of technology admitted his organisation had no idea the partner university was so strong in the area in question. The university partner in this case reported "*we knew we could do what they wanted straight away*" when the seeker explained the need. Service users made reference to the value of the facilitation of the first meetings where the intermediary made introductions, set the agenda, kept the discussion focused and within the bounds of the confidentiality agreements.

In the technological services group the main intermediary activity is around IP advice. Service users reported assistance with sharing IP for mutual benefit and managing expectations concerning the way in which parties were planning to take IP ownership forward.

Beyond the functions carried out by the intermediary, service users recognised a range of other skills that were valued. A technology seeker pointed to the intermediary's breadth of knowledge and the ability to communicate with the engineers and technicians on equal terms allowing a close relationship between the organisation and the intermediary. Similarly a technology provider identified the intermediary had the ability to understand the technology and identify how it can be applied beyond its initial application; the service user recognised that the methodology went beyond "looking at keywords".

Service users reported that intermediary functions can often be done by their own organisations but the way in which they are done differs. Corporate partners are aware of benefits in early identification of the technologies, the assumption being that the corporate would find the technology in the end but by utilising the intermediary organisation the process is accelerated and quality improvements are seen. On the academic side the university's commercial team may be able to identify potential partners but in the experience of the interviewee factoring in demand intelligence has accelerated the rate of technology uptake when compared to a conventional technology push approach.

During interviews with the technology providers it emerged that both are managing the ongoing opportunity by forming spin out companies around the technology. The deals with established organisations are perceived as lending market confidence to the company and facilitating potential and actual venture capital investment. In addition to using the connection as a platform to spin out companies the universities are also benefiting from deepening relationships with the corporate partner.

Beyond the initial connection made by the intermediary other factors for success were also identified. One such factor was keeping in touch and communicating on the project, as one user remarked *"Openness is a key part of open innovation"*. The universities value the existence of an open innovation team at their corporate partners to facilitate technology transfer and open up further projects.

6 Discussion

The profile of the InnovationXchange shows a focus on brokering supported by technology scanning and foresight along with assistance with managing intellectual property. The intermediary organisation does not engage in commercialisation activities, testing, regulation or accreditation.

From the seven activities identified by service users scanning / information processing was most commonly valued by service users. Management of technology intelligence has also been identified by others as a key factor in the open innovation process (Kerr *et al.* 2006; Lichtenthaler 2004). InnovationXchange benefits from its position in both academia and industry to *"introduce and solve a problem in one swoop"*. Harder aspects to quantify and formalise are the way that InnovationXchange Intermediaries were able to:

  i. Transfer technology from one field to another.
  ii. Communicate with and understand the needs of both academics and industrialists.
  iii. Create an environment where both sides feel free to communicate with each other and take their relationship forward.

This combination results in a good technology fit being supported by a good cultural fit.

A further unquantifiable benefit of InnovationXchange's positioning was that the intermediary organisation had nothing inherently to gain from any connection. This meant that the intermediaries were viewed as apolitical and pragmatic. This position engenders trust and a positive relationship. Although, clearly, a track record of failure in connections would lead to the conclusion of the working relationship.

This paper has focused on the activities of UK hub of the InnovationXchange. Further research into the collective impact of the three operating hubs may prove to be of value.

The innovation intermediary industry is still in its nascent phase and research has shown that a gap commonly exists between expectations and delivery in the area (Lichtenthaler and Ernst 2008). Further profiling of successful projects could provide insight into the positioning of intermediary organisations and assist in the selection of an appropriate intermediary organisation based on the service user's needs.

Acknowledgments

Many thanks to the interviewees JM, GH, TC and CN and their organisations for their time and thoughts that were invaluable in conducting this research.

The research was funded by InnovationXchange UK to help the company better understand its role and how it can deliver value to its clients.

References

Burt, R.: Structural Holes Versus Network Closure as Social Capital. In: Lin, N., Cook, K., Burt, R.S. (eds.) Social Capital: Theory and Research. Sociology and Economics: Controversy and Integration series, Aldine de Gruyter, New York (2001)

Chesbrough, H.: Open Innovation: The new imperative for Creating and Profiting from Technology. Harvard Business School Press, Boston (2003)

Chesbrough, H., Vanhaverbeke, W., West, J.: Open Innovation: Researching a new paradigm. Oxford University Press, Oxford (2006)

Fincham, R., Clark, T., Handley, K., Sturdy, A.: Configuring expert knowledge: the consultant as sector specialist. J. Organ. Behav. 29, 1145–1160 (2008)

Howells, J.: Intermediation and the role of intermediaries in innovation. Res. Policy 35, 715–728 (2006)

Kerr, C., Mortara, L., Phaal, R., Probert, D.: A conceptual model for technology intelligence. Int. J. Technol. Intell. Plan 1, 73–93 (2006)

Lichtenthaler, E.: Technological change and the technology intelligence process: a case study. J. Eng. Technol. Manag. 21, 331–348 (2004)

Lichtenthaler, U., Ernst, H.: Innovation Intermediaries: Why Internet Marketplaces for Technology Have Not Yet Met the Expectations. Creat. Innov. Manag. 17, 14–25 (2008)

Lopez, H., Vanhaverbeke, W., Wareham, J.: Connecting open and closed innovation markets: A typology intermediaries (2010),
http://www.eiasm.org/documents/JMW/jmp/1181.pdf
(accessed September 8, 2010)

Mortara, L.: Partners to help with open innovation: the role of intermediaries (2010) (manuscript in preparation)

NESTA , Open Innovation From Marginal to Mainstream (2010),
http://www.nesta.org.uk/library/documents/
Open-Innovation-v10.pdf (accessed September 6, 2010)

Obstfeld, D.: Social Networks, the Tertius Iungens Orientation and Involvement in Innovation. Adm. Sci. Q 50, 100–130 (2005)

Smith, C.: The Wholesale and retail markets of London 1660-1840. Econ. Hist. Rev. 55 (2002)

Technology Strategy Board, Connect and Catalyse. HMSO, London (2008)

Wolpert, J.: Breaking out of the innovation box. Harvard Business Review 80, 76–83 (2002)

Wolpert, J.: Adding intention to innovation. Fast Thinking (2006),
http://www.fastthinking.com.au/the-magazine/
autumn-2006/adding-intention-to-invention.aspx
(accessed September 8, 2010)

Abbreviations

IP	Intellectual Property
KTN	Knowledge Transfer Network
NESTA	National Endowment for Science Technology and the Arts

Appendix

Detail of functions of intermediaries. After Howells (2006) and Lopez *et al.* (2010)

Group	Activity	Functions
Connection Group	Gatekeeping and brokering	(a) Matchmaking and brokering
		(b) Contractual advice
	Middlemen between science policy and industry	
	Demand articulation	
Collaboration and support group	Foresight and diagnostics	(a) Technology foresight and forecasting
		(b) Articulation of needs and requirements
	Scanning and information processing	(a) Scanning and technology intelligence
		(b) Scoping and filtering
	Knowledge processing, generation and combination	(a) Combinatorial
		(b) Generation and recombination
	Commercialisation	(a) Marketing support and planning
		(b) Sales network and selling
		(c) Finding and organising potential capital funding or offerings
		(d) Venture capital
		(e) Initial public offering
Technological services group	Testing, validation and training	(a) Testing, diagnostics, analysis and inspection
		(b) Prototyping and pilot facilities
		(c) Scale-up
		(d) Validation
		(e) Training
	Accreditation and standards	(a) Specification setter or providing standards advice
		(b) Formal standards setting and verification
		(c) Voluntary and de facto standards setter
	Regulation and arbitration	(a) Regulation
		(b) Self-regulation
		(c) Informal regulation and arbitration
	Intellectual Property	(a) IP rights and advice
		(b) IP management for the client
	Assessment and evaluation	(a) Technology assessment
		(b) Technology evaluation

The ISSUES Project: An Example of Knowledge Brokering at the Research Programme Level

Katarzyna Przybycien, Katherine Beckmann, Kimberley Pratt, Annabel Cooper, Naeeda Crishna, and Paul Jowitt

Abstract. This paper examines knowledge brokering as a method of knowledge transfer at the research programme level using the ISSUES project as an example. The analy-sis is undertaken in the context of existing theories of knowledge bro-kering, fo-cussing on the three roles of knowledge brokers: knowledge managers, linking agents and capacity builders. To illustrate the nature of brokering at the pro-gramme level, the authors propose two models: the 'one-to-one' model, where brokers work with individual producers and users of knowledge, often supporting the transactional side of knowledge transfer; and the 'many-to-many' model, where brokers mediate between multiple producers and users of knowledge to en-courage the formation of individual relationships. This latter model has particular relevance to large and complex research programmes. Using examples from the ISSUES project, the paper recommends that future applications of this approach may benefits from embedding knowledge brokering into the work of the research programme as well as coordinating it with other knowledge transfer schemes.

1 Introduction

Recent governmental reviews such as Lambert Review (2003), Warry Report (RCUK 2006), Excellence with Impact (RCUK 2007a) consistently highlight the importance of knowledge transfer in academic research. Yet in the past funding for knowledge transfer available from research councils has appeared inconsistent and uncoordinated, causing confusion for researchers and end users alike (House of Commons Science and Technology Committee 2006). To overcome this, Re-search Councils conducted a review where groups of knowledge transfer schemes have been identified (RCUK 2007). They included: Dissemination, Placements, Secondments, Exchanges, Collaborative Research Fellowships and Research Bro-kering. This paper focuses on the latter grouping: 'Research Brokering'.

For Research Councils, brokering is achieved by employing '...a specialist who can identify opportunities for the commercialisation and exploitation of re-search and broker the linkages' (RCUK 2007). However, this definition ignores other dimensions of brokering that may also be undertaken by collective bodies. In the policy context these might include 'science advisory committees, Govern-mental research institutes, learned societies, consultancy firms, and think tanks...' (Holmes and Clark 2008). Large research programmes funded by research

R.J. Howlett (Ed.): Innovation through Knowledge Transfer 2010, SIST 9, pp. 297–307.
springerlink.com   © Springer-Verlag Berlin Heidelberg 2011

coun-cils may also benefit from tailored research brokering that extend the tradi-tional dissemination and networking approaches.

This paper explores how the 'many-to-many' model may help to contextualise brokering activities performed at research programme level, and puts this in con-text with examples from ISSUES.

1.1 Knowledge Brokering

Knowledge brokering refers to the '...processes used by intermediaries (knowl-edge brokers) in mediating between sources of knowledge (usually in re-search) and users of knowledge' (Bielak et al. 2008). The general purpose of knowledge brokering is to improve knowledge exchange for the wider benefit of all (Bielak et al. 2008) but sometimes its aims can be much more specific and tan-gible, for ex-ample: '... to identify opportunities for commercialisation and exploi-tation of research' (RCUK 2007). In either case the pro-activity of the broker in making linkages between researchers and end users is implied. Brokering may occur be-tween researchers and end users with commercial aims or may be under-taken for the 'public good', then filtering down to policy makers and practitioners who act on the evidence. By understanding the needs of involved parties, broke-ring can improve two-way information flow and thus better match between know-ledge 'push' and knowledge 'pull' (Bielak et al. 2008). Knowledge brokers can be indi-viduals, projects, organisations or bigger organisational structures. They are the 'intermediaries [...] who link the producers and users of knowledge to strengthen the generation, dissemination and eventual use of [...] knowledge' (Bielak et al. 2008). These activities fall roughly into three roles performed by knowledge bro-kers that Meyer describes as 'knowledge managers', 'link builders' and 'capacity builders' (Meyer 2010). This categorisation provided a framework to analyse ac-tivities conducted by the ISSUES Project as a programme level knowledge broker. In the following sections this paper will describe the current role of knowledge brokers and introduce two new models for conceptualising bro-kering. This will be followed by an analysis of the ISSUES Project as a know-ledge broker, illus-trated with activities which cover its roles as a knowledge ma-nager, linking agent and capacity builder.

1.2 Models of Knowledge Brokering Activities

Analysis of the existing forms of knowledge brokering reveals two distinct mod-els: 'one-to-one' and 'many –to-many'. The distinction is based on: the role of the broker; the method and purpose of brokering; and measurability. Other combina-tion models such as 'one-to-many' and 'many-to-one' had limited relevance for our work and so they have not been included in this analysis.

1.3 The 'One-to-One' Model

'One-to-one' brokering aims to support a specific research finding or expertise being utilised in a given context and occurs between a specific researcher and end user. The involvement of the broker in this process is direct and practical. Their support covers both finding the right stakeholders for engagement, and leading both parties through the technical and transactional process of knowledge transfer. The purpose of their involvement is 'to make things happen' and make sure the engagement goes smoothly. Examples of this type of brokering are likely to produce immediate, tangible results and quantitative metrics. Such brokering is often applied when research findings are already defined, tangible and have potential to be commercialised.

Currently, many funded brokering activities fall into the category of the 'one-to-one' model. They either have specific aims for finding opportunities for, and facilitating the commercial application of tangible research findings or academic expertise (examples from UK Research Council funded projects include: the Technology Translators at Engineering and Physical Sciences Research Council or Research Translators at Medical Research Council) (RCUK 2007). Alternatively, end user driven brokers seek to find solutions for particular problems (2KT project). In both cases the activities lead to a specific research result, expertise or a specific end user problem becoming the driver for an interaction between the knowledge producer and the user.

Fig. 1 'One-to-one' model

1.4 The 'Many-to-Many' Model

'Many-to-many' brokering applies where many research findings are relevant to many, diverse end-user communities. This type of brokering is typical of a complex research programme concerned with multifaceted research questions. The findings are often less straightforward than a specific solution to a problem and they rather they support better informed decision making. The uptake of this knowledge can happen through gradual influence on end user practice and this in itself may eventually stimulate 'one-to-one' interactions and the resultant utilisation of research. It is difficult to measure the results of this type of brokering due to the more diffuse nature of its outcomes or lack of awareness of the longer term effects. Indicators of activity provide measures of impact and therefore aid the assessment (Meagher et al. 2008). Here brokers do not represent any individual findings, expertise or their application but look at them collectively, enabling and encouraging individual interactions but not getting involved in them. Therefore the brokering that occurs in the 'many-to-many' model involves creating an environment for engagement, rather than connecting specific stakeholders and leading them through the process of knowledge transfer.

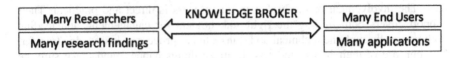

Fig. 2 'Many-to-many' model

1.5 The SUE Programme

The Sustainable Urban Environment (SUE) programme provides an example of a complex research programme where knowledge brokering through the ISSUES Project has followed the 'many-to-many' model. The EPSRC funded SUE Programme consists of eighteen consortia, involving researchers from 40 UK research institutions and 120 partners from the private and public sector[1]. The original twelve consortia were known collectively as SUE 1. The consortia were categorised into clusters around four themes: Urban and Built Environment; Waste, Water and Land Management; Transport; and Metrics, Knowledge Management and Decision Making. These were followed by six further consortia, known as SUE 2, which are currently at their mid-way point. Each consortium is cross-disciplinary, combining research groups from different backgrounds.

1.6 The ISSUES Project

The ISSUES Project[2] is the knowledge transfer arm of the SUE programme. It is a collaboration between Heriot-Watt University and the University of Cambridge. The ISSUES project facilitates the collective knowledge transfer of research outputs and outcomes from the SUE Programme to the potential end user communities from policy and practice. It began in January 2007 and will end in December 2010, spanning the end of the SUE 1 projects and ending half way through SUE 2.

2 The ISSUES Project as Programme Level Broker

Meyer's classification of knowledge brokering roles as 'knowledge managers, linking agents and capacity builders' has been used as a framework to put the activities undertaken by the ISSUES Project in the context of its role as a 'many-to-many' knowledge broker (see Table1). Each of these roles informs and justifies the other, and so it ensures a holistic approach to the knowledge transfer process, addressing the critique of many knowledge transfer and dissemination activities lacking strategic approach (Nutley et al. 2007).

---

[1] eg. city councils, architectural and engineering companies, charities, associations, etc.
[2] www.urbansustainability.co.uk

Table 1 Table of ISSUES Activities

ISSUES Activity	Knowledge managing	Linking	Capacity building
Information portal/search engine (SUE Gateway)	X	X	
Web directory of expertise (SUE Gallery of Experts)	X	X	
Online video portal (SUE Explains Vidiowiki)	X		X
Events (Ebbsfleet Challenge, Brave New City, etc.)		X	X
Movie poster tour	X		
Issues research based publications and reports	X		X
Feature articles about SUE in trade media	X	X	
Summarising research portfolios into plain language	X	X	
Newsletters, postcards, events	X	X	
SUE Exchange meetings for researchers		X	X
KTN network '_connect' platform		X	
Advocacy meetings	X	X	X
KT Guidebook for Researchers			X
Impact training for Researchers			X

2.1 Knowledge Managers: Searching, Categorising, Translating and Redistributing Knowledge

'Many-to-many' brokering requires achieving a balance between generalisation, ategorisation and a tailored approach.

2.2 Research and Categorisation

The initial stages of the ISSUES project focused on establishing knowledge foundations about SUE research, end-user audiences in urban environment and the processes of knowledge transfer itself. This exercise confirmed the existence of a gap between the supply and demand of knowledge in the urban environment (Crishna et al. 2010; Moncaster et al. 2009); the domination of a 'knowledge push' approach and use of printed formats in its execution; and the varying level of the perceived importance of knowledge transfer amongst researchers from relatively high to non-existent. It also revealed that academic jargon and an overload of information can be significant obstacles to knowledge uptake (Beckmann and Mason 2010).

2.3 Translation of Knowledge

The translation of knowledge in the 'many-to-many' context is an important and challenging task because the disciplines, language and information formats are varied and many. The translation of findings focused therefore on jargon free communication and use of common communication tools and formats. Knowledge translation occurred in various forms, from summarising research portfolios into plain language and distilling complex reports, to formatting the information using new media and online tools. Such translation activity can be exemplified by the SUE EXPLAINS Vidiowiki. This interactive forum consists of a network of linked video summaries by researchers, policy makers and practitioners who explain their work in a three minute synopsis. Throughout ISSUES, use of new media technologies and communication techniques was aided by employing a journalist who had experience in research communication.

2.4 Dissemination

The dissemination of multiple research findings into multiple destinations was a feature of 'many-to-many' brokering. It involved the creation of a coherent representation of the SUE programme as a whole, as well as providing tailored information from specific consortia to specific locations. The approach consisted yet again of managing the balance between 'pushing information out' through collective or tailored broadcasting of SUE and stimulating 'knowledge pull', by 'whetting the appetite' of stakeholders.

2.4.1 Effecting 'Knowledge Push' in the ISSUES Project

The ISSUES project conducted general dissemination, aimed at diffusing information and raising the profile of SUE research amongst urban environment stakeholders and the general public. Traditional dissemination strategies such as printed and internet based marketing materials, newsletters and organised events were complemented by less common approaches. Focus was placed on the creation of online tools, drawing on earlier research findings that practitioners preferred access to new knowledge via the internet and ideally through one portal (Beckmann and Mason 2010). These also helped to create a coherent public interface for the SUE Programme.[3]

The second part of the approach was targeted dissemination, which involved sending information to recognised sources of new knowledge for urban environment practitioners (Beckmann and Mason 2010). To date, over twenty articles have been published in trade journals and professional magazines. Finally,

---

[3] The web based tools included: a) the 'SUE Gateway' portal to search for evidence and tools in the SUE Programme from one single online space; b) the 'SUE Gallery of Experts' with summaries of and access to the expertise created by the SUE Programme, and c) the SUE EXPLAINS Vidiowiki, featuring summaries of SUE research presented by SUE researchers themselves (http://www.urbansustainabilityexchange.org.uk)

advo-cacy methods, such as meetings with government representatives (for example from Department for Environment, Food and Rural Affairs, Department for Transport, Department of Communities and Local Government), were undertaken to promote SUE as a sound and accessible source of evidence for policy making.

2.4.2 Generating Knowledge 'Pull' in the ISSUES Project

The ISSUES Project also used methods to specifically attract or 'pull' potential end users towards research produced by SUE. An example of this was the 'Brave New City' event series, held in Edinburgh and London. The invited panellists were asked to introduce a clip chosen from the ITN Source broadcast footage provider to spark off their discussion of the Brave New City. The stature of the speakers attracted interest in the event from the outset and ISSUES used it to attract two media partners: *The Architects' Journal* and *IKT magazine*, as well as a host of attendees who would be unlikely to attend a traditional academic dissemination event (The overwhelming majority (75%) of the audience were practitioners from both the private and public sector). This 'Trojan horse' approach to event dissemination ensured a more high profile panel, greater press coverage and attracted an audience not accustomed to attending 'academic' events or events about KT.

2.5 The ISSUES Project as a Linking Agent between Stakeholders

In the context of the 'many-to-many' model the linking process needs to occur within the research community itself, as well as between the researchers and the end user audience.

2.5.1 Creating Links Between the Knowledge Producers

The size and geographical spread of the SUE research programme was significant, and because of this, one of the first tasks of the ISSUES project was to create a coherent 'SUE identity'. This was achieved by publishing joint materials and creating a collective online presence such as the 'Gallery of Experts', where SUE researchers from across the programme feature as a collective body of knowledge on the urban environment. As well as creating these 'virtual' interfaces, ISSUES project provided opportunities for SUE researchers to meet, exchange experience and undertake joint actions by organising events.

2.5.2 Creating Links with End-User Communities

Due to the nature of the research, defining specific end users for research findings was challenging and resulted in a widespread approach to link building. End users included policy makers and practitioners from the fields of urban and transport planning, engineering, architecture, building and so on, each with their own modes and methods of practice. This diversity of practice across the public and private spheres is characteristic of the 'many-to-many' model. Linking SUE researchers with their diverse end user groups was achieved via several mechanisms and creating an environment for knowledge exchange and networking on both real and virtual platforms. Contemporary studies have suggested that the creation of informal contact and networks are the enabling factors for successful knowledge transfer (Holi et al. 2008; Meagher et al. 2008) and therefore ISSUES sought to provide various opportunities for these to occur. The physical spaces, where researchers could meet the potential end-users of their research face to face, provided opportunities for pitching research findings, for discussing the varying communication channels and techniques, and establishing new contacts and networks.

2.6 The ISSUES Project as Capacity Builders

Earlier findings from the ISSUES Project suggest there is a lack of shared language and capacity on both sides to bridge the gap between research and end user communities. The ISSUES project focused predominantly on overcoming this problem by building the capacity of knowledge producers to 'push' the knowledge out in a way that would stimulate knowledge 'pull' by end-users. This often included using tools reaching beyond media traditionally used in academia.

The ISSUES project also aimed to contribute to a culture change which promotes embedding knowledge transfer principles within research agendas. In order to achieve this the ISSUES project delivered workshops designed to give researchers practical tools for achieving effective knowledge transfer and organised events where end-users were invited to explain their knowledge seeking habits, or where researchers exchanged their good practices in regard to dissemination (SUE exchange meetings). The capacity building activities will go towards the production of 'a Knowledge Exchange Guidebook for researchers. We will also produce some materials relevant for funders and end users.

3 Barriers to ISSUES Project and Measures to Address Them

The ISSUES project encountered some challenges in its role as knowledge broker for the SUE programme. These were mostly related to the setting of the project within SUE (its scheduling, role, legitimacy and expectations). Firstly, its role as programme level broker was never embedded in the knowledge transfer strategy at the consortia level. Therefore, building legitimacy in the minds of the researchers took time and may have caused a loss or delay to the impacts of brokering, which has been recognised by other studies of knowledge brokering (Lomas 2007). This may have been exacerbated by the late start and early ending to the ISSUES project in relation to the lifespan of the SUE consortia. These challenges could be overcome by embedding the role of knowledge broker from the outset and more cohesive scheduling. Additionally because the ISSUES project was detached from the individual consortia in its role as broker for the whole SUE programme, it could not participate in the transactional part of the knowledge transfer. This programme level engagement may have stimulated the interest and action of the researchers and end-users to undertake the detailed one-to-one interactions but this process could be performed by the programme level broker themselves. Therefore closer integration of the ISSUES project with other knowledge transfer schemes, such as Follow-on Funding (RCUK) might have improved signposting and extended the direct impacts of brokering. Finally, a particular challenge for the project was the measurement of its impacts. Due to the long term and often intangible/conceptual nature of project impacts (e.g. increased awareness, changed behaviour) proxy indicators of input activities were used which allowed the team to assess what has been done to make an impact but not the impact itself. Realistic definition of the outputs and outcomes of the project on the outset approved by all stakeholders might have enhanced measurability of the impact.

4 Discussion and Conclusions

This paper has investigated the role of knowledge brokering in the knowledge transfer process, using examples from the ISSUES Project. Traditional brokering fits into the 'one-to-one' model, where the relationship between end-user and researcher is facilitated by the broker. However, programme level brokering may occur among many researchers and many end users. In this case, the three roles of the broker (knowledge manager, linking agent and capacity builder) must be adapted. 'Many-to-many' brokers aim to prepare and empower both researchers and end users to become involved in knowledge transfer, rather than conducting it for them. The paper identifies five features of knowledge brokering that distinguish it from other programme level knowledge transfer approaches.

Knowledge Push and Pull

Creating a 'pull' from end users can be increased by improving 'push' from researchers. Therefore capacity building activities focusing on researchers' ability to better disseminate and engage in knowledge transfer are crucial. Finding the right balance between knowledge 'push' and 'pull' can be ensured by ongoing research and formal and informal feedback loops.

Complementary approach

The ISSUES Project created networking environments that encouraged the creation of one-to-one relationships, where instrumental use of knowledge could occur. In this way, the 'one-to-one' and 'many-to-many' models are not mutually exclusive. Rather, in order to create an environment for effective knowledge transfer, all these approaches must be adopted in concert.

Ensuring Legitimacy

Establishing the ISSUES Project as legitimate part of the SUE Programme and even establishing the importance of knowledge transfer itself in the minds of the researchers took time. This delay could have been avoided if the ISSUES Project had been embedded in the structure of the programme and its role made clearer to researchers from the outset.

Being Heard in the Modern World

The ISSUES Project used modern communication technologies and references to popular culture to extend and regenerate traditional dissemination techniques. This has been due, in a large part, to the inclusion of a journalist on the ISSUES team

Creating Programme Level Identity

Part of the ISSUES Project's remit was the creation of a single identity for the SUE research programme, which helped end users distinguish the programme and its remit from other research and access researchers and findings more easily. It may also have strengthened links within the SUE, allowing synergies amongst SUE researchers and was in the interest of the funders who seek public recognition of the research they fund.

In conclusion the analysis of the brokering activities conducted by the ISSUES Project has illustrated that 'many-to-many' broker offers a holistic and strategic approach to research programme level knowledge transfer. Embedding the broker into the research programme and enabling links with other knowledge transfer schemes may help maximise the impact of many-to-many brokering.

References

2KT Project, Transferring Knowledge to Industry (2008), http://www.2kt.org.uk (last Accessed October 5, 2010)

Beckmann, K., Mason, C.: An Investigation of the Knowledge Exchange Practices of the End-Users of Sustainability Research. A Project Report for the ISSUES Project, Edinburgh (2010), http://www.urbansustainabilityexchange.org.uk/ (last Accessed: October 1, 2010)

Bielak, A.T., et al.: From science communication to knowledge brokering: the shift from 'science push' to 'policy pull. In: Cheng, D. (ed.) Communicating Science in Social Contexts, pp. 201–226. Springer, New York (2008)

Crishna, N., et al.: Bridging the gap - understanding the barriers to the exchange of knowledge between urban sustainability research and practice. The Environmentalist (106), 14–18 (2010)

Holi, M., Wickramasinghe, R., van Leeuwen, M.: Metrics for the Evaluation of Knowledge Transfer Activities at Universities. Library House, Cambridge (2008)

Holmes, J., Clark, R.: Enhancing the use of science in environmental policy-making and regulation. Environmental Science & Policy 11(8), 702–711 (2008)

House of Commons Science and Technology Committee Research Council Support for Knowledge Transfer: Third Report of Session 2005-06: Report, together with formal minutes, oral and written evidence, p.1.The Stationery Office. London (2006)

Lambert Review : Lambert Review of Business-University Collaboration (2003), http://www.hm-treasury.gov.uk/ lambert_review_business_university_collab.htm (last accessed: October 5, 2010)

Lomas, J.: The in-between world of knowledge brokering. British Medical Journal 334(7585), 129–132 (2007)

Meagher, L., Lyall, C., Nutley, S.: Flows of knowledge, expertise and influence: a method for assessing policy and practice impacts from social science research. Research Evaluation 17, 163–173 (2008)

Meyer, M.: 'The Rise of the Knowledge Broker'. Science Communication 32(1), 118–127 (2010)

Moncaster, A., et al.: The big issue: knowledge exchange practices between academia and industry. Engineering Sustainability 163(3), 167–174 (2009)

Nutley, S., Walter, I., Davies, H.: Using evidence: How research can inform public services. Policy Press, Bristol (2007)

RCUK, Research Councils UK, Knowledge Transfer Categorisation And Harmonisation Project (2007)

RCUK, Research Councils UK, Increasing the Economic Impact of the Research Councils (the Warry Report) (July 2006)

RCUK, Research Councils UK, Excellence with impact: Progress in implementing the recommendations of the Warry Report on the economic impact of the Research Councils, RCUK, Swindon (2007a)

RCUK, Research Councils UK Follow-on funding,
http://www.rcuk.ac.uk/innovation/ktportal/followon.htm
(last accessed October 5, 2010)

Student Supported Consultancy to Address Market Needs: Leeds Source-IT, a Case Study

Alison Marshall, Royce Neagle, and Roger Boyle

School of Computing, University of Leeds, Leeds, LS2 9JT, United Kingdom

Abstract. As universities become more commercially oriented, the demand by local businesses for small scale consultancy increases. Although there is a growing interest from within the academic community to undertake consultancy, as part of a portfolio of commercial activity, there are also real tensions with the core activities of research and teaching. A pilot software services initiative, using students as consultants, is described. The challenges faced are discussed in the context of benefits to the consultants, the host university and to the regional economy.

1 Background and Context

Many Universities now actively promote themselves as consultancy providers, often specifically to regional SMEs and sometimes in partnership with local business support agencies. There is growing recognition that consultancy can be an important income generating activity for a University and a perception that it may have additional, non-financial benefits for those taking part. An analysis of university linkages with industry in mid-range universities in UK, Belgium, Germany and Sweden [1], the value of 'consultancy and reach-out' is described in terms of transfer of tacit knowledge. The authors stress its importance as a knowledge transfer mechanism and superiority over licensing for some types of interactions. Interviewees are also reported as suggesting that consultancy is a better way to understand industry's problems and hence inform research. For smaller scale projects, it can be a more flexible mechanism. In a survey of over 4000 academic staff in UK universities, who had received funding from the Engineering & Physical Sciences Research Council (EPSRC), the evidence is that consultancy supports more direct commercialisation activities and vice versa [2]. Indeed, their findings show that the 'star researchers' tend to participate in university-industry linkages through multiple channels. Analysis of entrepreneurship activity in Sweden and Ireland [3] through structured questionnaire responses from 1857 academics, indicated that 51% of Swedish respondents and 68% of Irish respondents undertook consultancy and that this was a much higher level of involvement than was seen in patenting, licensing and spin off activity. The study found that the level of consultancy activity very often took place in spite of limited institutional support (and in a small minority of cases, institutional hindrance) for the practice of consultancy.

R.J. Howlett (Ed.): Innovation through Knowledge Transfer 2010, SIST 9, pp. 309–318.
springerlink.com   © Springer-Verlag Berlin Heidelberg 2011

Many academics were wary of the commitment required for spin off companies and found consultancy a more manageable way to generate extra income. In spite of this evidence, it still appears that in many cases consultancy assignments do not fit naturally into the academic structure. Unlike research contracts, which have a long lead time and are designed to generate publications as well as commercial deliverables, consultancy assignments have to be delivered to meet the client's needs and the client's timescales. Often there is a prohibition on publication and the nature of the task can be seen as uninteresting and diversionary. Even when the university allows a significant personal remuneration to the academic consultant (as does the University of Leeds), the incentive to take on a consultancy assignment, rather than do something more directly related to the 'day job', is not judged to be sufficient by many individuals. Academic staff are still rewarded and promoted largely on the basis of their research and teaching outputs, not their commercial work. The issue of 'ambidexterity' is discussed in an in-depth study of the commercialisation potential of 207 EPSRC funded research projects [4]. This paper builds on organisational ambidexterity literature to examine the conflicts between academic and commercially oriented activities, for a research-oriented university. The authors report that university structures have adapted over recent years to this tension, but individual academics find this transition harder. The study notes the development of 'dual structures' within institutions and the ability of 'star scientists' to negotiate this dichotomy. These 'star' individuals often have had career experience in industry.

Meanwhile, the market for technical consultancy work is strong, enhanced (at least until very recent times) by national and regional government agency drivers to facilitate interaction between industry and universities. Large and small companies have been encouraged to contact their local university for assistance with, amongst other things, product development, technology selection, technical troubleshooting, testing and analysis. Individual entrepreneurs also seek to build relationships with an organisation that has an appropriate technical infrastructure to keep costs low in the early stages of business start up. Yorkshire Forward [5], the regional development agency for Yorkshire & the Humber has established 10 Centres for Industrial Collaboration (CICs), each covering addressing a different industrial sector and involving one or more of the nine regional universities [6]. More recently, Yorkshire Forward and other regional development agencies have provided the Innovation Voucher Scheme [7], which subsidises consultancy contracts of up to £3000 for work undertaken by universities or other registered 'knowledge providers'. The hypothesis that investment in research universities can improve the regional economy is tested in a statistical analysis of data from universities rate as top research institutions by the USA National Science Foundation [8]. The author introduces the concept of university 'products', which include contracted research, trained labour, technology diffusion, new knowledge, new products & industries and proposes a model for how these elements interact with a technology-based economy. The importance of 'knowledge spillover' is highlighted, in which companies and universities interact closely on a number of levels.

A direct evaluation of a particular initiative has been undertaken by Lockett et al [9]. Interviews were conducted with small business clients and staff (academic and commercial) of Lancaster University's InfoLab21 [10]. InfoLab21 provides a range of consultancy, contract research and business incubation facilities, in partnership with the North West Development Agency under a model not dissimilar to the Yorkshire Forward CICs. Interviewees were asked to identify barriers to knowledge transfer and noted a) lack of time and different perceptions of time scales; b) intellectual property rights issues and incentivisation of university staff towards publishing research; c) the perception that SME problems are not likely to lead to 'cutting edge' research. Knowledge transfer is still seen as the 'third arm' after teaching and research. The study suggests that the 'market pull' for technology and knowledge transfer may be overestimated by public policy, although this is possibly due to a lack of appreciation of the benefits on both sides.

The tension of market pull that cannot be satisfied easily by universities is therefore partly due to historical and cultural attitudes, but also stems from real structural difficulties. At the University of Leeds, efforts have been made to remove barriers to consultancy work. A dedicated structure has been established within a subsidiary company (Consulting Leeds Ltd). Academic consultants and their parent School are directly remunerated. Processes, supported by specialist staff, are in place to deal with contractual issues, to help resolve any conflicts during delivery and to collect payment [11]. However, the fundamental barrier remains - that of non-compatibility with research and teaching roles. Where a strategic and ongoing market need is identified, such as with the CICs or InfoLab21, dedicated consultancy staff can be employed, who can respond quickly to enquiries and carry out both technical and commercial work with minimal involvement from academic staff. In some cases, this activity grows into a self-contained unit, which funds itself independently and exists within the university structure. Wright et al [1] and Lockett et al [9] independently note the importance of the role of intermediaries, both within and outside the university structure. These intermediaries are most effective if they have, or can develop, technical knowledge and credibility. This enables them to negotiate very directly on behalf of the academic unit – and in some cases to even deliver the consultancy themselves. However, ad hoc (or indeed early stage strategic) work is harder to resource and to justify. The result is that many enquirers are simply turned away as there is no internal structure to meet their needs, although there may be suitable expertise within the organisation.

Leeds Source-IT ('Source-IT') was formed to address the tension described, to service a 'market pull' and provide solutions to local companies. The School of Computing often receives enquiries for small scale programming and web development work. The growth of on-line social networking and wider access to the Internet has created increasing interest in businesses based on web applications and Web 2.0. These enquiries are very difficult to service. They are generally too small scale and too low level to be of interest to academics. Students could work on an external task as part of a project, but in practice the lead time is too long and there is no control by the company over the quality and approach. There are formal mechanisms (such as Knowledge Transfer Partnerships, EPSRC Industrial CASE Awards) to employ a postgraduate or postdoctoral project worker, but the tasks involved are often not suitable and the company generally wants a quick delivery. We therefore established Source-IT as a formal mechanism to meet this market need. The extra ingredient that enabled this to happen was the driver for personal development, work experience and financial remuneration by students and new graduates. This gave access to a community of potential consultants who could benefit directly.

In this paper we seek to examine whether the Source-IT model, in particular, and the structural model implied by it, in general, is successful in delivering value to its stakeholders. The paper is structured as follows. The next section describes how Source-IT was structured and launched. Section 3 presents factual details of the eight projects undertaken during the pilot year, which are analysed and discussed in Section 4. Finally, in Section 5, some conclusions are drawn and suggestions presented on the lessons learned and relevance to future policy.

2 Overview of Leeds Source-IT

Leeds Source-IT was established in direct response to enquiries from regional SMEs and individual entrepreneurs for small scale, low level programming and web application development work. The nature and size of the tasks made them largely unsuitable for any existing formal mechanisms (such as academic-led consultancy or Knowledge Transfer Partnerships – Shorter or Classic). Source-IT was established to enable small scale business to be serviced by students, within a formal consultancy structure. The key goals were:

- To work principally on prototypes and early stage development tasks for SMEs and individual entrepreneurs, at a level and scale that could be serviced by part-time programmers;
- To deliver within time and budget and at a quality consistent with the University of Leeds' brand and reputation;
- To provide students with valuable work experience and income, within a 'safe environment'.

Source-IT was launched as a pilot from September 2008 to November 2009, with management staff deployed on a part time basis. Source-IT was co-funded by two different sponsors with different goals in educational development (WRCETLE) and knowledge transfer (HEIF4), respectively. The pilot therefore sought to meet both of the sponsors' goals and to benefit the students as well as the University.

Source-IT was commercially administered through the University's existing (staff) consultancy channels, by the wholly owned subsidiary company Consulting Leeds Ltd (CLL). CLL provided indemnity insurance, credit checks, billing and financial management. Project management was handled locally by the Source-IT Technical Manager (20% FTE), using an online project management and document sharing system, configured specifically for this purpose. Students were expected to work an average 5 hours per week. Contracts were therefore structured into tasks and sub tasks so that a team of students could deliver a typical contract of 50-100 hours within an acceptable time period for the client. Business planning business development was undertaken by the Source-IT Commercial Manager (10% FTE), a member of the Faculty of Engineering Research & Innovation Centre. She led the sales, marketing and contract negotiation activity. The 'consultants pool' were drawn from undergraduate, postgraduate and recent alumni. In September 2008, a formal application process (which included a number of compulsory training workshops) recruited the first 'pool' of consultants. Some of these consultants left the pool the following summer (eg. because they completed their studies and went to full time work). The recruitment exercise was repeated in September 2009, with further students joining. Once recruited to the pool, consultants were given the opportunity to join contract teams, as work came in. Not all of the consultants worked on projects, often because they had coursework or other commitments at the time the work was needed.

Initially Source-IT benefited from securing a small number of 'softer' internal contracts, which provided a way to test project management and operational systems, while responding opportunistically to incoming enquiries. The first external contract was not secured until November 2008, when 2 internal contracts had already successfully started. At the start of 2009, Yorkshire Forward, the Regional Development Agency, launched their Innovation Voucher scheme. The Innovation Voucher scheme offered a mechanism for SMEs (including micro-companies and sole traders) to receive a 100% subsidy for a £3000 consultancy contract with a local University. This provided Source-IT with a useful stream of 1-5 enquiries per month, which fitted the size and scope of contract that it could easily service. As a result, sales and marketing activity was ramped up significantly from April 2009, with an increased rate of both quotations and contracts. In June-July 2009, a telesales company was engaged to undertake a sales campaign and generate a qualified prospects list. As a result of the increased prospects, two students were employed full time on summer internships and were able to be fully deployed on contracted work.

In the autumn of 2009, Source-IT had completed its first 12 month pilot and was assessed by the managers and internal sponsors. Year 1 financial accounts showed a 'book loss' of around 10% of turnover (when internal management costs were included). The break-even point was projected to be in Year 3, once turnover could be increased to effect economies of scale. Improvement in profitability was anticipated through building up libraries of code to be re-used on different jobs and through more repeat business contributing to lower management time demands. The telesales campaign had generated an impressive sales pipeline and word of mouth through Yorkshire Forward and Business Link partners were beginning to increase enquiry rate. However, there was a projected requirement for internal underwriting for a further 1-2 years, which was not available. Source-IT was therefore wound up and all outstanding contracts completed by March 2010.

3 Source-IT Projects

In its pilot period, Source-IT delivered a total of eight projects. All were contracted on a commercial basis, although four of them were to internal clients (in different parts of the University and using their own project funding). The external clients were all sole trader entrepreneurs, of which two were not yet trading. Source-IT was priced at a day rate comparable with competitor software engineering contractors, but at the lower end of the range. The main facts about the projects are summarised in Table 1.

Table 1 Key facts on the eight completed Source-IT projects

Project No.	Consultants	Consultant hours	Nature of project	Issues
1	3	105	Prototype web application	Project overrun due to mismatch of time availability between client and consultants. Client later had deployment issues which have limited usage, but otherwise successfully completed
2	2	162	Prototype web and mobile application	Successfully completed
3	2	15	Technical specification for web application	Successfully completed. Client did not have funding for next stage.
4	1	45	Management information tool	Successfully completed, deployed and piloted.
5	3	75	Management information tool	Successfully completed. Client did not have funding for next stage.
6	1	66	Management information tool	Successfully completed, deployed and piloted.
7	6	151	Prototype web application (parts)	Specification not clear enough and hence mismatch between client expectations and delivery. Problems arose, the project did not proceed to later stages. Project was loss making.
8	1	75	Technical specification and part prototype for web/mobile application	Successfully completed.

Source-IT consultants delivered a total of 694 billed hours, which included weekly project reporting to the Technical Manager and client demonstrations/meetings. In the first academic year 2008-09, 10 consultants from a total pool of 20 were employed on projects. These included undergraduates, postgraduates and recent alumni. The pool increased to 26 in 2009-10, of which 11 were new recruits (second and third year undergraduates). All the postgraduates were retained from the previous year, three of the third year undergraduates continued as postgraduate or alumni consultants. Of these 26 consultants, 9 were deployed on projects before the initiative was wound up.

The educational benefits to the participating consultants were surveyed and have been reported [12]. Students interviewed noted that they had benefited from the team working experience and the opportunity to work on real world problems. At least one student later gained his first employment partly as an outcome of participating in Source-IT. Consultants were paid a fixed fee per task, agreed in advance, but based on a relatively generous hourly rate. If the task took longer than had been anticipated, they would have to absorb the extra cost, but overall the fee was generally thought to be attractive.

Feedback from clients was also positive. Most projects were completed to the client's satisfaction and the client was pleased to note that the work was done efficiently and to a high standard. However, in many cases the results of the projects were not used due to budgetary issues or because the client decided not to proceed with the project for another reason, which was sometimes disappointing to the students. Only one project caused Source-IT real delivery problems (Project 7) and made a loss. Even in this case the client noted on several occasions how much she enjoyed working as part of the University environment and sharing the energy and enthusiasm of the student team. Project 7 was the most ambitious contract undertaken and was to be a prototype of a quite large web application for a start up business. The client was working separately with a team of graphic designers and integration between the two teams proved challenging. The specification that the Source-IT consultants prepared and worked to was not well understood by either the client or the graphic designers – and with hindsight was not sufficiently detailed. This led to a divergence in vision and expectations, resulting in project overrun. The client could not afford to delay her business launch and eventually it was (amicably) agreed that Source-IT would not complete the remaining sections of the task. This project was a great learning experience for all concerned, but ultimately was a factor in the decision to close down the initiative.

4   Discussion

The Source-IT pilot can be evaluated at a number of levels. Firstly, could this model be self-sustaining and operate within a university environment without institutional support? We believe that the answer to this is 'yes'. The small loss in the first year was due to largely to inexperience in negotiating and running contracts and consequently underestimating the work involved and hence the price. The marketing and sales activity demonstrated clearly that there was a strong and consistent market demand. Although there may be a temporary shortage of public subsidy for SMEs, in the long term there continues to be an appetite for exactly these kinds of services. However, establishing and piloting of Source-IT required a number of critical factors, including the availability of institutional KT funding and of deployable management staff and an openness by School and Faculty leadership to innovation and risk-taking.

The second question is more subtle. Does a self sustaining and independent consultancy unit generate value for a research-led university? Clearly there were strong educational benefits to the participating students, although we were conscious of the need to balance these carefully with their primary work priorities. Time management difficulties must certainly be one of the major management challenges and ultimately may limit the scale of a Source-IT-like unit. Other universities have addressed this problem by integrating consultancy to external clients with taught modules [13,14], which immediately gives much more management control over the consultants' time. However, in all these examples, as well as in more conventional consultancy by academics, there is still a niggling concern as to whether the work is really of a standard and level to justify the attention of an internationally leading research group. This issue may be the crux of the matter. The assumption that consultancy is part of a portfolio [1,2,9] and that transfer of tacit knowledge enhance the transfer of codified knowledge needs to be further explored. There is evidence that this is so for a minority of 'star' researchers, but these individuals may choose carefully which businesses they work with and what kind of work they do. The constraints of public funding to support SMEs often means that it is necessary to take on work that may not be so relevant to research and may be with an organisation that does not have the resource to carry through the project to a higher level. If a sustainable consultancy business could be established in which companies came in at 'entry level' with small scale, low level work and then progressed to projects that were of more academic interest, and if this business also provided skills development for students and early career scientists or engineers, it would certainly be of value to a research-led university. An analysis of the level and nature of repeat business from existing consultancy initiatives would be useful and interesting.

5 Conclusion

The Source-IT case study illustrates the tension between market pull for small scale, low level consultancy services from universities and the ability and motivation for the university to deliver. Whilst larger scale consultancy contracts can be very attractive to academics, smaller assignments are seen as a diversion. Source-IT overcame this problem by adding the dimension of the student consultants' motivation. Source-IT could provide valuable work experience and payment within a 'safe', managed environment, so that students benefited as well as the client. The pilot highlighted some of the practical challenges, but suggests that such a model could be self-sustaining with some adjustments. It is unlikely that an initiative such as Source-IT could be started, nurtured and grown to sustainability without adequate resourcing, with dedicated staff for a sufficient period of time. Within a research-led university and under the current economic climate, this kind of financial and organisational underwriting is difficult. More significantly, it is hard to demonstrate real strategic value unless the benefits can be translated into outputs by which academic staff and universities are judged. Leaving the educational issues aside, which have been evaluated elsewhere, a research-led university needs to see non-financial benefits to one or more research programmes. This could be, for example, from repeat business gradually leading to more strategic engagement with the company. It was not possible to demonstrate such value from the relatively short pilot for Source-IT and further work in this area will look at longer term iniatives.

Acknowledgements

Funding for launching the pilot was provided by the Higher Education Innovation Fund (HEIF4) and the White Rose Centre for Excellence in the Teaching and Learning for Enterprise (WRCETLE), with internal support for staff management costs underwritten by the School of Computing, Active Learning in Computing (ALiC) Centre of Excellence in Teaching and Learning (CETL) and the Faculty of Engineering Research & Innovation Centre. Valuable support and advice from colleagues in the Faculty of Engineering, the White Rose CETLE and the Enterprise & Innovation Office are acknowledged with thanks.

References

1. Wright, M., Claryssen, B., Lockett, A., Knockaert, M.: Mid-range universities' linkages with industry: Knowledge types and the role of intermediaries. Research Policy 37, 1205–1223 (2008)
2. D'Este, P., Patel, P.: University – industry linkages in the UK: What are the factors underlying the variety of interactions with industry? Research Policy 36, 1295–1313 (2007)
3. Klofsten, M., Jones-Evans, D.: Comparing Academic Entrepreneurship in Europe – The Case of Sweden and Ireland. Small Business Economics 14, 299–309 (2000)
4. Ambos, T., Mäkelä, K., Birkinshaw, J., D'Este, P.: When does university research get commercialized? Creating ambidexterity in research institutions. Journal of Management Studies 45 (December 8, 2008)
5. Yorkshire Forward, http://www.yorkshire-forward.com (accessed 24.08.10)
6. Access the region's Centres of Industrial Collaboration, http://www.yorkshire-forward.com/helping-businesses/improve-your-business/innovate/collaborating-for-success/collaboration-centres (accessed 24.08.10)
7. How would YOU spend an Innovation Voucher? http://www.yorkshire-forward.com/news-events/inform/How-would-YOU-spend-an-Innovation-Voucher (accessed 26.08.10)
8. Lendel, I.: The Impact of Research Universities on Regional Economies: The Concept of University Product. Economic Development Quarterly 24(3), 210–230 (2010)

9. Lockett, N., Kerr, R., Robinson, S.: Multiple Perspectives on the Challenges for Knowledge Transfer between Higher Education Institutions and Industry. International Small Business Journal 26(6), 661–681 (2008)
10. InfoLab21 – Lancaster University's Centre of Excellence for ICT, http://www.infolab21.lancs.ac.uk/ (accessed 26.08.10)
11. Consulting Leeds Ltd, a wholly owned subsidiary of the University of Leeds, http://www.consultingleeds.co.uk (accessed 24.08.10)
12. Neagle, R., Marshall, A., Boyle, R.: Skills and Knowledge for Hire: Leeds Source-IT. In: Proceedings of the 15th Annual Conference on Innovation and Technology in Computer Science Education, Bilkent, Ankara, Turkey (June 2010)
13. epiGenesys – a University of Sheffield company, http://www.epigenesys.co.uk/ (accessed 26.08.10)
14. The Kent IT Clinic – the graduates of tomorrow supporting the businesses of today, http://www.kitclinic.co.uk/kent-it-consultancy.php (accessed 26.08.10)

Session H
Knowledge Transfer with the Third and Public Sectors

Using KTP to Enhance Neighbourhood Sustainability – A Case Study of Wulvern Housing Association's Sustainability Indicators (WINS)

Paudie O'Shea

BA MSc LMRTPI AMInstKT CMI
KTP Associate and WINS Coordinator
Department of Interdisciplinary Studies
MMU Cheshire, Crewe Road, Crewe CW1 5DU

Abstract. Knowledge Transfer Partnerships (KTPs) are widely regarded as one of Europe's leading programmes to help businesses improve their competitiveness through incorporating knowledge, technology and skills from universities. While the KTP is still very popular with businesses around the UK, more and more public bodies and not for profit organisations are also recognising the benefits of such partnerships. Wulvern Housing Association is one of these organisations. They are seeking to expand their neighbourhood sustainability assessment tool (WINS) in order to identify 'sink' neighbourhoods, put initiatives in place to reduce the number of void properties, rent arrears and anti social behaviour (ASB) and ultimately to guide future investment strategies. The importance of sustainable neighbourhoods for housing associations across the UK obvious. Although the achievement of a universal decent homes standard has been long outlined as a priority area by successive Governments, organisations such as the National Housing Federation and the Homes and Communities Agency now demand a more expanded approach to asset management and indeed the concept of the 'community'.

This paper examines the role of KTPs in facilitating the development of the company's practical assessment instrument known as the WINS (Wulvern Indicators of Neighbourhood Sustainability) tool. It explains how the company determined that this type of assessment is important and why their sustainability indicators tool is innovative. Combining both quantitative and qualitative research data, the paper evaluates the relationship between neighbourhoods, public consultation and sustainability indicators within the Crewe area. It discusses and highlights the successful and less successful elements of the project so far. It details what the project is about, what has and what has not worked and can be used as a guide for future 'social' KTP projects in the UK. The paper also analyses some of the work carried out nationally which forms the basis or rationale for developing such a neighbourhood sustainability tool. The paper also outlines some of the problems facing Wulvern in Cheshire and housing associations nationally. It

R.J. Howlett (Ed.): Innovation through Knowledge Transfer 2010, SIST 9, pp. 321–333.
springerlink.com   © Springer-Verlag Berlin Heidelberg 2011

discusses Wulvern's attempts to improve housing stock and neighbourhoods by achieving decent home standards, energy efficiency ratings and progress towards measuring the condition of their neighbourhoods. Initial findings suggest that, in many instances, some of the main issues facing Wulvern customers and their neighbourhoods are less transparent. *Key terms: sustainability neighbourhoods, indicators, KTP, WINS, housing.*

1 Introduction

The WINS tool has been in existence in Wulvern for over five years. It is effectively an assessment tool which measures a selection of variables for individual neighbourhoods, currently using indicators ranging from crime levels, rent arrears and the number of void properties. While theoretically it seems natural that a company would want to know how their neighbourhoods are performing, in practice it is difficult to determine without an in depth examination requiring substantial resource investment. Although the concept of sustainability stands on the three pillars of economic, social and environmental well being, at Wulvern, it is only recently that there has been real recognition of the social requirements of this concept, towards the 'promised land' of the sustainable community. While sustainable communities are often discussed in this context, accurate and validated indicators are seldom used to measure the success of sustainability strategies. Wulvern recognised that both the methods of data collection and the resultant evidence base needed to change, in order to improve the status, validity and reliability of the WINS tool. While indicators relating to the number of void properties and average rent arrears are easily quantified data and information, what might be considered 'fluffy' indicators such as reputation and community cohesion, although meaningful, are less easily identified or measured. As a result, the company decided to join in partnership with MMU to try and develop the WINS tool further, into a predictive and diagnostic system capable of improving monitoring and to steer future investment. A major part of the company's strategy is to significantly reduce voids and rent losses and to improve the quality of the housing stock, through more rigorous monitoring and evaluation. Therefore the aim of the KTP is to develop the WINS tool which Wulvern will commercialise and use to enhance core operations, focusing actions and investments in construction, repair and modernisation to improve demand, customer satisfaction and neighbourhood improvement. The company's aims for growth and expansion are supported by improved data capture and analysis, steering future investments in planning, housing developments and maintenance, supporting neighbourhood regeneration and wider green issues for future investments.

2 The Rationale and Policy Context for Developing WINS

Since the change in Government earlier this year, local and national budget cuts in health, education and importantly housing have been the major talking points of the coalition government's radical debt reduction agenda. Housing related organisations across the UK are waiting anxiously for announcements of further cuts over the next few years. Nevertheless, for Wulvern and its tenants, major issues still exist in the neighbourhoods and they need to be addressed. Whether funding for future regeneration of Wulvern neighbourhoods exists or not, the company needs to find smarter ways of improving its neighbourhoods with a 'value for money' as well as a 'customer satisfaction' mindset in place. The WINS tool can facilitate this 'smarter' and 'local' way of working, because it can help the company make evidence based decisions about its neighbourhoods and any future regeneration strategy. Wulvern's quality assessors, as well as the Audit Commission continue to demand improvements in service provision and enhanced efficiencies in their business operations.

Nationally, it is accepted that attempts to measure the sustainability of communities are justified; economically, environmentally and socially. Thus, in terms of establishing a rationale for developing an effective and holistic assessment tool, the WINS project builds upon pieces of research using some of the recommendations for measuring and analysing sustainability within neighbourhoods from Long (2000), 'A Toolkit of Sustainability Indicators'; Williamson and Legg (2001), 'Investors in Communities' and Jozsa & Brown (2005), 'Neighbourhood Sustainability Indicators'. Neighbourhood assessment tools can assist local authorities and all housing providers to respond to Governments' sustainability policies, including urban regeneration initiatives to improve deprived and 'run down' areas and services. Locally and nationally, these themes are outlined in policies such the English House Condition Survey Annual Report 2007, the Cheshire Sub Regional Housing Strategy 2009-2012, the Housing Green Paper 2007 and the 2001 National Strategy for Neighbourhood Renewal Action Plan. However, it is apparent that, presumably owing to the current economic climate, significant proportions of the Government's housing and regeneration strategy have been delayed. Fortunately for some housing associations, there is still some money in the pot to conduct such an analysis of housing stock and neighbourhoods. And by using some of the recommended 'big society' initiatives promoted by the coalition Government, neighbourhood assessments and the associated techniques can be looked upon favourably by local and national policy makers.

3 Issues in Cheshire

Like many housing associations around Britain, Wulvern suffers from rent arrears, void properties, anti-social behaviour and crime related incidents. Consequently, a major part of the company's strategy is to significantly reduce these voids and rent losses and to improve the quality of life for residents, by identifying the major underlying issues in their neighbourhoods. Wulvern is also attempting, through its Lean System Thinking interventions[1], to understand the actual reasons for void

---

[1] Lean Systems Thinking, for Wulvern, is a process of shaping services to meet the tenant's needs and demands. It is basically listening to demand to understand what is important to customers and designing services and equipping staff to meet that demand.

properties and rent arrears and to gain a better understanding of its customers in general. Wulvern realises that by undertaking such an analysis of neighbourhoods, it will have a positive impact on both the neighbourhoods and the company itself by:

1. Reducing the number/ cost of void, vandalised and 'lost rent' properties;
2. Identifying areas in need of improvement such that the condition of assets and stock is enhanced;
3. Improving neighbourhood sustainability and quality of the local environment;
4. Recommending targeted investment in particular housing types and other urban facilities around Cheshire.

The above interventions should result in the following key impacts:

1. An increase in main business turnover (in rents and building condition);
2. Enhanced bargaining power of the company with creditors (by identifying areas in need of improvement and thus directing funds more efficiently);
3. Reduction of social disturbance and nuisance, thus improve the quality of life in these neighbourhoods;
4. Enhancement of the local environment thus improving a sense of community and customer satisfaction;
5. Increase the standing of the company in the housing association sector.

Improvement in neighbourhood sustainability will have positive impacts on Wulvern's expenditure and losses and will save time and money in repair and management costs amongst others. Establishing this type of neighbourhood assessment is also especially important in the current financial and political situation as it can result in a clear and defined audit trail, value for money and a justification to the coalition Government that more capital is needed in areas of multiple deprivation. But apart from making sense on a national scale, assessing neighbourhoods also presents the company with a real chance to reach the 'promised land' of the sustainable community. Progress in Wulvern neighbourhoods has so far been achieved through interventions such as new kitchens and bathrooms and external environmental works such as new fences and driveways, yet rent losses seem to be consistent in these areas of investment. The consequence of maintaining properties in undesirable areas and socially deprived neighbourhoods, according to Ciniglio (2005), results in hard-to-let properties and reduced rental income for the company. This in turn makes property financially unsustainable and arguably not worthwhile maintaining. Housing associations that hold a high proportion of unsustainable housing will thus be put in a position of risk, undermining the overall strategic objectives of the company. Initially, my research has found that most housing associations, local authorities and private companies acknowledge the importance of linking asset management strategies with neighbourhood assessments but in most cases an integrated monitoring system or a suitable neighbourhood assessment process has not yet been established owing primarily to a lack of resources. Nonetheless, a housing association's asset management strategy, whether weighted towards an aggressive disposal policy or otherwise, should be informed by hard evidence, quantified and measured as has been seen in a growing number of cases in the UK, Europe and the USA (USA: Cascadia Scorecard, Pacific North West and Germany: the HQE²R approach).

4 Neighbourhood Assessments in the UK, Europe and the USA

Developing a theoretical sustainability measurement tool in the United Kingdom is not new, however in practice there still lacks a real commitment towards implementation of this kind of diagnostic instrument. Neighbourhood assessment indicators have been developed by Housing Associations across the country, from Drum Housing in Southern England, Sunderland Gentoo in the North West and Knowsley Housing Trust in the Merseyside area. Attempts to measure sustainability have also been made in the United States, Canada and in Europe. In almost every case, the development of a sustainability appraisal tool is set within the wider context of a company's asset management strategy. The emergence of sustainable development and its importance to government bodies in a social housing context has resulted in a more holistic approach to sustainability agenda. It is anticipated that assessment tools such as WINS will become examples of good practice in the asset management arena in the future. According to Ciniglio (2005), there is currently a significant push on this particular agenda as over two thousand Housing Associations are currently in the process of writing or reviewing their approach to asset management and it is believed that greater representation of sustainability within housing associations strategies is required to reflect current concerns.

In October 2004, Sunderland Housing Group (Now Sunderland Gentoo) was employed on a consultancy level by Wulvern Housing Association to provide recommendations for a prospective sustainability appraisal tool for the company. These recommendations were followed by meetings to share and develop ideas in this area between Wulvern and the Sunderland consultants. Twenty-one sustainability indicators were agreed upon. Vital indicators such as demand, turnover and stock condition were analysed. Within Sunderland Gentoo, the Neighbourhood Assessment Matrix (NAM) model has been operating effectively and was considered a suitable strategy for Wulvern to adopt. In terms of seeking international guidance, Wulvern can look to Europe and the USA for good examples of measurement techniques and public engagement methods. Although social housing exists in some form in many parts of the globe – whether it is a large sector, as in the Netherlands, or a small part of the stock as in the United States, evidence of operational assessment tools similar to WINS is minimal.

However, one aspect of the WINS tool – greater participation of residents in strategy making – seems to be increasingly accepted as an important aspect of strategy across Europe and the United States. In most cases the social sectors now try to take account of tenant views and levels of customer satisfaction. In England, the Tenant Services Authority is responsible for ensuring that the social sector responds to tenants' views and that its standards reflect tenant needs. The

Chartered Institute of Housing in partnership with the Matrix Housing Partnership carried out a brief international study to compare emerging proposals in England with the mechanisms for taking account of tenant views in several other countries (Perry & Lupton, 2009). This study found that national-level surveys of tenant satisfaction are surprisingly common, and that they typically show that about three-quarters of tenants are satisfied with their housing.

Nevertheless, there are some interesting differences in approaches to how tenants' views are taken into account in regulating the social housing sector. In Holland for example, housing associations have developed their own 'rented housing label' (in the UK it would be called a kite mark) based on tenant experiences with different landlords. The New Zealand housing corporation also carries out quarterly surveys of tenant views, which give it a continuous picture of customer satisfaction and enable the association to monitor opinions about particular issues. A similar survey is carried out by the Northern Ireland housing executive. From these examples, it is clear that in order for a well-managed and sustainable housing stock to be realised, housing associations will need to develop more creative, holistic and active approaches to asset management, with particular emphasis put on the views of the residents in the neighbourhoods.

A key challenge to the WINS project is applying contemporary principles and practices in social geography, community psychology and environmental sciences to construct a predictive investment planning tool. In the UK, examples of neighbourhood assessment mechanisms seem to be popular, however the quality of data is variable. This is a significant problem and explains why other housing associations are shying away from such assessments. However, Karol and Brunner's 2009 paper on 'Tools for Measuring Progress towards Sustainable Neighbourhood Environments' is a good starting point for identifying a variety of themes and sub-themes that support assessment tools at both the project design phase and the project operational phase currently in operation in the UK. They provide a good insight into other neighbourhood sustainability project that have been initiated here - One Planet Living (OPL), the South East England Development Agency (SEEDA) checklist and SPeAR (Sustainable Project Appraisal Routine). However, developments at Wulvern are different, mainly because the data is current, clearly presented, concise, evidence based and gathered directly by the company, through the employment of a specific post (KTP Associate) and other dedicated resources. Thus, these 'fluffy' indicators are no longer unreliable. Their performance is based on evidence and validated by established academic methods, derived from resources from the relevant Higher Education Institution (HEI). Furr-Holden et. al. (2008) argue that there is a limited range of validated quantitative assessment methods for measuring features of the built and social environment that might form the basis for preventive interventions. Therefore, this presents Wulvern with a unique opportunity to develop a specialised form of assessment to fill that gap. Wulvern has, in the past, learned about relevant 'housing' issues such as void properties, rent arrears and reducing anti-social behaviour from other housing associations in the UK. However, in the particular field of sustainable neighbourhood regeneration and strategy, Wulvern can become an industry leader in the methods and techniques of finding the true measure of a neighbourhood.

5 The Innovative Route – Expanding and Shaping New Indicators for Wulvern

Indicators enable Wulvern to measure how their neighbourhoods are progressing towards becoming sustainable communities. Currently, the tool is based on primary research (door to door primary data collection through surveying) and does reveal in significant detail some of the major and underlying issues affecting satisfaction and value for money in Wulvern neighbourhoods. In the past, general crime was measured but now, the important crime indicators are broken into separate entities (emphasising their importance). Consequently, the WINS tool can now address some more searching questions such as:

- What are the major underlying issues affecting Wulvern tenants and private residents satisfaction levels in their neighbourhoods?
- Regarding sustainability, what directions are Wulvern communities moving in?
- Is the number of void properties increasing, and why?
- What is the reason (if any) for the increase in crime and anti social behaviour?
- What is the environmental quality standard of Wulvern communities?

The answers to these questions will enable Wulvern to establish a set of benchmarks from which to measure neighbourhoods correctly by defining the 'triple bottom line' (worst case scenario) and the truly sustainable neighbourhood (best case scenario) (Hornsbyshire Council, 2010). The 'triple bottom line' trigger effectively identifies the point at which a neighbourhood reaches such an unsustainable level that the only economically viable option is disposal. Indicators themselves can be broken down into three key dimensions of sustainability: environmental, economic and social. Although it is useful to talk of the 'triple bottom line' of economic, social and environmental sustainability, more detail is required to develop effective targets. Although social and environmental targets are of paramount importance, economic sustainability is probably the most important factor for Wulvern to focus on because not until communities can meet their vital Maslowian needs of shelter, food and sanitation will they become aware of the social and environmental problems. When residents cannot afford their basic living expenses, they usually do not address the associated social and environmental decline in their neighbourhood. Economic prosperity stabilises families and enhances revenues to pay for public services (Di Cosmo, 2009).

In Crewe especially, while economic stability in some areas is being maintained physically, too many places are neither cohesive, connected, well-designed nor well-planned. Some neighbourhoods, for various reasons, have lost the essential glue, the community cohesion that binds them together. Newer areas, often big estates, are sometimes soulless places, disconnected and car-dependent, wasteful of energy and built with little recognition of the wider environment (Homes and Communities Agency, 2010). Wulvern needs to reappraise these areas, to reinvigorate older neighbourhoods and to create new places where people want to live – carbon-efficient, socially cohesive and well-connected to local services.

Currently, Wulvern are involved in gathering their primary data for both the WINS tool and the neighbourhood plans. This is a continuous exercise as both the plans and the WINS tool need constant updating. The neighbourhood level survey is designed to gather information about the major issues affecting Wulvern tenants and indeed private residents in their neighbourhoods. Initial findings, gathered on the pilot Selworthy Drive neighbourhood during January 2010 – suggest some interesting results. Sixteen streets were surveyed overall. Here, it seems that some issues are more prevalent in certain streets than in others. For example poor street lighting and an abundance of resident generated litter were identified as problematic to residents in Wheelman Road and Rigby Avenue; however in Frank Bott Avenue the untidy or damaged external condition of private properties was identified as a major issue. The maintenance and utilisation of the green spaces in the neighbourhood was identified by nearly all residents surveyed as another major issue that needs attention and investment. From this, it is obvious that some issues are important on a neighbourhood level, but probably the most significant problems that need to be addressed are at a street level. Being able to identify street level issues and focus resources accordingly confirms that Wulvern is leading the way in finding out what makes a neighbourhood sustainable.

Neighbourhood level information has also been gathered in some of the rural areas in South Cheshire. Social housing in rural areas does not fit the stereotypical view of the 'council' house. Some of the properties are located in affluent locations and the problems experienced by urban areas seem a lifetime away. However, initial findings of the WINS tool suggest that in a small proportion of neighbourhoods tenant arrears are a major problem for the company and this needs to be addressed. Reputation of Wulvern neighbourhoods is also something that needs to be considered for improvement. For example, from the findings of the neighbourhood level survey and subsequently the WINS tool, the reputation of the Cronkinson neighbourhood in Nantwich would seem to be relatively healthy (from the residents' point of view). However, external opinion of that area is less complementary and consequently, the neighbourhood is associated with negativity and has, locally, a bad reputation. Peoples perceptions built up over years of hear say and rumours can have a hugely damaging effect on the local perceptions of a neighbourhood, and this makes it more difficult for Wulvern to let properties in this neighbourhood.

Blanket survey coverage, including interviewing all non-Wulvern tenants, was required to learn the extent, nature and severity of major issues troubling residents in these neighbourhoods. Once these have been identified, the WINS tool can be employed to measure the indicators, and used to explain why sustainability is improving or declining on a street-by-street level. The Selworthy Drive neighbourhood itself was selected or a pilot survey primarily because the company has recently invested over £3.5 million pounds on this neighbourhood including internal and external improvements such as new kitchens, heating systems, bathrooms, doors, roofs and driveways. To meet decent homes standards and fulfil the promises made by Wulvern when they received the housing stock from the local council in 2003, the improvement programme has covered practically every element of the stock in that neighbourhood. There is no doubt that these changes have had a positive affect on the neighbourhoods however anti-social behaviour is still occurring, rent arrears are still evident and the reputation of the area remains tarnished. Housing conditions internally and externally have improved significantly, but it is clear that the path to true sustainability needs to involve more than physical regeneration. It is easy to throw money at a neighbourhood, but achieving a sense of place and community cohesion is slightly more difficult.

So, through this neighbourhood-level customer consultation (quantitative: questionnaires and qualitative: focus groups), Wulvern neighbourhood plans, fed by WINS, can assist the company with strategic decision-making; serve as part of an early detection system that assists in identifying risk areas that threaten the health of the community and present a snapshot of the community's progress towards its sustainability vision. Following consultation and engagement with Wulvern residents (January - October 2010), the next step of the programme will involve:

- Continuing the monitoring system to see if any common trends are developing;
- Creating a structured outcome process where an intervention is undertaken in a neighbourhood or street which is constantly performing poorly. If a street is consistently scoring poorly with rent arrears for example, Wulvern will be looking at investigating if there are underling financial management problems being encountered in the neighbourhood;
- Rolling out the actions identified in the neighbourhood plan action plans.

Thresholds, established through a process of minimum standards are used where available, to evaluate the indicators. So the question now is: what is the difference between Wulvern's approach to sustainable assessment and other projects elsewhere in the UK? The answer is simple: quality of data and the techniques for acquiring this data. Valid and accurate data is essential to attain accurate interpretation of issues and situations in neighbourhoods. Data gathered by most other housing organisations have been acquired elsewhere from private consultancies or other government reports, is frequently outdated or lacks specific relevance at street level and generally omits resident consultation and feedback. Wulvern, where necessary, uses its own data to feed its sustainability assessment. This provides a 'true' indication of the condition of neighbourhoods from which the company can then put forward an evidence based and justified asset management strategy. The WINS tool will eventually be linked to the company's Housing Quality Indicators (HQIs) which measure, for example, accessibility to services and schools. Bringing the information to life through GIS is something that will also be developed in line with WINS. So although major lessons can be learned from other attempts at this kind of assessment, research has shown that 'one size doesn't fit all' and housing associations must begin to tap directly into what makes their neighbourhoods tick if they are to realise the suitable indicators from which to measure their progress (or indeed lack of progress).

6 Initial Findings from the WINS Tool

At the time of writing, WINS2 has been monitoring Wulvern neighbourhoods for over ten months. The process has worked well, and has highlighted a number of key issues that need to be address if our neighbourhoods are to be fully sustainable for the company and for the tenants. Some of these issues such as high tenant arrears in rural areas, fear of crime in some urban neighbourhoods and increasing 'end to end' time for repairing properties will need to be monitored closely and if certain streets continue on a downward trajectory, a form of intervention or further investigation needs to be initiated in order to stop the decline. A better idea of how the neighbourhood is really performing can be assessed after a number of consecutive assessments. Following on from this, qualitative research techniques such as in-depth interviews and focus groups will be exercised to dig even deeper into some of the underlying issues or indicators which have scored poorly.

From these sustainability scores, early findings suggest that there are financial management problems in our neighbourhoods and some kind of a financial inclusion or financial management awareness campaign should be rolled out to investigate 'why' people are having difficulty managing their rent. End to End time for repairing empty properties is also a concern, and perhaps a stricter monitoring of tenants and maintenance of their properties should be a priority for the company going forward. These actions will reduce the amount of rent owed to the company and should also reduce the number of void properties which meets the overall project aims and objectives. However, as with any attempt at trying to measure performance of neighbourhoods, there are a number of areas of the project which need more focus and probably the main stumbling block is the process of the primary data collection. While this is vital to gaining an exact insight into different issues in the neighbourhood, and while it does compliment the 'Con Dems' localism agenda, the fact is that primary data is hard to acquire. It is expensive, time consuming and difficult to execute properly. In trying to combat this, Wulvern have taken the innovative approach and employed volunteers from MMU (another benefit from the KTP partnership) to help with the neighbourhood consultation. We are also looking at using the new census information which will be made available after 2011. Overall, the process is working, however momentum needs to be maintained after the KTP project period is completed in order for the company (and the neighbourhoods) to reap the benefits of sustainability measurement.

7 From Decent Homes to Decent Neighbourhoods

Based on the indices of multiple deprivation and other relevant government publications, some of Wulvern's neighbourhoods might indeed seem a "symbol of the apparently intractable web of problems faced" by some of the most deprived neighbourhoods in the country (Redwood, 2009). Initial reading of the indices of multiple of deprivation and 'OnePlace' do not bode well for inhabitants in some of Crewe's (and Nantwich to a lesser extent) neighbourhoods which fall within the top ten percent most multiply deprived neighbourhoods in England. Issues such as low levels of educational attainment, low measures for standards of living, below average life expectancy and higher than average unemployment rates make for uneasy reading for local residents. However, the situation is surely not beyond salvage; material improvements are occurring. But, as mentioned earlier, the key is creating the pride in community, creating a place where people want to stay and live in a 'decent' environment. For too long, sustainability has been associated predominantly with the housing stock and/or assets, rather than with the community and neighbourhood needs as defined by residents. It's not just a house, or a home in fact but a community and neighbourhood that makes an area sustainable.

For Wulvern customers and private residents alike, the WINS tool can identify potential hotspots in specific neighbourhoods, bringing these to the attention of the company which can then intervene in an attempt to bring their sustainability rating up to standard. This new approach will create more sustainable communities and enhance the quality of life for Wulvern customers. The aim for Wulvern is to make their assets desirable places in which to live both now and in the future and thus preserve their value. Properties cannot however be viewed in isolation as the importance of creating and maintaining a sense of community is essential to successful housing management. Ciniglio (2005) argues that this is dependent on a range of externalities over which many housing associations may have little or no control. Housing associations, after all, have the ability to affect the lives of many of Britain's residents, so it is important that they act responsibly. Wulvern has already aligned its asset management strategy to link in with the WINS tool and the neighbourhood plans so by adopting and applying this strategy, the process of sustainability will contribute to the end goal of sustainable development. Unsustainable housing presents problems for housing associations from loss of rental income and ongoing management costs. Unsustainable housing and neighbourhoods for residents present much wider problems (diminished quality of life, reduced life expectancy among many other issues).

The WINS tool can, given further investment, facilitate profit growth through reduced voids, bad debt losses and reduced rent arrears. It will assist the company in meeting improved environmental and sustainability standards and will consolidate the company's position as a market leader in the housing association sector. In practice, WINS combines the use of a series of indicators with objective assessments from external statistical data sources, quantitative and qualitative internal data, including in-house surveys and questionnaires, interviews and focus groups with tenants, residents and other relevant stakeholders.

8 Conclusion – 'WINS' – The Third Way

In conclusion, this paper presents Wulvern Housing's WINS tool as a new departure; an innovative strategy and 'a third way' in measuring the sustainability of neighbourhoods in the UK. In theory, the WINS tool is not innovative as examples of neighbourhood assessment tools can be found in housing associations around the UK. In practice however, the tool is probably the closest any housing association has come to really knowing what makes a neighbourhood tick. This is the result of community-level real and raw data. The primary data which feeds the WINS tool is the reason why it will become one of the most accurate and evidence based neighbourhood assessment tools in the UK. The use of primary information, (and some secondary information where necessary) will ensure that the WINS tool will be able to pinpoint the major underlying issues currently faced by Cheshire neighbourhoods. From this information, justified and evidence based decisions on the future of these neighbourhoods can be made. Although this type of data collection might not be possible on a large scale (in major cities for example), Wulvern's attempts at achieving a sustainable 'decent neighbourhood standard' for their residents should not go unnoticed.

Overall, the project, already at the half way stage, has so far been a success. A better idea of how the neighbourhoods are really performing can be assessed after a number of consecutive assessments. Following on from this, qualitative research techniques such as in-depth interviews and focus groups will be exercised to dig even deeper into some of the underlying issues or indicators which have scored poorly. From the early sustainability scores, findings suggest that there are financial management problems in our neighbourhoods and some kind of a financial inclusion or financial management awareness campaign should be rolled out to investigate 'why' people are having difficulty managing their rent. Work is already in motion to deal with this problem. End to End time for repairing empty properties is also a concern, and perhaps a stricter monitoring of tenants and maintenance of their properties should be a priority for the company going forward.

These actions will reduce the amount of rent owed to the company and should also reduce the number of void properties which meets the overall project aims and objectives. Because of the success of the project so far, confidence in the program has been shown by Wulvern, MMU and KTP by looking favourably on two further KTPs to be based at Wulvern, one looking at the impact of environmental change 'the green agenda' on social housing including energy affordability, and a KTP that looks at raising youth aspirations in Crewe. If these projects are approved, they will place Wulvern on the national stage as a centre of excellence re knowledge transfer and have the potential to attract significant resources into the organisation, invigorating existing teams and enabling the company to make interventions that are indisputably 'evidence led'.

References

1. Audit Commission. Local authority property: A management handbook, pp. 6–14. HMSO, London (1988)
2. Ciniglio, P.: Developing a sustainable asset management index for affordable housing (SAMi). MA - Sustainable Futures, 5–46 (2005)

3. Cosmo, D., Green, F.: Communities Assistance Kit. United States Environmental Protection Agency, 8–12 (2009)
4. Furr-Holden, C., Smart, M.J., Pokorni, J., Ialongo, N., Leaf, P.J.,Holder, H., Anthony, J.C.: The NIfETy Method for Environmental Assessment of Neighbourhood-level Indicators Society for Prevention Research. Prev Sci. 9, 245–255 (2008)
5. Hornsbyshire Council 'Sustainability & triple bottom line'. Sustainable Energy Strategy 2006-2010, p.14 (2010)
6. Jozsa, A., Brown, D.: Neighbourhood Sustainability Indicators – Report on Best Practice Workshop, pp. 1–20. School of Urban Planning, McGill University and the Urban Ecology Centre (2005)
7. Karol, E., Brunner, J.: Tools for Measuring Progress towards Sustainable Neighbourhood Environments. Sustainability 1(3), 612–627 (2009)
8. Long, D., Hutchins, M.: A Toolkit of Indicators of Sustainable Communities, pp. 1–9. The Housing Corporation and the European Institute for Urban Affairs, London (2003)
9. Perry, J., Lupton, M.: What Tenants Want– Globally!, p. 9. Chartered Institute of Housing publication (2009)
10. Sheltair Group.: Indicators for Sustainable Communities - A Case Study Scan of Performance Indicator Initiatives, pp.5–35 (2007)
11. Sunderland Housing Group. The Neighbourhood Assessment Matrix. Sunderland Gentoo consultation at Wulvern Housing (August 2003)
12. Redwood, M.: Slum dog Sustainability, p. 16. The International Development Research Centre (2009)
13. Tait, K.: Managing the Assets, pp. 34–40. National Housing Federation, London (2003)

Knowledge Transformation in the Third Sector: Plotting Practical Ways to Have an Impact

Razia Shariff

Head, Knowledge Exchange Team, Third Sector Research Centre
2nd International Conference on Innovation through Knowledge Transfer
7th & 8th December 2010, Coventry

Abstract. The Third Sector Research Centre (TSRC) in the UK is a unique initiative established in 2008 by the Economic and Social Research Council (ESRC), Office of the Third Sector (recently changed to the Office for Civil Society) and Barrow Cadbury Trust. Initially over a five year period it aims to develop a solid evidence and knowledge base about the third sector to inform policy and practice. TSRC, hosted by the Universities of Birmingham and Southampton has established mechanisms for knowledge transfer in the design and delivery of the research process, offering virtual as well as physical participatory spaces for knowledge exchange to occur. This paper explains TSRC's policy approach to knowledge exchange based on current definitions, theories and models of knowledge transfer. The paper details some of the initial reflections from the approach used by the Centre to engage with, and involve non-academic stakeholders in knowledge exchange through its formal structures and the activities of the knowledge exchange team. The paper presents, a Knowledge Exchange Impact Matrix (KEIM) (adapted from Arnstein's 'Ladder of participation') which plots different types of TSRC knowledge exchange activities based on: the extent of meaningful knowledge exchange; and the number of stakeholders engaged. The paper ends by exploring TSRC's planned methods for monitoring and evaluating TSRC's knowledge exchange activities and how these support research reach and impact.

1 Introduction

There is a growing demand by higher education and research funders to demonstrate the impact of their research investment[1]. In the policy world, evidence-based and informed policy is being promoted (see Nutley 2009). In the

---

[1] ESRC Strategic Plan 2009-2014 Delivering Impact through Social Sciences, http://www.esrcsocietytoday.ac.uk/ESRCInfoCentre/strategicplan/ Accessed 10th June 2010 and HEFCE (2009) Research Excellence Framework (REF), http://www.hefce.ac.uk/research/ref/ Accessed 4th June 2010.

R.J. Howlett (Ed.): Innovation through Knowledge Transfer 2010, SIST 9, pp. 335–348.
springerlink.com   © Springer-Verlag Berlin Heidelberg 2011

world of the practitioner, commissioners and funders want to know what evidence and theoretical models are being used for project implementation. Knowledge exchange is becoming a core component of researcher projects to ensure that the knowledge that has been gained through research is disseminated and used to inform policy and practice and have an impact on society. Recent studies[2] by HEFCE also indicate that researchers benefit from knowledge exchange, informing their teaching practice and providing a reality check on their research work. The traditional approach to knowledge exchange in the world of academia has typically been limited to a written paper which is presented at a conference and then hopefully published in academic or professional journals for a wider audience. The establishment of TSRC has created the opportunity to offer a 'step change' in the linear model of 'first the research is published then it is disseminated' to a more participatory and interactive model that engages key stakeholders throughout the research process. This paper explores the initial learning from this innovative approach to knowledge exchange between researchers and the third sector.

2 TSRC Model

The approach of TSRC has been to develop: a strong formal partnerships between the research, policy and practice communities; a commitment to incorporating knowledge exchange throughout the research process; and a mechanism to monitor and evaluate impact. The intention is to achieve the long term vision of TSRC as a sustainable research centre providing a resource which is valuable and influential in the UK and abroad.

The investment, design and delivery mechanism of the Third Sector Research Centre is innovative and unique in that it integrates capacity building and stakeholder engagement throughout the Centre's approach and includes staff teams based within the user community. TSRC anticipates 'process impacts' through its Capacity Building Clusters (CBCs), and the methods used to undertake research e.g. engaging stakeholders and undertaking action research. TSRC promotes 'instrumental impacts' from its research on policy makers, decision makers and practitioners through knowledge exchange activities e.g. policy symposiums and partnership impact events. TSRC will have 'conceptual impacts' on thinkers and academics interested in the third sector through publications and conference presentations. The TSRC is not just establishing itself as a national centre of research on the third sector but ensuring that through its Capacity Building Clusters and Knowledge Exchange Team (KET), working collaboratively with other ESRC Centres, and developing an international academic reputation, it will have a major impact on UK third sector policy and practice.

---

[2] HEFCE Reports (March 2010)'*Knowledge Exchange and the generation of civic and community impacts*' & (February 2010) '*Synergies and trade-offs between research, teaching and knowledge exchange*' produced by Centre for Business Research (CBR) and Public Corporate Economic Consultants (PCEC).

The establishment of a high level Advisory Board made up of key partners in the third sector, academia and Government, and individual specialist Reference Groups for each research stream and administration within the UK, provides a unique opportunity to develop new ways of engaging and producing research knowledge and its application in the wider sector. TSRC's Knowledge Exchange, Communications and Impact Strategy (KECIS) supports this by using a blended approach of creating an off- and on-line presence within third sector communities and includes an interactive website with videos, podcasts, discussion boards and blogs, as well as partnership seminars, workshops and events to explore the implications of the research and its implementation. The CBC's offer PhD Case studentships, Knowledge Transfer Partnerships, voucher and placement schemes in partnership with third sector organisations. The three CBC's are led and themed as follows: Middlesex University, Social Enterprise; Lincoln University, community engagement; University of Bristol, economic impact.

During the initial five years of its contract the Third Centre Research Centre has seven key priorities: (1) to establish a sustainable resource of a robust database on the sector, and key subsectors, in the UK; (2) to establish longitudinal analysis of the sector and organisational dynamics to secure a base for ongoing analysis into the future; (3) to undertake robust analysis of the impact and value of the sector; (4) to develop models which can be used by policy makers and practitioners; (5) to establish a framework of action research which engages all key stakeholders in the development and dissemination of the activity of the Centre; (6) to enhance considerably the capacity for research on the sector and work closely with sector agencies to ensure a sustainable programme of knowledge exchange; (7) to extend theoretical and conceptual analysis of the sector to broaden and deepen understanding of its scope and diversity, and the differing impact of policy interventions across these dimensions.

The overall vision of the Knowledge Exchange Team (KET) is to demonstrate the value of robust and relevant research by creating a platform and infrastructure for knowledge interactions between third sector organisations, policy makers and researchers. TSRC researchers will ensure the production of high quality academic research, while KET aims to ensure through stakeholder interactions that the research is relevant to the third sector, and can be readily used to inform policy and action in order to have an impact. The overall aim of the KET is to ensure that TSRC's research is fully accessible to those for whom it is relevant, both in terms of reach and understanding, and to ensure that policy-makers and practitioners have the opportunity to engage in the research process.

Our activities as the Knowledge Exchange Team will have a number of baseline principles: systematically capturing and storing knowledge, by bringing together research findings, evidence and analysis and making it accessible to all those interested in the third sector (through the internet, searchable database of research papers, working paper e-alerts, publications, articles in relevant media, presentations and stands at events); maximising research uptake and impact through the knowledge sharing and learning process and ensuring this is monitored and refined on a regular basis; being proactive in developing positive and productive collaborative mechanisms with other stakeholders in the collation of knowledge and its effective use for and with the sector; supporting engagement from different interests (government, policy and decision makers, third sector funders and practitioners, academic institutions and research bodies); making research available in a range of formats depending on access requirements. Ultimately we want to establish an infrastructure and network of reciprocal knowledge flows between the third sector stakeholders and academia to inform future research, policy and decision-making and practice.

Figure 1, TSRC's model of knowledge exchange, is based on the Third Sector Research Centre's policy strategy (KECIS). There will be a two-way exchange of knowledge between stakeholders and researchers during the research process. Stakeholders will be engaged as active partners in developing the priorities and focus of the research streams through the formal Reference Groups and Advisory Board. KET will develop a diverse range of mechanisms for informal engagement through the website with dedicated discussion boards and blogs for research streams, partnership events, workshops and other activities once initial evidence reviews have been undertaken to inform the next phase of the research. Once the research is produced there will be opportunities to explore the research implications and the ways in which it might be used to inform the work of different stakeholders through policy circle seminars and impact events.

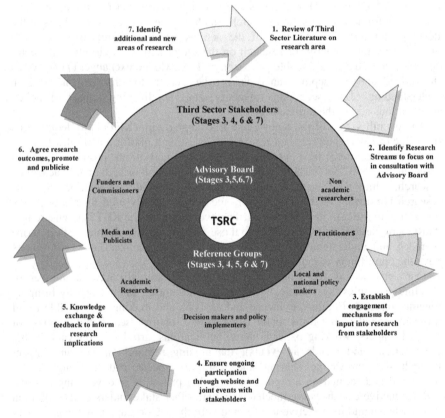

Fig. 1 TSRC's Model of Knowledge Exchange

Figure 1 outlines TSRC's framework for stakeholder engagement and knowledge exchange: the **inner core** represents the TSRC research teams based in the Universities of Birmingham, Southampton, and Middlesex; the **dark ring** represents the formal engagement structures, and the **light ring** the stakeholders that we will offer engagement opportunities to; the **outer arrows** are indicative of the stages of the research cycle and relate to engagement opportunities for all stakeholders.

The website has dedicated pages for each research stream outlining the key areas of focus and opportunities for those interested to find out more and to reflect by responding through discussion boards and blog comments. Once initial findings from the research emerge these will be disseminated and discussed through knowledge exchange activities which will consider the implications for the sector. This will also be an opportunity to identify new areas of research. We believe this iterative process will both ensure the academic quality of the research and its relevance to the sector.

3 KET Activities Analysis

TSRC, by considering current definitions, methods and theories of knowledge exchange, has developed its own policy model of knowledge exchange which prioritises exchange, as opposed to knowledge transfer through producer-push or user-pull models (Lomas 2007, Lavis 2007, Sudsawada 2007, Jackson-Bowers 2006). The approach offers the development of knowledge in action, valuing experiential and tacit knowledge as part of the research knowledge process so as to maximise relevance for impact (Graham 2006). Most research on knowledge exchange has explored this in sectors other than the third sector e.g. in science, international development and health[3]. The process of knowledge exchange has been researched theoretically and has raised key challenges for effective knowledge exchange which the TSRC model aims to address (see Dobbins 2004 and Canadian Health Service Research Foundation Digest Series). In the UK research has explored evidence based policy and practice, how to encourage knowledge exchange, and the way in which knowledge is translated suggesting conditions for successful knowledge exchange (see Nutley 2008, Eppler 2007, Nutley 2000). However it has been suggested by some authors including Robeson et al (2008) that there is a "...lack of guidance available for planning and evaluating knowledge brokering interventions." Others suggest limited research knowledge on how knowledge exchange works in different contexts (Robeson 2008, Ward 2009a, Jackson-Bowers 2006). In reviewing current knowledge exchange literature, there also seems to be limited literature comparing knowledge exchange activities and their effectiveness. These are issues that TSRC's policy

---

[3] See for example the Joint Information Systems Committee UK, to bring together higher education establishments with science and business http://www.jisc.ac.uk/aboutus/ partnerships/knowledgeexchange.aspx; Overseas Development Institute website with tools for knowledge and learning, www.odi.org.uk/rapid/tools; and the Canadian Health Service Research Foundation digests www.chsrf.ca

approach is trying to address by developing a model for knowledge exchange, its planning, implementation, monitoring and evaluation and is detailed in our Knowledge Exchange Communications and Impact Strategy (KECIS). This paper begins to address this by assessing activities TSRC has undertaken to date, and by developing a Knowledge Exchange Impact Matrix (KEIM) to plot and compare activities.

Much of the current literature focuses on the need for knowledge exchange activities to bridge the divide between the worlds of policy, practice and academia (Lomas 2000, Ward 2009b, Blinder 2006, Nutley 2000). TSRC's strategy and implementation addresses this by ensuring that the research process engages other stakeholders through KET activities, and in the formal structure of TSRC through the Advisory Board, Reference Groups for each research stream and devolved administrations. Recognising the need to bridge the gap between different stakeholders TSRC has a dedicated Knowledge Exchange Team, who have sound experience and strong networks in third sector stakeholder communities and are based in London at the National Council of Voluntary Organisations (NCVO) offices, the leading umbrella body in the sector.

An initial review of our activities since September 2009 has informed the development of the Knowledge Exchange Impact Matrix (KEIM) which plots the different types of knowledge exchange activities we have undertaken to date in relation to: the extent of meaningful knowledge exchange; and the number of stakeholders the activity has had an impact on. The matrix is adapted from Arnstein's 'Ladder of Participation' (Arnstein 1969), from a bottom rung representing the dissemination of information through to a the top rung representing meaningful participatory knowledge exchange which has an impact. The ladder is complemented with a horizontal dimension to plot the number of stakeholders engaged in an activity. Figure 2 plots different knowledge exchange activities in different quadrants, primarily focusing on non academic stakeholders.

Figure 2, TSRC's Knowledge Exchange Impact Matrix is split into four quadrants A, B, C, and D:

Quarter A and C offer optimum meaningful knowledge exchange the former to the maximum number of stakeholders the latter to fewer stakeholders (targeting specific stakeholder groups);

Quarters B and D offers the building blocks for successful knowledge exchange: the former, by profiling TSRC and its research to a larger audience; the latter, by raising and promoting TSRC activities and research to specialist groups of stakeholders within the sector. These activities predominantly offer the dissemination of research with limited opportunity or space for interaction and knowledge exchange unless knowledge recipients take further initiatives.

Although quadrant A offers the optimum type of activity to maximise meaningful knowledge exchange while having an impact on the largest number of stakeholders, we would argue that quadrant C is probably the most effective for intense quality knowledge exchange with smaller groups of stakeholders. However, in order to achieve meaningful knowledge exchange, other types of activities (in quadrant B & D) are needed to develop the groundwork so that stakeholders know about the activities and access the research available. Each activity has resource implications that also need to be considered when deciding different mechanisms to use for effective knowledge exchange. To date we have experienced the result of a number of knowledge exchange activities from each of the matrix quadrants and below we illustrate, based on our experience of knowledge exchange: the implications of different types of activities in each quadrant; reflect on the impact of the activity; its limitations; and cost implications.

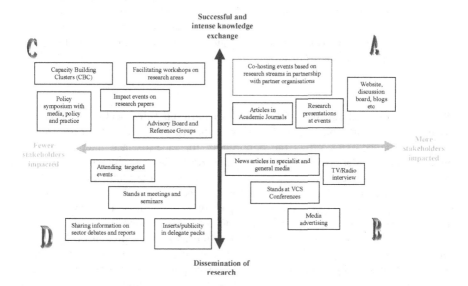

Fig. 2 TSRC Knowledge Exchange Impact Matrix

Quarter A

We have had **75 research presentations at events** accessing an audience of 5000 stakeholders but this has been mainly in conferences targeting academics, researchers and Government policy makers (over 60%) and to a lesser extent practitioners, third sector and local public sector stakeholders. This activity delivers on key elements of successful knowledge exchange, to influence the thinking and understanding of participants but in a formal large scale participatory space which limits the interactive nature of the exchange while offering greater reach. This activity has minor cost implications as researchers are usually invited to present by non academic event organizers.

We have established a dedicated **website, with discussion boards, blogs, and vodcasts etc,** which have had a high volume of hits (given that it was launched in September 2009) but limited interactions. In an average month over 1000 visitors, spend on average three minutes on the website, and look at four pages per visits, 58% access the site through search engines (mainly from google), 22% direct traffic, 20% from a referring site which has a link to our website. After the home page (our landing page), the most visited pages are our publications, staff pages, research pages, followed by the 'about page'. The type of visitors to the website include: 10% from within TSRC and University of Birmingham and Southampton, 10% from other known university networks, the remaining 80% from non academic sources. Of those who access the website, 60% are visiting it for the first time. Although the website has the potential to offer a high volume of stakeholder engagement, there is currently minimal interaction with our platforms for exchange such as commenting on blogs, discussion boards and contacting us directly (this is not uncommon even with high profile websites). As yet it is difficult to know if the information accessed is used, but we plan to undertake online evaluations and ask those downloading papers from the website their purpose in doing so. This activity has substantial cost implications to build, manage and maintain the website.

Quarter B

KET have attended and hosted **stands at 13 national third sector conferences,** with 300 – 600 participants at each raising the profile of TSRC and our research to a wide audience. This offers the opportunity for face-to- face interaction with potential stakeholders of TSRC, and with the KET who can tailor information to meet the needs of the enquirer. Most enquirers sign up to our e-bulletin so that they will receive ongoing information about TSRC's research, but there is limited scope at these events for detailed exchange on particular research areas. Also, there are normally cost implications for having a stand and for someone to manage it.

We have had a number of **press releases that have been picked up by social media** (websites, blogs, and twitters). One press release led to: nine online media articles; increased hits on the research papers web page from 10 – 50 hits in a day; a headline piece in a major trade magazine with 115,000 subscribers; which led to over a dozen people posting comments; which led to it being picked up and mentioned in popular sector blogs with approximately 500 followers; and being tweeted by 8 people reaching 4,182 followers. The article put an angle on the news story which didn't include the detail, most people commented on the article rather than reading the research, and the researchers had to monitor and respond to comments made to clarify misinterpretations of their research. The researchers also had a meeting with one of the commentators as a result. This activity raised the profile of TSRC, and created the opportunity for a wide audience to learn about the research but led to limited further knowledge exchange.

Quarter C

Our **Advisory Board and six Reference Groups** have met between one and three times with key stakeholders over the past year, and their comments and suggestions have been used to direct the research priorities, and the content of our research. This form of ongoing formal engagement and knowledge exchange has a cost implication but offers the development of trust, understanding and the ability to build relationships between researchers and stakeholders, a vital element of effective knowledge exchange. It also offers 'buy in' and ownership of the research outcomes to inform stakeholders policy and practice, although the number of stakeholders (on average 15 members per Reference Group and 30 members on our Advisory Board)) are limited and there is no guarantee that research knowledge will be cascaded through the stakeholders.

Our three **Capacity Building Clusters** have, in the last 12 months, agreed twenty PhD studentships, nine Knowledge Transfer Partnerships, twelve voucher schemes and five placements. These are primarily targeted at academics and knowledge exchanges with a partner third sector organisation. These activities take place over a set period of time and knowledge exchange is limited to between key individuals on a given research topic.

We have **facilitated a workshop** using an action learning set method on social enterprise. Although the number of participants is small, and the time commitment (over three day sessions) and therefore cost more extensive, there is intense and meaningful knowledge exchange between participants from different stakeholder communities. The report produced includes reflections on actions which have been agreed and implemented by the learning sets, and is widely circulated and has been used to inform our research.

Quarter D

TSRC researchers and KET have attended **35 targeted events** as participants. These are smaller scale events which offer participants an opportunity to contribute to the debate using the research knowledge of TSRC. However, impact is inconsistent and depends on the event, the issues that are raised, and the opportunity to contribute to them.

We have held **stands at 8 meetings and seminars.** These offer the opportunity to promote our work to selective, strategic stakeholders such as policy and decision makers and funders although knowledge exchange is primarily introductory rather than detailed.

KET regularly sends **information on sector debates and reports to the research teams** to inform their research process. This offers scope for further knowledge exchange initiated by the researcher.

We have also had **inserts in delegate packs** but this is limited to targeted dissemination (unless the delegate initiates contact) and has a cost implication.

Through our KET activities we have had over 800 people signing up to our e-bulletin over the past six months. Those registered receive a regular update on our research. Our e-newsletter usually increases hits on our website from 100 to over 250 a day when it is sent out. We have also disseminated 500 hard copies of each of our first 15 briefing papers through our stands. We can assume that this has in some part stimulated 60% of the new hits on our website on a monthly basis - with 1000 people accessing our website every month. This can be seen as an indication that the building blocks for more intense knowledge exchange have been laid.

The second phase of work for the KET, from 2010, is to **organise partnership events** and start engaging proactively with the **media**. The latter activity is in Quarter B, although accessing large numbers of people, it does not necessarily guarantee a match between the angle of the research being reported and the reader/viewers' research needs. Although largely a 'free profiling' activity (unless through media advertising) there are risks of knowledge being lost in translation where the reporting of the research is not accurate. **Organising partnership events, workshops and other activities** (Quarter C) such as impact events and policy circle seminar will develop working relations with other stakeholders in the third sector and ensure that knowledge exchange is accurate and meaningful. Depending on the size of the group and the extent of participation, these will have cost implications. These costs could be shared with the partner, who are also more likely to be able to guarantee attendance at the event through their networks and contacts.

4 TSRC Monitoring and Evaluating Impact

In addition to the ESRC reporting and monitoring mechanisms, TSRC has agreed additional evaluation mechanisms for the knowledge exchange element of the Centre. This is due in part to the unique nature and opportunity created by establishing a dedicated knowledge exchange resource as part of Third Sector Research Centre; the widely acknowledged lack of guidance available for planning and evaluating knowledge broker interventions (Robeson et al (2008), Ward et al (2009a) & Jackson-Bowers et al (2006)); and a lack of knowledge about how it works, what contextual factors influence it and how effective it is (Conklin et al (2008) & Ward et al (2009a).

By planning and designing a framework for the monitoring and evaluation of our knowledge exchange activities, initially over the next four years we would hope to contribute to the knowledge base of what works in knowledge exchange. KET's Critical Circle of Friends (a panel of knowledge exchange experts from across sectors and administration) plays a role in advising and guiding the team in this and helps us to reflect, learn and develop our approach to knowledge exchange. Our conceptual framework is adapted (from Sullivan et al (2007) to provide an overview of our knowledge exchange work. The framework offers a flow mechanism from **inputs** of human and institutional resources, to **processes** of product and service development and dissemination, which lead on to **outputs** of the information products and service, so as to enable **reach** through initial distribution, secondary distribution and referrals, to **audiences** in policy, practice and academia, with **initial outcomes** including usefulness, user satisfaction and quality, **intermediate outcomes** of use in being more informed, enhancing practice, adding to research knowledge and collaboration and **intended long term outcomes** based on our vision and goals.

We have ensured that data collection is integrated into the work of the Knowledge Exchange Team on a regular basis by undertaking **routine recording** of our activities e.g. dissemination of our research papers at events; through e-bulletins downloaded from the web; a record of requests for information and outcomes; web statistics; citations in other publications and reports; number of, and attendance at, events; contacts made; circulation lists; media coverage; and number of collaborative ventures. We plan **user surveys** from those who download working papers, online feedback surveys, focus group discussions and event evaluations; and consultations with our Advisory Board and Reference Groups. We will assess **research use** through content analysis of: publications;

bibliographical references; citations in academic journals and other papers and reports; and case study examples.

We will **continually monitor** the results of our Knowledge Exchange, Communication and Impact Strategy, evaluating success and will revise and make changes to increase its impact. We monitor the numbers of people and the varieties of audience we are reaching, and the types and level of action taken by audiences in a number of different ways:

Monitoring: web activity – hits, downloads etc; monitoring size and growth of our database – numbers of people receiving newsletters, journals, other communication; sectors/ types of audience receiving information – i.e. third sector organisations/ academics/ policy-makers, what types of third sector organisations, how many community groups, below the radar groups, social enterprises etc;

Recording: media coverage; citation in academic and policy documents; actions taken as result of communication – i.e. requests for more information, signing up to mailing lists, joining groups, giving feedback; actions taken as a direct result of specific communications – i.e. how many people took action they were directed to take by each communication;

Evaluating: media coverage; audience experience of communication through qualitative feedback, including feedback forms on website, face-to- face feedback in meetings and through stakeholder groups.

We plan to **annually evaluate** our work based on the implementation plan of our Knowledge Exchange, Communications and Impact Strategy. This will inform the work of the following year and the priorities we address. We will undertake a Performance Management (PM) evaluation every two years with stakeholders of our work as the Knowledge Exchange Team. We will share our learning and report on the main activities in an Annual Review of our work. As part of our second annual review we have established the following standardised indicators to measure reach, usefulness and use of our research, collaborative ventures and capacity building initiatives. This mechanism will help us understand the knowledge exchange pathways that inform policy and practice.

Monitoring **reach** through: primary distribution lists size and representation (e-bulletins, RSS feeds, stall distribution); secondary distribution lists (web counts and downloads, media coverage); referrals (number of citations, referencing, posts and links on other websites).

Monitoring **usefulness** by: user satisfaction (focus groups on satisfaction, format and presentation, content, knowledge gained and changing views); quality assurance (online feedback, e-mail responses to queries, Journal Impact factor).

Monitoring **use** by: online usage (pop up poll when downloading on sector, purpose and relevance); feedback questionnaire (need to adapt information, use to inform policy/decision making or improve practice).

Monitoring **collaboration and capacity building** initiatives through: activities, partners involved, numbers participating and their feedback.

As part of our Performance Management Evaluation we will undertake a more **intensive qualitative evaluation** of our knowledge exchange activities every two years. This will be the most resource intensive element of our evaluation, in addition to the evaluation of all our monitoring information. It will be based on the Realistic Evaluation Model (Pawson and Tilley (1997)) which suggests that all programmes are theories and through their model they offer a way of differentiating what works in which context and with whom. Vicky Ward et al (2009a, and forthcoming publication) have been developing a framework for gathering evidence using this type of evaluation for knowledge exchange interventions in health.

The process to be used will be to agree the programme theories and mechanisms we have used to bring about change and impact including the target audiences and desired changes and outcomes from our knowledge exchange work. We will then codify and map these mechanisms in relation to user motives, outcomes and contexts through stakeholder and recipient views, administrative data and cross tabulating responses. Using this we will identify outcome patterns in the analysis and suggest the different outcomes that occur in different contexts (policy, practice and academia).

The ongoing monitoring, Annual Reviews, Performance Management Evaluations and qualitative realistic evaluations will all feed into the 5 year evaluation of TSRC and our knowledge exchange work. Our ongoing commitment to evaluation will hopefully offer reflective learning for future research centres, knowledge exchange activities, and the TSRC.

5 Implications of a Matrix Approach to TSRC Understanding and Practice

The matrix works on the assumption that effective knowledge exchange can occur by bringing different stakeholders together in a participatory space for knowledge exchange where research knowledge is shared and interaction welcomed to inform the development of the research and clarify inconsistencies between the stakeholders reality and the research. Using this matrix approach means that TSRC will not equate high numbers of citations or media coverage as key indicators of successful knowledge exchange but will want to highlight the quality of the knowledge exchange activity and how meaningful the interaction is to influence further research, policy and practice.

By using a matrix to plot knowledge exchange activities we cross reference our activities with the opportunity to engage in meaningful knowledge exchange as apposed to simply disseminating research findings, and the number of stakeholders we access. The matrix does not explore the different stakeholder sector's cultures and ways of working or their existing understanding of the research topic before they engage in the activity. Nor does it consider the actual impact of the interaction after the activity i.e. if the participant of the activity re-engages with TSRC in the future. It also does not consider the additionality created by the activity through the cascading of research information to other potential users. But these should be picked up through the other mechanisms we are using to monitor and evaluate our activities. This matrix does not assume that wide coverage and dissemination of research leads to meaningful knowledge exchange.

6 Conclusion

TSRC has created the scope to develop a model that offers a step change in the way academic institutions can integrate knowledge exchange as part of a more participatory research process. The formal engagement mechanisms, along with a dedicated knowledge exchange team has created an insight into plotting different activities and their potentials and limitations on having an impact on society. Initial key lessons are that knowledge exchange should not be an add on, it should not be supplementing activity once research has been undertaken but needs to be integral to the research process. In an ideal world effective knowledge exchange can only be realised over time once the building blocks have been established which raise the profile of academics and build their relationships with stakeholders. When thinking about knowledge exchange we need to consider the resource implications if it is to succeed, as it involves intense facilitation of flows of knowledge and ongoing interaction between researchers and other stakeholders to establish greater awareness and understanding between researchers, decision makers and practitioners. Based on the Knowledge Exchange Impact Matrix analysis there is not necessarily a correlation between accessing a large number of stakeholders and effective knowledge exchange, as it seems that the cumulative effect of more intensive and meaningful knowledge exchange with a smaller group of people is more likely to have an impact. Although accessing a large number of stakeholders offers solid building blocks for subsequent more successful knowledge exchange activities.

References

Arnstein, S.: A Ladder of Citizen Participation. JAIP 35(4) (1969)

Blinder, A.: Stinglers Lament. Eastern Economic Journal 32(3) (2006)

Canadian Health Service Research Foundation, Digest Issues (1,11,21,22,24,25,28,35, 38,40,45, 46 and 52) relating to knowledge transfer

Conklin, et al.: Briefing on linkage and exchange: facilitating diffusion of innovation in health services Occasional Paper 231, Cambridge RAND Europe (2008)

Dobbins, M.: A knowledge transfer strategy for public health decision making. Worldviews on Evidence Based Nursing 1(2) (2004)

Eppler, M.: Knowledge communication problems between experts and decision makers: an overview and classification. The Electronic Journal of Knowledge Management 5(3) (2007)

Graham, I., et al.: Lost in knowledge translation. Time for a map? Journal of Continuing Education in the Health Proffessions 26(1) (2006)

Jackson-Bowers, E.: Focus on: knowledge brokering, Primary Health Care Research and Information Services. vol. 4. Flinders University, Australia (2006)

Lomas, J.: The in-between world of knowledge brokering ,334 (2007), http://www.BMJ.com

Lavis, J.: Research into practice: a knowledge transfer planning guide, Institute for work and health, Canada (2006)

Pawson, R., Tilley, N.: Realistic Evaluation (1997)

Nutley,et al.: Flows of knowledge, expertise and influence: methods for assessing policy and practice impacts from social science research, Innogem Working Paper 55 (2008)

Nutley, et al.: What works? Evidence based policy and practice in public services (2000)

Robeson, et al.: Life as a knowledge broker in public health. Journal of the Canadian Health Libraries Association 29(3) (2008)

Sudsawada, P.: Knowledge translation: introduction to models, strategies and measures. National Centre for the dissemination of disability (2007)

Sullivan, T., et al.: Guide to Monitoring and evaluating Health Information Products and Services. Centre for Communications Program, Johns Hopkins Bloomberg School of Public Health, Baltimore, Maryland (2007)

Ward, et al.: Developing a framework for transferring knowledge into action: a thematic analysis of the literature. JHSRP 14(3) (2009a)

Ward, et al.: Knowledge brokering: the missing link in evidence to action chain. Evidence and Policy 5(3) (2009b); unpublished paper Knowledge brokering: exploring the process of transferring knowledge into action based on a study at the University of Leeds supported by the Medical Research Council (2009c)

Author Index